High-Pressure
Microbiology

High-Pressure Microbiology

Edited by

Chris Michiels
Katholieke Universiteit Leuven,
Leuven, Belgium

Douglas H. Bartlett
Scripps Institution of Oceanography,
University of California, San Diego,
La Jolla, California

Abram Aertsen
Katholieke Universiteit Leuven,
Leuven, Belgium

ASM
PRESS

Washington, DC

Address editorial correspondence to ASM Press, 1752 N St. NW, Washington, DC 20036-2904, USA

Send orders to ASM Press, P.O. Box 605, Herndon, VA 20172, USA
Phone: 800-546-2416; 703-661-1593
Fax: 703-661-1501
E-mail: books@asmusa.org
Online: estore.asm.org

Library of Congress Cataloging-in-Publication Data

High-pressure microbiology / edited by Chris Michiels, Douglas H. Bartlett, Abram Aertsen.
    p. ; cm.
    Includes bibliographical references and index.
    ISBN 978-1-55581-423-6 (hardcover)
    1. Water microbiology. 2. Hydrostatic pressure—Physiological effect. I. Michiels, Chris. II. Bartlett, Douglas Hoyt. III. Aertsen, Abram.
    [DNLM: 1. Water Microbiology. 2. Adaptation, Physiological. 3. Hydrostatic Pressure. QW 80 H638 2008]

    QR105.H54 2008
    579—dc22

                                                                    2007044218

10 9 8 7 6 5 4 3 2 1

Cover figures: (Top left) Phase-contrast microscopy of exponential-phase cells of *Escherichia coli* K-12 grown overnight at 40 MPa and 37°C. Due to inhibition of cell division, cells grow as long filaments. (Courtesy of A. Aertsen and C. Michiels.) (Top right) Epifluorescence microscopy of exponential-phase cells of *E. coli* K-12 containing a reporter plasmid with the promoter of *sulA* fused to the green fluorescent protein gene *(gfp),* 3 h after high-pressure shock (100 MPa, 15 min, 20°C). Cells are bright green and slightly elongated as a result of SOS induction. (From chapter 5, Fig. 1. Courtesy of A. Aertsen and C. Michiels.) (Bottom left) Racks of pressure vessels being used for the high-pressure storage of marine microorganisms collected from various deep-sea locations around the world. They belong to A. Aristides Yayanos (Scripps Institution of Oceanography). (Courtesy of D. Bartlett.) (Bottom right) Transmission electron micrograph of thin sections showing morphological changes in exponential-phase cells of *E. coli* pressure treated at 200 MPa for 2 min (30% survival). Prominent are the unusual conformation of the nucleoid and its fibrillar DNA and the appearance of dark condensed regions that are presumably protein. (From chapter 4, Fig. 6. Courtesy of P. Chilton and B. M. Mackey.) (Background) Some pressure gauges, pumps, and high-pressure tubing from the marine microbiology laboratory of A. Aristides Yayanos (Scripps Institution of Oceanography). (Courtesy of D. Bartlett.)

CONTENTS

CONTRIBUTORS

Fumiyoshi Abe • Extremobiosphere Research Center, Japan Agency for Marine-Earth Science and Technology (JAMSTEC), 2-15 Natsushima-cho, Yokosuka 237-0061, Japan

Tjakko Abee • Top Institute for Food and Nutrition, P.O. Box 557, 6700 AN, and Laboratory of Food Microbiology, Wageningen University, Bomenweg 2, 6700 EV, Wageningen, The Netherlands

Abram Aertsen • Centre for Food and Microbial Technology, Department of Microbial and Molecular Systems (M^2S),vii Faculty of Bioscience Engineering, K.U.Leuven, Kasteelpark Arenberg 22, B-3001 Leuven, Belgium

Shizuka Arakawa • Extremobiosphere Research Center, Japan Agency for Marine-Earth Science and Technology, Yokosuka 237-0061, Japan

Douglas H. Bartlett • Mail code 0202, Marine Biology Research Division, Center for Marine Biotechnology and Biomedicine, Scripps Institution of Oceanography, University of California, San Diego, La Jolla, CA 92093-0202

Dennis A. Bazylinski • School of Life Sciences, University of Nevada, Las Vegas, Las Vegas, NV 89154-4004

Boonchai B. Boonyaratanakornkit • Department of Chemical Engineering, University of California at Berkeley, Berkeley, CA 94720

Douglas S. Clark • Department of Chemical Engineering, University of California at Berkeley, Berkeley, CA 94720

Juliana R. Cortines • Programa de Biologia Estrutural, Instituto de Bioquímica Médica and Centro Nacional de Ressonância Magnética Nuclear Jiri Jonas, Universidade Federal do Rio de Janeiro, 21941-590, Rio de Janeiro, RJ, Brazil

Samir Damare • Institut für Biotechnologie, Biotechnikum, Walter Rathenau Strasse 49A, 17489 Greifswald, Germany

Matthias A. Ehrmann • Technische Mikrobiologie, Technische Universität München, 85350 Freising, Germany

Jiasong Fang • Department of Geological and Atmospheric Sciences, Iowa State University, Ames, IA 50011-0001

Gail Ferguson • School of Medicine, Department of Medicine and Therapeutics, Institute of Medical Sciences, University of Aberdeen, Foresterhill, Aberdeen AB25 2ZD, United Kingdom

Patricia M. B. Fernandes • Biotechnology Core, Federal University of Espirito Santo, Vitória, ES Brazil, 29040-090

Andre M. O. Gomes • Programa de Biologia Estrutural, Instituto de Bioquímica Médica and Centro Nacional de Ressonância Magnética Nuclear Jiri Jonas, Universidade Federal do Rio de Janeiro, 21941-590, Rio de Janeiro, RJ, Brazil

Rafael B. Gonçalves • Centro de Pesquisa em Ciência e Tecnologia do Leite, Universidade Norte do Paraná, Fazenda Experimental da UNOPAR, 86125-000, Tamarana, PR, Brazil

Karel Heremans • Department of Chemistry, Katholieke Universiteit Leuven, B-3001 Leuven, Belgium

Kimon A. Karatzas • Top Institute for Food and Nutrition, P.O. Box 557, 6700 AN, Wageningen, The Netherlands, and School of Clinical Veterinary Science, University of Bristol, Langford House, Langford, Bristol, BS40 5DU, United Kingdom

Chiaki Kato • Extremobiosphere Research Center, Japan Agency for Marine-Earth Science and Technology, Yokosuka 237-0061, Japan

Sheila M. B. Lima • Fundação Oswaldo Cruz, Instituto de Tecnologia em Imunobiológicos, Programa de Vacinas Virais, Rio de Janeiro, RJ 21045-900, Brazil

Mark Linton • Agriculture, Food and Environmental Science Division (Food Microbiology Branch), Agri-food and Biosciences Institute, Newforge Lane, Belfast BT9 5PX, Northern Ireland, United Kingdom

Bernard M. Mackey • Department of Food Biosciences, The University of Reading, Whiteknights, Reading RG6 6AP, United Kingdom

Pilar Mañas • Departamento de Producción Animal y Ciencia de los Alimentos, Facultad de Veterinaria, Miguel Servet 177, 50013, Zaragoza, Spain

Filip Meersman • Department of Chemistry, Katholieke Universiteit Leuven, B-3001 Leuven, Belgium

Chris W. Michiels • Centre for Food and Microbial Technology, Department of Microbial and Molecular Systems (M^2S), Faculty of Bioscience Engineering, K.U.Leuven, Kasteelpark Arenberg 22, B-3001 Leuven, Belgium

Roy Moezelaar • Top Institute for Food and Nutrition, P.O. Box 557, 6700 AN, and WUR, Wageningen University and Research Centre, Food Technology Centre, Bornsesteeg 59, 6708 PD, Wageningen, The Netherlands

Kaoru Nakasone • Department of Chemistry and Environmental Technology, School of Engineering, Kinki University, Higashi-Hiroshima 739-2116, Japan

Yuichi Nogi • Extremobiosphere Research Center, Japan Agency for Marine-Earth Science and Technology, Yokosuka 237-0061, Japan

Andréa C. Oliveira • Programa de Biologia Estrutural, Instituto de Bioquímica Médica and Centro Nacional de Ressonância Magnética Nuclear Jiri Jonas, Universidade Federal do Rio de Janeiro, 21941-590, Rio de Janeiro, RJ, Brazil

Margaret F. Patterson • Agriculture, Food and Environmental Science Division (Food Microbiology Branch), Agri-food and Biosciences Institute, and Department of Food Science, Queen's University, Belfast, Newforge Lane, Belfast BT9 5PX, Northern Ireland, United Kingdom

Chandralata Raghukumar • National Institute of Oceanography, Dona Paula, Goa 403 004, India

Takako Sato • Extremobiosphere Research Center, Japan Agency for Marine-Earth Science and Technology, Yokosuka 237-0061, Japan

Waleska D. Schwarcz • Centro de Pesquisa em Ciência e Tecnologia do Leite, Universidade Norte do Paraná, Fazenda Experimental da UNOPAR, 86125-000, Tamarana, PR, Brazil

Peter Setlow • Department of Molecular, Microbial and Structural Biology, University of Connecticut Health Center, Farmington, CT 06030-3305

Ana Cristina B. Silva • Programa de Biologia Estrutural, Instituto de Bioquímica Médica and Centro Nacional de Ressonância Magnética Nuclear Jiri Jonas, Universidade Federal do Rio de Janeiro, 21941-590, Rio de Janeiro, RJ, Brazil

Jerson L. Silva • Programa de Biologia Estrutural, Instituto de Bioquímica Médica and Centro Nacional de Ressonância Magnética Nuclear Jiri Jonas, Universidade Federal do Rio de Janeiro, 21941-590, Rio de Janeiro, RJ, Brazil

Hideyuki Tamegai • Department of Chemistry, College of Humanities and Sciences, Nihon University, Setagaya-ku, Tokyo 156-8550, Japan

Giorgio Valle • Department of Biology, University of Padua, 35131 Padua, Italy

Rudi F. Vogel • Technische Mikrobiologie, Technische Universität München, 85350 Freising, Germany

Marjon Wells-Bennik • Top Institute for Food and Nutrition, P.O. Box 557, 6700 AN, Wageningen, and NIZO Food Research, Kernhemseweg 2, 6718 ZB Ede, The Netherlands

Xiang Xiao • Key Laboratory of Marine Biogenetic Resources, Third Institute of Oceanography, State Oceanic Administration, Xiamen 361005, People's Republic of China

PREFACE

Unicellular microorganisms, and in particular prokaryotes, thrive in extremely diverse environments. In fact, the range of physicochemical conditions under which microbial life has been observed continues to expand as microbiologists explore more remote and hostile environments. Today, microorganisms have been described to occur in habitats spanning an extraordinary range of more than 120°C, 10 pH units, and millimolar to molar concentrations of solutes like NaCl. To grow under these extreme conditions, these microorganisms have evolved specific and unique adaptations, and the study of these provides interesting insights not only into microbial physiology but also into the most fundamental properties of living systems as compared to nonliving systems. Temperature, pH, and osmotic pressure are among the most-studied environmental parameters in microbiology, probably because they are easy to manipulate and because they are relevant in our daily life, for example, in food preservation. By comparison, much less is known about microbial adaptation to high pressure, although high-pressure environments are more widespread in nature than high-temperature, high- or low-pH, and high-osmolarity environments. The compartment of the oceans 200 m below sea level constitutes more than 95% of the volume and represents 55% of the prokaryotic cells of all aquatic habitats on earth. Pressure levels vary 3 orders of magnitude, from 0.1 MPa at sea level to more than 110 MPa at the deepest point in the ocean, approximately 11,000 m deep. Besides the deep sea, the deep terrestrial subsurface is another enormous high-pressure habitat, probably the last large remaining unexplored habitat on the earth. Several piezophilic bacteria from the deep sea have been isolated and characterized today, but clearly, the full amplitude of microbial diversity in high-pressure habitats and the physiological adaptations in these organisms are only beginning to be unraveled.

Besides piezophiles, nonpiezophilic microorganisms are an object of study for investigators of high-pressure microbiology. The interest in studying the effects of high pressure on these organisms has been fueled by the development and commercial introduction since the 1990s of a food preservation technique based on high pressure without the need to apply heat. Large amounts of quantitative data on the high-pressure inactivation of food-borne pathogens and spoilage organisms have become available since then. The pressures used in these inactivation studies are typically in the range of 200 to 800 MPa, much higher than the pressures at any depth in the ocean. We have learned from these studies that sensitivity to heat and to high pressure are not necessarily linked and, most remarkably, that vegetative bacteria can acquire extreme resistance to high-pressure inactivation. The findings have, in turn, raised interesting fundamental questions about the actual molecular perception of high pressure, the nature of the cellular damage it causes, and more specifically the adaptations that render these cells high-pressure resistant. Adaptation to stress has always enjoyed a wide interest among microbial physiologists, and several bacterial stress responses are known in great molecular detail. High-pressure stress has been largely neglected in this field, but in view of its unique effects on biomolecules and biomolecular assemblies, which are different and sometimes even opposed to the effects of heat, high pressure is likely to become a very useful addition to the "toolbox" available to microbial-stress investigators. High pressure has indeed been shown to induce some unique responses in nonpiezophiles, like *Escherichia coli,* and may thus help to uncover novel stress response pathways and adaptational mechanisms in bacteria that have thus far remained cryptic.

It is clear from the above that the field of high-pressure microbiology has evolved along two tracks and that both in the piezophile and in the nonpiezophile tracks, significant progress has been made in the description and understanding of high-pressure effects. Against this background, we endeavored to compile this book, which is probably the first to be entirely devoted to high-pressure microbiology. The objectives are to give an update on the progress in the field and to stimulate a cross-fertilization between the piezophile and the nonpiezophile high-pressure research fields. This approach is reflected in the structure of the book. While the first chapter introduces elementary thermodynamic principles related to high pressure and focuses on biomolecules and biochemical reactions, chapters 2 and 3 look at viruses and bacterial spores, respectively, and form the bridge to the cellular response and adaptation strategies of nonpiezophilic microorganisms, discussed in chapters 4 to 9. Chapter 10 closes the first section with an overview of food- and microbe-related features that affect the efficiency of high-pressure processing. The second section of the book (chapters 11 to 18) highlights different aspects of deep-sea microorganisms and deals both with their isolation, diversity, and ecology and with their physiological adaptations. We hope that this book will appeal to a large readership of microbiologists, not only those actively involved in high-pressure research but also those interested in microbial stress responses or more generally in microbial physiology.

We are grateful to all coauthors for their contributions and to Greg Payne and Susan Birch from ASM for their help in bringing this book to press.

High-Pressure Microbiology
Edited by C. Michiels, D. H. Bartlett, and A. Aertsen
© 2008 ASM Press, Washington, DC

Chapter 1

High Hydrostatic Pressure Effects in the Biosphere: from Molecules to Microbiology

Filip Meersman and Karel Heremans

PRESSURE AS A VARIABLE IN THE BIOSPHERE AND THE BIOSCIENCES

The fundamental transformation in the ideas on the possible effects of pressure on living systems can be traced back to the scientific expeditions that were undertaken during the second half of the 19th century. Before 1850 it was assumed that the deep sea would not be suitable for life in general. In 1872 HMS *Challenger* sailed around the globe for about 4 years, and the findings of this expedition were considered as "the greatest advance in the knowledge of our planet since the celebrated discoveries of the fifteenth and sixteenth centuries." During 1882–1883 the French expedition with the *Talisman* recovered a large amount of organisms from a depth of 6,000 m. This drastically changed the opinions on the role of pressure in living systems. It is now well established that a large part of the biosphere is exposed to extremes of temperature and hydrostatic pressure. Approximately 70% of Earth's surface is covered with water. The average depth of the oceans is 3,800 m. Consequently the average hydrostatic pressure is about 38 MPa, with a maximum of about 100 MPa at the deepest point, the Mariana trench. At these depths the temperature is as low as 2°C, except in the vicinity of hydrothermal vents. Even under these conditions thriving microbial communities and invertebrates have been found. Understanding the adaptation of these organisms to this extreme environment requires the knowledge of temperature and pressure effects on the molecules of which they are composed and on their metabolic reactions.

A second, more fundamental reason for studying the effect of pressure on biomolecules is related to the fact that one needs to consider pressure as a variable in order to obtain a full thermodynamic description of a molecular system, i.e., a biomolecule and the solvent. In particular, pressure experiments provide information on the volume changes of a system composed of, e.g., a protein in solution. This can be inferred from the following thermodynamic relationships. The Gibbs free energy *(G)* of a system is defined as

$$G = H - TS = E + pV - TS \tag{1}$$

Filip Meersman and Karel Heremans • Department of Chemistry, Katholieke Universiteit Leuven, B-3001 Leuven, Belgium.

where E, V, and S represent the internal energy, the volume, and the entropy of the system, respectively, p is pressure, and T is temperature. The change in free energy as a function of pressure and temperature is given by

$$dG = Vdp - SdT \qquad (2)$$

At constant temperature ($dT = 0$) and

$$(\partial G/\partial p)_T = V \qquad (3)$$

Thus, for a reversible process, the change of the Gibbs free energy (ΔG) with pressure is given by the volume change of the process (ΔV):

$$(\partial \Delta G/\partial p)_T = \Delta V \qquad (4)$$

It is clear from this equation that, according to the Le Châtelier-Braun principle, the system will react on an increase in pressure by shifting towards the state that occupies the smallest volume. A similar quantitative statement can be made for temperature, where a temperature increase will shift the equilibrium towards the state of the largest heat content. Furthermore, equation 4 also shows that pressure affects primarily the volume of the system under study. This is a great advantage of pressure over temperature, which changes both the volume and the internal energy of the system.

The aim of this chapter is to give an outline of the effect of high hydrostatic pressure on proteins, lipids, nucleic acids, and their interactions and to provide a thermodynamic and kinetic framework to describe these effects. The pressure effects of single-component systems (e.g., a protein in solution) are then related to the viability of microorganisms under extremes of high hydrostatic pressure. On the basis of the similarity between the stability curves of proteins and the viability diagrams of microorganisms (in the pressure-temperature plane) (Fig. 1 and 2), it is argued that proteins, and in particular protein-protein interactions, are the main target in the pressure-induced inactivation of microorganisms.

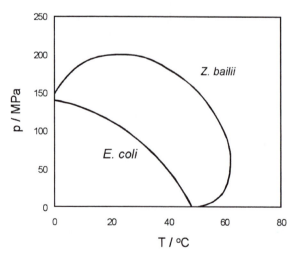

Figure 1. Isokineticity profiles of the inactivation of a bacterium *(Escherichia coli)* and a yeast *(Zygosaccharomyces bailii)* as a function of a combined pressure and temperature treatment. For first-order reactions the decimal reduction time *(D)* is inversely proportional to the inactivation rate *(k)*. Similar differences in stability have been observed for proteins (30, 46). (Redrawn after references 24 and 38.)

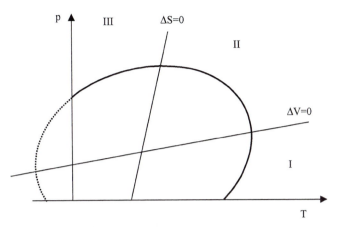

Figure 2. Pressure-temperature phase diagram of proteins. In zone I ΔV and ΔS are positive, in zone II ΔV is negative and ΔS is positive, and in zone III both ΔV and ΔS are negative.

PRESSURE VERSUS TEMPERATURE EFFECTS: FROM WATER TO MACROMOLECULES

A temperature increase will cause a volume expansion, and an increase in pressure will cause a reduction in volume. If, however, as in the case of water, the forces are strong, then an increase in temperature might actually decrease the volume, as is observed between 0 and 4°C, where it reaches its maximum density under ambient pressure conditions.

Under compression all liquids will invariably decrease their volume, although the effect is much smaller in the case of water than in the case of hydrocarbons. How can a system shrink under compression? This can only happen by pushing the molecules closer together. In the pressure range that is of interest here, one can safely neglect the pressure effects on covalent bonds. When strong intermolecular forces are also highly directional, as is the case for hydrogen bonds in water, then these forces will oppose a further closer approach of the molecules under compression.

The above-mentioned general principles become even clearer when we consider the effect of pressure on the melting temperature (dT_m/dp) of solid hydrocarbons and ice. Since the volume of the solid phase of hydrocarbons is smaller than that of the liquid phase, pressure will increase the melting point of hydrocarbons. The dT_m/dp value is about 15°C/100 MPa and is almost independent of the length of the hydrocarbon chain. The value is higher for unsaturated hydrocarbons. From the Clapeyron equation one can estimate the volume and the enthalpy change:

$$dT_m/dp = T_m \, \Delta V/\Delta H \tag{5}$$

As melting is an endothermic process ($\Delta H > 0$), the volume change on melting is positive. The opposite is true for the melting of ice, where the T_m decreases with increasing pressure down to about -20°C ($dT_m/dp = -20$°C/100 MPa). In contrast, the T_m of other ice phases increases with increasing pressure, as can be seen from the phase diagram of water (46).

It is of particular interest to consider briefly the effect of pressure and temperature on mixtures of small molecules, lipids, and proteins in aqueous solutions. The liquid-liquid phase separation that occurs at ambient pressure in the case of water-alcohol mixtures disappears at high pressure. On the other hand, pressure can also induce phase separation in water-soluble-synthetic polymer systems. Its antagonistic effect to temperature has been nicely demonstrated in these systems (32). In the case of aqueous suspensions of lipid vesicles, formed by the insolubility of the lipid molecules in water, the general outcome is a phase diagram which shows pressure dependencies of the phase transition temperature of the order of that found in pure hydrocarbons. The Clapeyron equation (equation 5) can be used to describe the pressure dependence of the phase transition temperature.

In contrast to the rather simple thermodynamic description of the phase diagrams for small molecules and lipids, the stability diagrams of proteins dissolved in water cannot be described by simple thermodynamic considerations. Kinetics may become very important in the unfolding and aggregation processes. The latter is often the cause of the irreversibility of protein denaturation. The molecular details of the denaturation of a protein are also much more complex than the melting of water or hydrocarbons. Experiments and computer simulations make it clear that water-water, water-solute, and solute-solute interactions are equally important in the behavior of the system.

PROTEINS AND ENZYMES: FOLDED, UNFOLDED, AND AGGREGATED

The pioneering work in high-pressure protein research is that of Bridgman, who observed that a pressure of 600 MPa will give egg white an appearance similar, but not identical, to that of a cooked egg (3). Here we describe the mechanism of pressure-induced denaturation, the role of water in this process, and how the pressure-denatured state relates to other denatured states as induced by heat or chemical denaturants. Finally, the effect of pressure on protein assemblies and the pressure-temperature phase diagram of proteins are discussed.

Mechanism of Pressure Denaturation

Suzuki (48) found that at temperatures below 30°C the kinetics of the pressure denaturation of carbonylhemoglobin and ovalbumin were characterized by a negative activation enthalpy. Such negative activation energies have also been observed for the urea-induced denaturation of proteins. To explain his observations, he proposed the following mechanism:

$$P + nH_2O \leftrightarrow P(H_2O)_n \rightarrow P_D \qquad (6)$$

where P is the native protein, $P(H_2O)_n$ is the hydrated protein, and P_D is the denatured protein. Thus, he suggested that pressure induces the penetration of water into the protein in a strongly exothermic step, thereby causing denaturation. A similar conclusion was reached by Silva and Weber (45) on the basis of experimental data. Computer simulations on the association of methane in water (19) and on bovine pancreatic trypsin inhibitor (54) have provided further evidence for this water penetration mechanism. The former simulation demonstrated that high hydrostatic pressure disrupts hydrophobic contacts in favor of the solvent separated apolar partners, whereas the latter showed that under pressure, protein-protein hydrogen bonds are replaced by protein-water hydrogen bonds. Pressure also affects other noncovalent interactions, mainly electrostatic interactions (Table 1). Note that the volume change associated with hydrogen bond formation is close to zero, implying that pressure will not strongly affect this interaction.

Protein denaturation is associated with volume changes on the order of -10 to -100 ml·mol^{-1}. What is the origin of the volume decrease? The volume of a protein in solution, V_i, is the sum of

$$V_i = V_{atom} + V_{cavities} + \Delta V_{hydration} \qquad (7)$$

where V_{atom} and $V_{cavities}$ are the volumes of the atoms and the cavities (that originate from imperfect packing of the atoms in the native conformation), respectively, and $\Delta V_{hydration}$ is the volume change resulting from the interactions of the protein with the solvent (17). Upon protein denaturation the volume of the atoms will not change, so the volume change accompanying the denaturation can be written as

$$\Delta V = \Delta V_{cavities} + \Delta \Delta V_{hydration} \qquad (8)$$

Contributions to $\Delta \Delta V_{hydration}$ are summarized as follows. Exposure of charged and hydrophobic groups to water will cause a volume decrease (Table 1). The former is due to a phenomenon called electrostriction: upon the formation of an ion in solution, the nearby water dipoles will be strongly attracted by the Coulombic field of the ion. The solvation of a monovalent ion is accompanied by a volume decrease of ~ 10 ml·mol^{-1}. In the case of hydrophobic groups, the mechanism underlying the volume change is not fully understood. Presumably the contribution to the volume change arising from the compressibility of the hydrophobic hydration layer plays an important role (28). Solvation of polar groups, on the other hand, also results in a volume decrease (Table 1).

The elimination of cavities upon denaturation is also expected to contribute to the observed negative volume change. This has been confirmed experimentally. Mutants of

Table 1. ΔV associated with specific biochemical reactions (25°C)

Reaction	Example	ΔV (ml·mol^{-1})
Protonation	$H^+ + OH^- \rightarrow H_2O$	21.3
	Imidazole $+ H^+ \rightarrow$ imidazole·H^+	-1.1
	Tris $+ H^+ \rightarrow$ Tris·H^+	-1.1^a
	$HPO_4^{2-} + H^+ \rightarrow H_2PO_4^-$	24
Hydrogen bond formation	Poly(L-lysine) (helix formation)	~ 0
Hydrophobic hydration	$C_6H_6 \rightarrow (C_6H_6)$water	-6.2
	$(CH_4)_{hexane} \rightarrow (CH_4)_{water}$	-22.7
Hydration of polar groups	n-Propanol$\rightarrow$(n-propanol)$_{water}$	-4.5
Protein dissociation	Lactate dehydrogenase ($M_4 \rightarrow 4M$)	-500
Protein denaturation	RNase A (at pH 2)	-46

^{a}The small ΔV for Tris-HCl indicates that the pH of this buffer is pressure insensitive. It is therefore the ideal buffer for pressure studies near physiological pH. A phosphate buffer, in contrast, will have a pH shift of approximately 0.4/100 MPa. In practice the effect may be less pronounced, since the ΔV of ionization becomes smaller at high pressure. It should also be pointed out that pressure-insensitive buffers often show a large temperature dependence (large ΔH) and vice versa.

RNase A in which the mutations created additional cavities are characterized by a larger negative ΔV upon denaturation (50).

Summation of the above contributions would result in a large and negative ΔV. However, experimentally only small, negative volume changes are observed. This suggests that there is also a positive contribution that, at least in part, compensates for the above negative contributions. The origin of this contribution is still the subject of debate (40).

Water as a Reaction Partner

From the above-described mechanism it is clear that water should be considered a reaction partner rather than an inert background. This is supported by other observations. For instance, the presence of water is required for enzyme activity, and it strongly affects the temperature stability of proteins. Fujita and Noda observed a decrease in the denaturation temperature from $\sim137°C$ for dry lysozyme to $\sim67°C$ at a hydration level of ~0.4 g of H_2O/g of protein (10). This observation has been made for several proteins, and it applies also to the pressure denaturation (12, 35). The resistance of bacterial spores and small organisms, such as tardigrades, and the stability of amyloid fibrils are other illustrations of the importance of water in the effects of pressure on organisms and molecules (27, 43, 47).

Water availability is also one of the crucial parameters for the growth of microorganisms. The effect of water activity on bacterial growth is more pronounced than the effect of water activity on protein denaturation. Hence, other parameters, such as osmotic pressure, play a role in the water stress response of bacteria.

Nature of the Pressure-Denatured State

It is now well recognized that the denatured state of a protein is not a random coil but that there are still some persistent native-like long-range contacts. However, the degree of conformational change depends on the denaturation method. Pressure-denatured proteins are often considered to be molten globule-like structures (49). The molten globule state of a protein is generally characterized by a loss of tertiary contacts, whereas the bulk of the secondary structure is maintained. This results in a more hydrated and more expanded conformation. In contrast, the heat-denatured state often shows a more disordered conformation of the protein in which both tertiary and secondary structures are lost (28). The above-described pressure denaturation mechanism provides a basis for the observed differences between the heat- and pressure-denatured states (Fig. 3) (30).

The different characteristics of the heat- and pressure-denatured states affect properties such as the aggregation propensity and the gelation of proteins. The latter is being exploited in the food industry to produce foodstuffs with novel properties using high-pressure treatment.

Pressure Effect on Multimeric Proteins and Aggregates

So far we have considered the effect of pressure on monomeric proteins, which generally become denatured between 400 and 800 MPa. Moderate pressures, of 200 to 300 MPa, are also known to dissociate protein oligomers into their monomers (45). The latter can maintain their native conformation or may denature in this process. Of particular interest is the pressure-induced depolymerization of larger protein assemblies, such as cytoskeletal

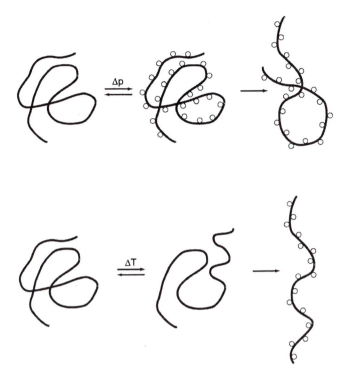

Figure 3. Highly schematic representation of the pressure-induced (top) and temperature-induced (bottom) denaturation of a protein. The circles represent water molecules. The first step of the pressure-induced denaturation is the insertion of water without much change in the conformation. For the temperature-induced denaturation the first step is a change in conformation of the protein.

proteins, which has been shown to result in morphological changes in both eukaryotic and bacterial cells (20, 52). Note that these changes occur under low pressure (~50 MPa) and that, under certain conditions, the original cell morphology is restored after the pressure is returned to ambient.

More recently, the ability of high pressure to dissociate protein assemblies has been applied to larger amorphous and fibrillar aggregates. It has been found that aggregates, such as inclusion bodies, can be dissociated by pressure under solution conditions where the monomeric protein is in its native conformation. Hence, it is possible to rescue these proteins from an aggregate and refold them, an observation that has interesting industrial and biotechnological applications. Amyloid fibrils, involved in a number of debilitating diseases, such as Alzheimer's disease and transmissible spongiform encephalopathies, have also been shown to be sensitive to pressure, at least in the early stages of their formation. However, as aggregates are left to mature, they become more and more pressure insensitive. The mature aggregates are also extremely temperature stable ($T_m > 100°C$), which has been attributed to the anhydrous nature of the fibrillar aggregates. In this respect they resemble bacterial spores. The pressure stability of amyloid fibrils has been discussed at length by Meersman and Dobson (27). In addition, an increasing number of examples suggest that aggregation may also occur at high pressure, but the aggregate presumably has different characteristics from the conventionally heat-induced aggregates (31).

Not all protein aggregates cause disease. Proteins similar to the prion proteins involved in bovine spongiform encephalopathy and Creutzfeldt-Jakob's disease also occur, for instance, in yeasts. Here, its presence is not related to disease, but it represents a protein-based element of inheritance involved in a non-Mendelian transfer of phenotypical traits (23). It is worthwhile to consider that if pressure affects the structure of the prion protein in its aggregated state, as has been shown in homogenates of hamster brain (9), it is, at least in principle, possible to induce a phenotypic switch in a particular strain by high-pressure treatment. However, the pressure and temperature required for optimal dissociation are 500 MPa and 60°C, which are lethal to most microorganisms.

Pressure-Temperature Phase Diagram of Proteins

Following the seminal observation on egg white by Bridgman (3), Suzuki (48) studied the denaturation kinetics of the proteins ovalbumin and carbonylhemoglobin as a function of temperature and pressure and found an elliptic curve when connecting the points of iso-kineticity. Similarly, Hawley investigated the equilibrium denaturation of chymotrypsinogen and RNase A and observed the re-entrant behavior (14). Elliptic diagrams have also been observed in the case of starch gelation (42) and the inactivation of microorganisms (Fig. 1) (13, 51). These data suggest that proteins may be the primary target in the pressure and temperature inactivation of organisms. In the following sections thermodynamic and kinetic descriptions of the phase diagram are given.

Thermodynamic Description of the Phase Diagram

The general equation for the temperature and pressure dependence of phase transitions was applied to the denaturation of proteins by Hawley (14). Mathematically the elliptic shape originates from the fact that in a Taylor expansion of the free-energy difference between the denatured (G_D) and the native (G_N) states, $\Delta G = G_D - G_N$, as a function of temperature and pressure, the second-order terms give a significant contribution (46). Consider the following reversible, two-state folding/denaturation process:

$$N \text{ (native)} \leftrightarrow D \text{ (denatured)} \qquad (9)$$

Note that this is a simplified representation of equation 6. The pressure and temperature dependence of ΔG is given by

$$d(\Delta G) = -\Delta S \, dT + \Delta V \, dp \qquad (10)$$

Integration, assuming that both ΔS and ΔV are temperature and pressure dependent, gives the following expression:

$$\Delta G(p,T) = \Delta G_0 - \Delta S_0(T - T_0) + \Delta C_p[(T - T_0) - T \ln(T/T_0)] \qquad (11)$$
$$+ \Delta V_0(p - p_0) - (\Delta\beta/2)(p - p_0)^2 + \Delta\alpha \, (T - T_0)(p - p_0)$$

where ΔG_0, ΔV_0, and ΔS_0 refer to the reference conditions p_0 and T_0. In this equation the second-order terms, $\Delta\alpha$, $\Delta\beta$, and ΔC_p, are proportional to the differences in thermal expansion factor, compressibility factor, and heat capacity between the denatured and the native states of the protein, respectively. These parameters are defined as follows:

$$\Delta\alpha = (\partial\Delta V/\partial T)_p = -(\partial\Delta S/\partial p)_T \quad \Delta\beta = -(\partial\Delta V/\partial p)_T \quad \Delta C_p = T(\partial\Delta S/\partial T)_p \quad (12)$$

An important assumption in the derivation of equation 10 is the temperature and pressure independence of $\Delta\alpha$, $\Delta\beta$, and ΔC_p. The curve will have an elliptic shape only if

$$\Delta\alpha^2 > \Delta C_p\Delta\beta/T_0 \qquad (13)$$

This condition implies that $\Delta\beta$ and ΔC_p have opposite signs. The slope of the phase boundary can be deduced as

$$\frac{dT}{dp} = \frac{\Delta V_0 - \Delta\beta(p - p_0) + \Delta\alpha(T - T_0)}{\Delta S_0 - \Delta\alpha(p - p_0) + \Delta C_p[(T - T_0)/T_0]} \qquad (14)$$

Note that if $\Delta\beta$, $\Delta\alpha$, and ΔC_p are zero, this equation is reduced to the classical Clapeyron equation (equation 5). This clearly demonstrates the importance of the second-order terms in the elliptic nature of phase diagrams of proteins.

The main advantage of the Hawley equation (equation 11) over purely empirical equations lies in the fact that all parameters can be given a physical interpretation. However, using this equation one can only obtain the differences in compressibility, thermal expansion, and heat capacity between the denatured and the native states. Therefore, other techniques are required to determine β_T, C_p, and α_p. These thermodynamic quantities are of particular interest, as they can be related to fluctuations in volume, energy, and a cross-correlation of volume and energy, respectively (17). Such fluctuations underlie the dynamic behavior of proteins in water (5).

At the phase boundary the change in Gibbs free energy for denaturation ($\Delta G = G_D - G_N$) is zero. Within the elliptic contour the protein is in the native conformation ($\Delta G > 0$); outside the contour the protein is denatured ($\Delta G < 0$). At the highest pressure (p_{max}) where the native state is stable, the slope of the tangent on the ellipse is zero. At the highest temperature (T_{max}) the slope is infinite. At these points ΔS and ΔV, respectively, are equal to zero, and these can be represented by a straight line (Fig. 2). It can be seen that these lines divide the $\Delta G = 0$ contour in three regions. In the first region (I), where ΔS and ΔV are both negative, an increase in temperature will stabilize the protein against pressure denaturation. From the van't Hoff equation it can be derived that in this region the enthalpy change (ΔH) will be negative. In the second region (II), ΔS is positive and ΔV is negative. Here increasing temperature lowers the denaturation pressure, and vice versa. In the third region (III), ΔH, ΔV, and ΔS are all positive. It can be seen that increasing pressure stabilizes the protein against thermal denaturation (30). Note, however, that the pressure-induced thermal stabilization is not observed for all proteins. A similar observation has been made for the inactivation kinetics of microorganisms (Fig. 1). Moreover, some studies suggest that the sign of dT/dp in region III may also depend on the nature of the conformational transition, i.e., native-to-molten globule, native-to-denatured state, or molten globule-to-denatured state, and may even become negative. The signs of ΔV and ΔS provide a thermodynamic basis for the mechanistic and conformational differences between the pressure and heat denaturation of proteins and rationalize the similarities between the pressure and cold denaturation (29).

Finally, it is of interest to consider the effect of the solvent and the charge of the protein on the position and shape of the phase diagram. Zipp and Kauzmann (56) studied the phase diagram of myoglobin in a wide pH range. They observed that the shape of the diagram decreases at extreme pH values, bringing the cold- and heat-denatured states closer together and lowering the denaturation pressure. The presence of cosolutes such as urea and salts has a similar effect (36). Other factors such as macromolecular crowding have been virtually unexplored with respect to pressure stability. Moreover, food scientists investigating the inactivation of microorganisms in foodstuffs are well aware of the fact that the pressure sensitivity of vegetative bacteria depends on the composition of the food matrix. Van Opstal and coworkers demonstrated that when plotting the decimal reduction time (D) in the p,T plane the inactivation of *Escherichia coli* in carrot juice follows a linear pattern, whereas in HEPES buffer a typical elliptic outline can be observed (51). This example clearly shows that pressure sensitivity strongly depends on the nature of the experimental medium. For the prion protein, amyloid fibrils were found to be pressure sensitive in brain homogenates, but pressure insensitive when purified and dissolved in various buffers (9). Hence, one should be careful when extrapolating data from in vitro (buffer) systems to real-life systems, as the latter involves a large number of unknown factors that are presently beyond our grasp.

Kinetic Aspects of the Phase Diagram

In many instances, the rate of denaturation as a function of pressure and temperature is studied, yielding a p,T,k diagram, where k is the rate constant of inactivation or denaturation. A mathematical analysis of the isokineticity curves yields the activation parameters for the denaturation. The change of the free energy of activation ($\Delta G^{\#}$) as a function of pressure and temperature is now given by

$$d\Delta G^{\#} = \Delta V^{\#}dp - \Delta S^{\#}dT \qquad (15)$$

where $\Delta G^{\#} = -RT \ln k$, being the difference in free energy between the transition state and the native state. According to the transition state theory, the activation volume ($\Delta V^{\#}$) is defined as

$$\Delta V^{\#} = -RT \frac{\partial \ln k}{\partial p} \qquad (16)$$

Similar to the p,T phase diagram, the p,T,k diagram can be divided in three regions based on the signs of $\Delta V^{\#}$, $\Delta H^{\#}$, and $\Delta S^{\#}$. This can easily be understood from the thermodynamic background of the kinetic theory of the transition state. An important aspect that has to be taken into account is the irreversibility of the protein denaturation, which is generally due to protein aggregation. Such a phenomenon can be represented by the following mechanism (16):

$$N \underset{k_2}{\overset{k_1}{\longleftrightarrow}} D \xrightarrow{k_3} I \qquad (17)$$

where N and D are the native and reversibly denatured protein and I is the irreversibly denatured protein. One possible molecular interpretation for such a scheme is given in equation 6. From the viewpoint of the phase diagram, two conditions are worth considering. First, if $k_3 \ll k_1, k_2$, then there exists an equilibrium between N and D, while the transformation of D into I is slow. Under this condition the apparent rate constant, k_{obs}, can be defined as

$$k_{obs} = (k_1/k_2)k_3 = Kk_3 \qquad (18)$$

Thus, the overall rate of the reaction is mainly determined by the formation of I. Second, when $k_3 \gg k_1, k_2$, then all the reversibly denatured molecules will be immediately incorporated into an intermolecular network. In this case the denaturation of N into D is the rate-limiting step. Since the temperature dependence of the rate constants is given by

$$k_3 \approx \exp(-\Delta G^{\#}/RT) \qquad (19)$$

it can be seen that the first condition is most probable at low temperature and high pressure, whereas the second condition will likely take place at high temperature.

LIPIDS: WHY PIEZOPHILES LIKE UNSATURATED LIPIDS

Biological membranes very often are represented as a lamellar phospholipid bilayer matrix. It is, however, well known that many biological lipids can also form nonlamellar liquid-crystalline phases, such as the micellar, hexagonal, and cubic mesophases (7, 53). These structures are thought to be involved in processes such as membrane fusion, endo- and exocytosis, and fat digestion (22, 25).

Lamellar phases can undergo two thermotropic phase transitions, a gel-to-gel pretransition ($L_{\beta'}/P_{\beta'}$) and a gel-to-liquid-crystalline ($P_{\beta'}/L_{\alpha}$) transition, which occurs at a higher temperature. In the fluid-like L_{α} phase the acyl chains of the lipid bilayers are conformationally disordered, whereas in the gel phase ($L_{\beta'}$ or $P_{\beta'}$) the chains are more extended and highly ordered (53). Thus, one can envisage that under pressure (or at a low temperature) the equilibrium will be shifted from the L_{α} phase to the gel phase in order to reduce the volume of the lamellar bilayer, whereas a temperature increase will shift the equilibrium in the opposite direction, i.e., the melting of the lipid bilayer characterized by a transition T_m. For phospholipid bilayers with acyl chains of more than 16 carbon atoms, the formation of additional gel phases, such as the interdigitated gel phase, has been observed under pressure. Depending on the chemical nature of the lipid, high pressure can also induce lamellar-to-nonlamellar transitions. Moreover, in binary lipid mixtures of fatty acids and phosphatidylcholines, nonlamellar phases will be formed, and in phospholipid mixtures intermediate fluid-gel coexistence regions will occur. In the latter case the lamellar bilayer has undergone a two-dimensional phase separation. The rate and mechanism of the phase transition depend on the degree of hydration of the bilayer components involved and on the driving forces of the phase transition.

Lipid phase transitions are the most pressure sensitive of all biological systems, as can be seen from the phase diagram for several (single-component) phospholipid bilayer systems (Fig. 4). Typical values of dT_m/dp are ~22°C/100 MPa for saturated and mono-*cis*-unsaturated lipids. In contrast, the slopes of di-*cis*-unsaturated lipids are smaller (dT_m/dp, ~14°C/100 MPa). The latter can be rationalized from the fact that the *cis*-double bonds

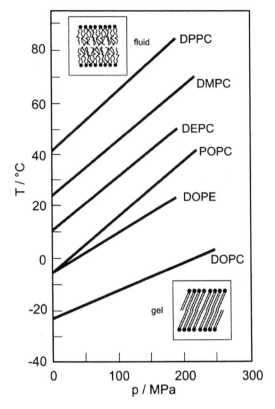

Figure 4. Pressure-temperature phase diagram of different phospholipid bilayer systems in aqueous suspension. Single component bilayers are considered. Note the different slopes of DOPE and DOPC which are both di-*cis* unsatured, compared to the mono-*cis*-unsatured POPC. DPPC, 1,2-dipalmitoyl-*sn*-glycero-3-phosphatidylcholine; DMPC, 1,2-dimyristoyl-*sn*-glycero-3-phosphatidylcholine; DEPC, 1,2-dielaidoyl-*sn*-glycero-3-phosphatidylcholine; POPC, 1-palmitoyl-2-oleoyl-*sn*-glycero-3-phosphatidylcholine; DOPE, 1,2-dioleoyl-*sn*-glycero-3-phosphatidylethanolamine; DOPC, 1,2-dioleoyl-*sn*-glycero-3-phosphatidylcholine. (Adapted from reference 53.)

impose kinks in the linear acyl chains, thereby creating voids that oppose the ordering effect induced by pressure. Note that the T_m of the melting transition depends on the acyl chain length as well as on the chemical nature of the hydrophilic head group, whereas dT_m/dp is not affected significantly by these parameters (53). The T_m of unsaturated lipids is particularly lower than that of saturated lipids, as illustrated in Fig. 4.

Organisms living in the deep sea are exposed to high hydrostatic pressure and low temperature. Among the many adaptations required for survival in such an environment is the maintenance of membrane fluidity in order to allow the transport of essential nutrients and gases (2). From the above it is clear that this can be achieved by regulating the phospholipid composition, i.e., by combining lipids differing in chain length, head group, methyl branching, and chain unsaturation. This is called homeoviscous adaptation (2). Incorporation of proteins and sterols, such as cholesterol, into the membrane also affects its phase behavior. The presence of both can induce two-phase coexistence regions, albeit their effects are concentration dependent.

MEMBRANES: PROTEIN-LIPID INTERACTIONS

In the cell, protein behavior is governed not just by hydration but also by the presence of other macromolecules, i.e., crowding, and small molecules like osmolytes. In addition, many proteins interact with lipid membranes, whether they are membrane bound or not

(7). It has been demonstrated that membranes can influence the folding-denaturation reactions of proteins and change the protein conformation compared to the solution state. Of particular interest is the possible role of lipid membranes in a number of human diseases, such as Creutzfeldt-Jakob disease, light chain amyloidosis, and Alzheimer's disease. Furthermore, it is known that when a protein is inserted into the membrane, the protein can affect the structural organization of the lipid or vice versa (7), but few experimental studies have addressed this issue with respect to changes in high hydrostatic pressure.

The effects of high pressure on membrane proteins have been explored in a number of cases (39). Pressure treatment of *Salmonella enterica* serovar Typhimurium resulted in the loss of a large fraction of the outer membrane proteins. This is presumably due to a weakening of the protein-lipid interactions, thereby causing the release of large integral and peripheral proteins from membranes (53). Others have investigated the effect of pressure (up to 90 MPa) on ion channels and observed that pressure affects the kinetics of these channels, but not their conductance (26). The altered kinetics is assumed to be due to the effect of pressure on the displacement of a non-lipid-interacting section of the channel protein. Hence, there is no direct influence of possible changes in membrane viscosity on this phenomenon. By contrast, in the case of Ca^{2+}-ATPase a correlation between thermotropic transitions in membrane lipids (see below) and changes in enzyme activity has been demonstrated on the basis of high-pressure studies (18). The pressure-induced decrease in activity of an Na^+/K^+-ATPase was found to correlate to a reduced fluidity of the membrane under pressure (11). Moreover, because thermotropic lipid transitions are more sensitive to pressure than protein conformational changes (dT/dp, ~ 17 to $22°C/100$ MPa versus 2 to $5°C/100$ MPa), it was suggested that pressure is a useful tool to discriminate between these two phenomena in complex systems (18). More recently, it was found that gramicidin, a channel-forming pentadecapeptide often used in model studies of protein-lipid interactions, was pressure stable up to 1 GPa. As not many proteins have such a high-pressure resistance, it was suggested that the presence of the lipid bilayer has a stabilizing effect (53). In addition, gramicidin can adopt different conformations of which the equilibrium populations are tightly correlated with the gel-fluid state of the lipid bilayer. Altered protein-lipid interactions under pressure have also been considered as the basis of membrane leakage, in which the action of ion pumps is compromised, and, by consequence, also the viability of the microorganism (34, 38).

GENE EXPRESSION: PROTEIN-NUCLEIC ACID INTERACTIONS

DNA occurs as a double-stranded helix in its native conformation and is dissociated into single-stranded coils by heating. The midpoint of the melting transition T_m (at atmospheric pressure) depends on the base pair composition and the sequence of the DNA, as well as on the salt concentration, indicating that the stability of DNA is intimately related to its hydration (6, 37).

After the observation of an elliptic phase diagram for proteins, Hawley (14) revisited the pressure-temperature stability of nucleic acids in order to investigate whether a similar diagram is associated with DNA (15). He found that dT_m/dp is linear and positive (up to 600 MPa) with slightly steeper slopes at higher salt concentrations. Because the melting enthalpy (ΔH_m) is positive at atmospheric pressure (0.1 MPa), one can derive from the Clapeyron equation that ΔV is positive, indicating that pressure stabilizes the helix conformation of *Clostridium perfringens* DNA (15). This is not unexpected, as base stacking and hydrogen bonds are stabilized by high pressure. By plotting the slope of the coexistence lines as a function of T_m, he observed that the slope changes sign at $T_m \approx 59°C$, implying that below this temperature ΔV would be negative and thus pressure would destabilize the DNA double helix. Note that 59°C is well below the T_m of natural chromosomes under physiological conditions. Indeed, recent work by Dubins and coworkers on synthetic DNA or RNA duplexes revealed that pressure destabilizes the double-stranded conformation at T_m values below approximately 50°C (6); in other words, dT_m/dp becomes negative at high salt concentrations and low temperatures. For instance, the midpoint for the pressure-induced melting of the poly(dA)·poly(rU) DNA/RNA duplex is 60 MPa at 25°C. Such a change in sign of dT_m/dp also has been observed for water-soluble synthetic polymers, depending on the nature and concentration of the added salt (21). The effect of salts has been attributed to changes in water structure and can be related to the Hofmeister series (4, 55). This again underscores the importance of hydration in the processes considered in this chapter. Dubins et al. (6) also calculated the phase diagram in an extended p,T region using equation 11. The calculated phase diagram reveals that as pressure is further increased dT_m/dp changes sign and above 600 MPa dT_m/dp is close to zero (regardless of T_m at 0.1 MPa). Unfortunately, these authors did not explore this pressure range experimentally in order to verify the correctness of their prediction. Also, Hawley and Macleod did not observe any transition in *C. perfringens* DNA under pressures in excess of 900 MPa (15).

Interestingly, when comparing the heat- and pressure-induced helix-to-coil transition it was found that the cooperative length, i.e., the number of base pairs that melt as one unit, of the pressure-induced transition is twofold greater than the one for the heat transition, suggesting that these processes are mechanistically different. Moreover, on the basis of the hypochromicity of several infrared bands and a comparison of thermodynamic variables (α_p, β_T, and ΔV) it was concluded that the pressure-induced single-stranded DNA is more structured than the heat-induced coil, due to the greater amount of stacking at high pressure (37). These findings are reminiscent of the differences in high-pressure versus high-temperature behavior of proteins.

Several cellular processes involving nucleic acids, such as replication, transcription, and recombination, depend on the correct recognition of protein and DNA binding partners. X-ray crystallography of protein-DNA complexes can identify the noncovalent interactions involved in the complex. Electrostatic and hydrophobic interactions are the primary forces involved, and these will be destabilized by pressure (see above). Therefore, pressure can provide information on the stoichiometry and thermodynamic parameters of the association. A typical dissociation constant for BamHI-DNA complex is 4.6 ± 0.4 nM at 50 MPa versus 0.7 ± 0.1 nM at 0.1 MPa, demonstrating a clear destabilization at high pressure. Molecular dynamics studies show that pressure forces water into the protein-DNA complex and that it is sequestered at the intermolecular interface, similar to the effect of pressure on protein oligomers. Moreover, as most DNA-interacting proteins are oligomers, high-hydrostatic-pressure studies can also reveal information on the effect of DNA binding on their stability. Depending on the protein involved, DNA has been found both to stabilize and to destabilize protein oligomers (44). For instance, the tetrameric LacI repressor protein is stabilized by the inducer but destabilized by DNA (41). In contrast, the dimeric LexA repressor, involved in the regulation of the transcription of the SOS system in *E. coli,* is stabilized upon DNA binding (33).

OUTLOOK

The comparison between the effects of temperature and pressure on biological macromolecules and microorganisms has revealed some interesting similarities. The similarities between the shapes of the protein stability diagrams and the viability diagrams of microorganisms (Fig. 1 and 2) strongly suggest that proteins are the main targets of temperature and pressure. There are, however, differences that are worth noting. Whereas the temperatures for the inactivation of microorganisms and the denaturation of proteins are in the same range, the influence of pressure is quite different. Since pressure influences protein-protein interactions more strongly than protein stability, it is natural to conclude that pressure affects these interactions in the first place. However, the alternative explanation, that viability is related to maintaining a functional membrane, cannot be ruled out at present and remains an important topic for further investigations (34, 38). Other important differences are related to the solvent conditions. Although one can now describe the effect of a number of cosolvents on the denaturation of proteins by temperature and pressure, it seems much more difficult to understand the medium effects on the viability of microorganisms (51). Since the metabolic response and gene expression, in short, the piezophysiology (1,8) of microorganisms, is a much more complex issue, our understanding of the effects of the medium composition is far from optimal.

Mapping structural features of biomolecules in the pressure-temperature plane is an important research topic for the molecular biologist to see which state of biomolecules is physiologically relevant. But before we can understand the external conditions (pressure, temperature, and chemical composition of the medium) that allowed the evolution of life, we should understand in much greater detail metabolic and genetic responses of microorganisms.

REFERENCES

1. **Abe, F.** 2004. Piezophysiology of yeast: occurrence and significance. *Cell. Mol. Biol.* **50:**437–445.
2. **Bartlett, D. H.** 2002. Pressure effects on in vivo microbial processes. *Biochim. Biophys. Acta* **1595:**367–381.
3. **Bridgman, P. W.** 1914. The coagulation of albumen by pressure. *J. Biol. Chem.* **19:**511–512.
4. **Cacace, M. G., E. M. Landau, and J. J. Ramsden.** 1997. The Hofmeister series: salt and solvent effects on interfacial phenomena. *Q. Rev. Biophys.* **30:**241–277.
5. **Chalikian, T. V.** 2003. Volumetric properties of proteins. *Annu. Rev. Biophys. Biomol. Struct.* **32:**207–235.
6. **Dubins, D. N., A. Lee, R. B. Macgregor, Jr., and T. V. Chalikian.** 2001. On the stability of double stranded nucleic acids. *J. Am. Chem. Soc.* **123:**9254–9259.
7. **Epand, R. M.** 1998. Lipid polymorphism and protein-lipid interactions. *Biochim. Biophys. Acta* **1376:**353–368.
8. **Fernandes, P. M. B., T. Domitrovic, C. M. Cao, and E. Kurtenbach.** 2004. Genomic expression pattern in *Saccharomyces cerevisiae* cells in response to high hydrostatic pressure. *FEBS Lett.* **556:**153–160.
9. **Fernández García, A., P. Heindl, H. Voigt, M. Büttner, P. Butz, N. Tauber, B. Tauscher, and E. Pfaff.** 2005. Dual nature of the infectious prion protein revealed by high pressure. *J. Biol. Chem.* **280:**9842–9847.
10. **Fujita, Y., and Y. Noda.** 1978. Effect of hydration on the thermal stability of protein as measured by differential scanning calorimetry. *Bull. Chem. Soc. Jpn.* **52:**2349–2352.
11. **Gibbs, A., and G. N. Somero.** 1990. Pressure adaptation of teleost gill Na^+/K^+-ATPase: role of lipid and protein moieties. *J. Comp. Physiol. B* **160:**431–439.
12. **Goossens, K., L. Smeller, J. Frank, and K. Heremans.** 1996. Pressure-tuning the conformation of bovine pancreatic trypsin inhibitor studied by Fourier-transform infrared spectroscopy. *Eur. J. Biochem.* **236:**254–262.
13. **Hashizume, C., K. Kimura, and R. Hayashi.** 1995. Kinetic analysis of yeast inactivation by high pressure treatment at low temperatures. *Biosci. Biotech. Biochem.* **59:**1455–1458.

14. **Hawley, S. A.** 1971. Reversible pressure-temperature unfolding of chymotrypsinogen. *Biochemistry* **10:** 2436–2442.

15. **Hawley, S. A., and R. M. Macleod.** 1974. Pressure-temperature stability of DNA in neutral salt solutions. *Biopolymers* **13:**1417–1426.

16. **Heremans, K., and L. Smeller.** 1997. Pressure versus temperature behaviour of proteins. *Eur. J. Solid State Inorg. Chem.* **34:**745–758.

17. **Heremans, K., and L. Smeller.** 1998. Protein structure and dynamics at high pressure. *Biochim. Biophys. Acta* **1386:**353–370.

18. **Heremans, K., and F. Wuytack.** 1980. Pressure effect on the Arrhenius discontinuity in Ca2+-ATPase from sarcoplasmic reticulum. *FEBS Lett.* **117:**161–163.

19. **Hummer, G., S. Garde, A. E. Garcia, M. E. Paulaitis, and L. R. Pratt.** 1998. The pressure dependence of hydrophobic interactions is consistent with the observed pressure unfolding of proteins. *Proc. Natl. Acad. Sci. USA* **95:**1552–1555.

20. **Ishii, A., T. Sato, M. Wachi, K. Nagai, and C. Kato.** 2004. Effect of high hydrostatic pressure on bacterial cytoskeleton FtsZ polymers in vivo and in vitro. *Microbiology* **150:**1965–1972.

21. **Kunugi, S., Y. Yamazaki, K. Takano, N. Tanaka, and M. Akashi.** 1999. Effects of ionic additives and ionic comonomers on the temperature and pressure responsive behavior of thermoresponsive polymers in aqueous solutions. *Langmuir* **15:**4056–4061.

22. **Landh, T.** 1995. From entangled membranes to eclectic morphologies: cubic membranes as subcellular space organizers. *FEBS Lett.* **369:**13–17.

23. **Lindquist, S., S. Krobitsch, L. Li, and N. Sondheimer.** 2001. Investigating protein conformation-based inheritance and disease in yeast. *Philos. Trans. R. Soc. Lond. B* **356:**169–176.

24. **Ludwig, H., W. Scigalla, and B. Sojka.** 1996. Pressure- and temperature-induced inactivation of microorganisms, p. 346–363. *In* J. L. Markley, C. Royer, and D. Northrup (ed.), *High Pressure Effects in Molecular Biophysics and Enzymology.* Oxford University Press, Oxford, United Kingdom.

25. **Luzzati, V.** 1997. Biological significance of lipid polymorphism: the cubic phases. *Curr. Opin. Struct. Biol.* **7:**661–668.

26. **Macdonald, A. G.** 2002. Ion channels under high pressure. *Comp. Biochem. Physiol. A* **131:**587–593.

27. **Meersman, F., and C. M. Dobson.** 2006. Probing the pressure-temperature stability of amyloid fibrils provides new insights into their molecular properties. *Biochim. Biophys. Acta* **1764:**452–460.

28. **Meersman, F., C. M. Dobson, and K. Heremans.** 2006. Protein unfolding, amyloid fibril formation and configurational energy landscapes under high pressure conditions. *Chem. Soc. Rev.* **35:**908–917.

29. **Meersman, F., L. Smeller, and K. Heremans.** 2002. Comparative Fourier transform infrared spectroscopy study of cold-, pressure-, and heat-induced unfolding and aggregation of myoglobin. *Biophys. J.* **82:**2635–2644.

30. **Meersman, F., L. Smeller, and K. Heremans.** 2005. Extending the pressure-temperature state diagram of myoglobin. *Helv. Chim. Acta* **88:**546–556.

31. **Meersman, F., L. Smeller, and K. Heremans.** 2006. Protein stability and dynamics in the pressure-temperature plane. *Biochim. Biophys. Acta* **1764:**346–354.

32. **Meersman, F., J. Wang, Y. Wu, and K. Heremans.** 2005. Pressure effect on the hydration properties of poly(*N*-isopropylacrylamide) in aqueous solution studied by FTIR spectroscopy. *Macromolecules* **38:**8923–8928.

33. **Mohana-Borges, R., A. B. F. Pacheco, F. J. R. Sousa, D. Foguel, D. F. Almeida, and J. L. Silva.** 2000. LexA repressor forms stable dimers in solution—the role of specific DNA in tightening protein-protein interactions. *J. Biol. Chem.* **275:**4708–4712.

34. **Molina-Höppner, A., W. Doster, R. F. Vogel, and M. G. Gänzle.** 2004. Protective effect of sucrose and sodium chloride for *Lactococcus lactis* during sublethal and lethal high-pressure treatments. *Appl. Environ. Microbiol.* **70:**2013–2020.

35. **Oliveira, A. C., L. P. Gaspar, A. T. Da Poian, and J. L. Silva.** 1994. Arc repressor will not denature under pressure in the absence of water. *J. Mol. Biol.* **240:**184–187.

36. **Randolph, T. W., M. Seefeldt, and J. F. Carpenter.** 2002. High hydrostatic pressure as a tool to study protein aggregation and amyloidosis. *Biochim. Biophys. Acta* **1595:**224–234.

37. **Rayan, G., and R. B. Macgregor, Jr.** 2005. Comparison of the heat- and pressure-induced helix-coil transition of two DNA copolymers. *J. Phys. Chem. B* **109:**15558–15565.

38. **Reyns, K. M. F. A., C. C. F. Soontjens, K. Cornelis, C. A. Weemaes, M. E. Hendrickx, and C. W. Michiels.** 2000. Kinetic analysis and modelling of combined high-pressure-temperature inactivation of the yeast *Zygosaccharomyces bailii.* *Int. J. Food Microbiol.* **56:**199–210.

39. Ritz, M., M. Freulet, N. Orange, and M. Federighi. 2000. Effects of high hydrostatic pressure on membrane proteins of *Salmonella typhimurium*. *Int. J. Food Microbiol.* **55:**115–119.

40. Royer, C. A. 2002. Revisiting volume changes in pressure-induced protein unfolding. *Biochim. Biophys. Acta* **1595:**201–209.

41. Royer, C. A., A. E. Chakerian, and K. S. Matthews. 1990. Macromolecular binding equilibria in the lac repressor system: studies using high-pressure fluorescence spectroscopy. *Biochemistry* **29:**4959–4966.

42. Rubens, P., and K. Heremans. 2000. Pressure-temperature gelatinization phase diagram of starch: an in situ Fourier transform infrared study. *Biopolymers* **54:**524–530.

43. Seki, K., and M. Toyoshima. 1998. Preserving tardigrades under pressure. *Nature* **395:**853–854.

44. Silva, J. L., A. C. Oliveira, A. M. O. Gomes, L. M. T. R. Lima, R. Mohana-Borges, A. B. F. Pacheco, and D. Foguel. 2002. Pressure induces folding intermediates that are crucial for protein-DNA recognition and virus assembly. *Biochim. Biophys. Acta* **1595:**250–265.

45. Silva, J. L., and G. Weber. 1993. Pressure stability of proteins. *Annu. Rev. Phys. Chem.* **44:**89–113.

46. Smeller, L. 2002. Pressure-temperature phase diagrams of biomolecules. *Biochim. Biophys. Acta* **1595:**11–29.

47. Sojka, B., and H. Ludwig. 1997. Effects of rapid pressure changes on the inactivation of *Bacillus subtilis* spores. *Pharm. Ind.* **59:**436–438.

48. Suzuki, K. 1960. Studies on the kinetics of protein denaturation under high pressure. *Rev. Phys. Chem. Jpn.* **29:**49–56.

49. Torrent, J., M. T. Alvarez-Martinez, F. Heitz, J.-P. Liautard, C. Balny, and R. Lange. 2003. Alternative prion structural changes revealed by high pressure. *Biochemistry* **42:**1318–1325.

50. Torrent, J., J. P. Connelly, M. G. Coll, M. Ribó, R. Lange, and M. Vilanova. 1999. Pressure versus heat-induced unfolding of ribonuclease A: the case of hydrophobic interactions within a chain-folding initiation site. *Biochemistry* **38:**15952–15961.

51. Van Opstal, I., S. C. M. Vanmuysen, E. Y. Wuytack, B. Masschalck, and C. W. Michiels. 2005. Inactivation of *Escherichia coli* by high hydrostatic pressure at different temperatures in buffer and carrot juice. *Int. J. Food Microbiol.* **98:**179–191.

52. Wilson, R. G., Jr., J. E. Trogadis, S. Zimmerman, and A. M. Zimmerman. 2001. Hydrostatic pressure induced changes in the cytoarchitecture of pheochromocytoma (PC-12) cells. *Cell Biol. Int.* **25:**649–665.

53. Winter, R. 2002. Synchrotron X-ray and neutron small-angle scattering of lyotropic lipid mesophases, model biomembranes and proteins in solution at high pressure. *Biochim. Biophys. Acta* **1595:**160–184.

54. Wroblowski, B., J. F. Diaz, K. Heremans, and Y. Engelborghs. 1996. Molecular mechanisms of pressure induced conformational changes in BPTI. *Proteins* **25:**446–455.

55. Zhang, Y., and P. S. Cremer. 2006. Interactions between macromolecules and ions: the Hofmeister series. *Curr. Opin. Chem. Biol.* **10:**658–663.

56. Zipp, A., and W. Kauzmann. 1973. Pressure unfolding of metmyoglobin. *Biochemistry* **12:**4217–4228.

High-Pressure Microbiology
Edited by C. Michiels, D. H. Bartlett, and A. Aertsen
© 2008 ASM Press, Washington, DC

Chapter 2

Effects of Hydrostatic Pressure on Viruses

Andréa C. Oliveira, Andre M. O. Gomes, Sheila M. B. Lima,
Rafael B. Gonçalves, Waleska D. Schwarcz, Ana Cristina B. Silva,
Juliana R. Cortines, and Jerson L. Silva

Virions have evolved to move their genome between cells of a susceptible host and between hosts. The virus particle is composed of either a membrane enveloped or nonenveloped protein shell and nucleic acid (20). Evolution has tailored the protein shell to assume multiple functions, including shielding of the nucleic acid, participation in chemical reactions for particle maturation, and ability to penetrate into the cell and undergo disassembly. The coat proteins are mostly arranged in a shell with an icosahedral shape. To pack an infectious genome, integral multiples of 60 subunits are required to form the shell, resulting in nonidentical contacts between subunits (20).

Hydrostatic pressure has been used to study the thermodynamics and dynamics of protein folding and protein-protein interactions (26, 37). One of the main advances obtained from using high pressure is the stabilization of folding intermediates such as molten-globule conformations, thus providing a unique opportunity for characterizing their structure and dynamics. Because the structure of a virus is highly dependent on protein-protein interactions, the use of pressure opens a large number of experimental assessments of the biology of viruses. For example, pressure has permitted study of how the plasticity required for successful assembly of a virus particle is coded into the native conformation of a capsid protein subunit (37). With the aid of high pressure, combined thermodynamic and structural approaches have been used to try to identify the general rules that govern virus assembly. In general, it has been found that the capsid coat proteins (monomers or dimers) are much less stable against pressure than the assembled icosahedral particles (37). The

---

Andréa C. Oliveira, Andre M. O. Gomes, Ana Cristina B. Silva, Juliana R. Cortines, and Jerson L. Silva
• Programa de Biologia Estrutural, Instituto de Bioquímica Médica and Centro Nacional de Ressonância Magnética Nuclear Jiri Jonas, Universidade Federal do Rio de Janeiro, 21941-590, Rio de Janeiro, RJ, Brazil. *Sheila M. B. Lima* • Fundação Oswaldo Cruz, Instituto de Tecnologia em Imunobiológicos, Programa de Vacinas Virais, Rio de Janeiro, RJ 21045-900, Brazil. *Rafael B. Gonçalves and Waleska D. Schwarcz* • Centro de Pesquisa em Ciência e Tecnologia do Leite, Universidade Norte do Paraná, Fazenda Experimental da UNOPAR, 86125-000, Tamarana, PR, Brazil.

isolated capsid and the assembly intermediates assume different partially folded states in the assembly pathway (8, 9, 15, 29, 32). Single-amino-acid substitutions in the hydrophobic core of the coat protein of icosahedral viruses produce large decreases in stability against pressure and chemical denaturants. High pressure has permitted us to trap a ribonucleoprotein intermediate (a potential target for antiviral drugs), where the coat protein is partially unfolded but bound to RNA (8, 9, 15, 29, 32). Intermediate states also appear in the dissociation of empty capsids, such as P22 procapsids (10, 32).

In several animal viruses, hydrostatic pressure causes inactivation with maintenance of immunogenicity (19, 29, 38). In picornaviruses, pressure-inactivated particles resemble the A-particle detected in poliovirus and rhinovirus upon interaction with cells, which lacks the internal capsid protein VP4 (29). The A-particle is substantially less infectious than natural virions and is often identified as an intermediate in uncoating. The "fusion intermediate states" have been found in many nonenveloped and enveloped viruses. The substantial evidence that high pressure traps viruses in the fusion intermediate states (found with alphaviruses, influenza viruses, retroviruses, etc.), not infectious but highly immunogenic, is promising for vaccine development (19, 29, 38).

In the subsequent sessions, we describe in detail how high pressure has been used to tackle basic and applied problems in virus biology.

EFFECTS OF PRESSURE ON THE ASSEMBLY OF PLANT VIRUSES

The contribution of protein folding and protein-nucleic acid interactions to virus assembly has been evaluated in many bacterial, plant, and animal viruses (8–10, 15, 23, 29, 32, 40). In the plant virus cowpea mosaic virus, the free-energy linkage was examined by comparing the pressure stabilities of empty capsids and ribonucleoprotein particles. As expected, empty capsids were much less stable than the ribonucleoprotein particles (8). For several viruses, we find that the isolated capsid proteins and the assembly intermediates are not fully folded, and that association of 60 or more subunits into an icosahedral particle is coupled to progressive folding of the coat protein and also to changes in interactions with the nucleic acid (37). Several viruses undergo cold denaturation under pressure, suggesting the entropic stabilization of the particles, which can be explained by the formation of protein-protein contacts at the expense of releasing molecules of water (9, 15, 29, 32).

Viruses with helical structure have also been studied by high pressure. The dissociation promoted by hydrostatic pressure of tobacco mosaic virus was first reported by Lauffer and Dow (22). More recently these studies were deepened by Bonafe et al. (2), who performed cold-denaturation studies under pressure.

A ribonucleoprotein intermediate has been detected in the pressure dissociation of several viruses (8, 9, 15, 29, 32), as shown in Color Plate 1 for cowpea severe mosaic virus. It has been postulated that the ribonucleoprotein intermediate would serve as a core for ready regeneration of the particle when the pressure is reduced. An unusual feature of this ribonucleoprotein intermediate is the presence of partially unfolded coat proteins bound to the RNA. RNA apparently plays a chaperone-like role during assembly of the capsid. In the absence of RNA the subunits drift to a disorganized structure and cannot renature when the perturbation is withdrawn. The ribonucleoprotein intermediate seems to have a condensed structure (15), which demonstrates the high degree of plasticity of the coat protein-RNA complex.

Another common feature found in the pressure studies of different viruses is a progressive decrease in folding structure in moving from assembled capsids to ribonucleoprotein intermediates (in the case of RNA viruses), free dissociated units (dimers or monomers), and finally unfolded monomers. A gradient of molten-globule states between the fully structured coat protein in the capsid and the unfolded monomers has been proposed (Color Plate 1) (8, 9, 12, 15, 32). High pressure would affect primarily the quaternary and tertiary structure of the capsid protein, leading to partially unfolded, molten-globule conformations. In contrast, high urea concentrations would primarily disrupt the secondary structure.

PRESSURE EFFECTS ON BACTERIOPHAGES

Correct assembly of macromolecular complexes such as viruses requires specific protein-protein and protein-nucleic acid interactions. Bacteriophage MS2 has been successfully used by us as a model for the study of such interactions. MS2 is a member of a large group of small RNA phages that infect *Escherichia coli* (11). Its icosahedral shell consists of 180 copies of coat protein (M_r, 13, 728) arranged in a T=3 quasiequivalent surface lattice surrounding the single-stranded RNA genome. T is the triangulation number or the number of protein subunits in each asymmetric subunit of the icosahedron. Since an icosahedron has 60 asymmetric subunits, a T=3 virus capsid presents 180 protein subunits. Each virion also contains one copy of the maturase (or A) protein, responsible for attachment of the virus to *E. coli* through the F-pilus. Coat protein folds as a dimer of identical subunits and consists of a 10-stranded antiparallel β-sheet facing the interior of the phage particle, with antiparallel, interdigitating α-helical segments on the virus' outer surface.

In recent years, we have investigated how the packing and stability of virus capsids are sensitive to single-amino-acid substitutions in the coat protein. Tryptophan fluorescence, bis-8-anilinonaphthalene-1-sulfonate fluorescence, circular dichroism, and light scattering were employed to measure urea- and pressure-induced effects on MS2 bacteriophage and temperature-sensitive mutants (24). M88V and T45S mutant particles were less stable than the wild-type (WT) forms and completely dissociated at 3.0×10^8 Pa of pressure. M88V and T45S mutants also had lower stability in the presence of urea. The results led to the conclusion that the lower stability of M88V particles is related to an increase in the cavity of the hydrophobic core (24).

Bis-8-anilinonaphthalene-1-sulfonate fluorescence increased for the pressure-dissociated mutants but not for the urea-denatured samples, indicating that the final products were different. The phage titer was dramatically reduced when particles were treated with a high concentration of urea. In contrast, the phage titer recovered after high-pressure treatment. Thus, after pressure-induced dissociation of the virus, information for correct reassembly was preserved. In contrast to M88V and T45S, the D11N mutant virus particle was more stable than the WT virus, despite also possessing a temperature-sensitive growth phenotype.

Overall, our data show how point substitutions in the capsid protein, which affect either the packing or the interaction at the protein-RNA interface, result in changes in virus stability. In this study it was possible to identify cavity-induced mutations, as, for example, M88V (24). M88V showed a large decrease in stability when exposed to conditions of high pressure and high concentrations of urea. This was explained by the large potential of the substitution to create a cavity, as well as to sterically interfere with side chain packing in the protein's interior. Using the program VMD (18) and a probe radius of 1.4 Å, we analyzed the pdb coordinates of WT bacteriophage (17) and the M88V mutant (by substitution using the same program) for the existence of internal cavities in this region in the structure. The molecular representation in Fig. 1 shows the presence of a significant cavity in the WT phage structure in the neighboring Met88 residue. After substitution of this residue with valine, the cavity volume increased to 43 Å^3, reflecting a reduction in the surface area of the residue (59 Å^2) and in the interactions occurring there (Fig. 1).

In a second study, to better investigate the roles of protein-protein and protein-nucleic acid interactions in virus assembly we compared the stabilities of native bacteriophage MS2, of virus-like particles (VLPs) containing nonviral RNAs, and of an assembly-defective coat protein mutant (dlFG) and its single-chain variant (sc-dlFG) (25). Again, physical (high pressure) and chemical (urea and guanidine hydrochloride) agents were used to promote virus disassembly and protein denaturation. In contrast with the effects of chemical agents, we found that high pressure was able to promote only small changes in the spectral center of mass of the MS2 virus. VLPs were actually more stable than the authentic virus, suggesting the possibility of additional contacts between coat protein subunits and the heterologous RNA.

The dlFG mutant dimerizes correctly. However, it fails to assemble into capsid particles because of the absence of the FG loop (15 amino acid residues) implicated in interdimer

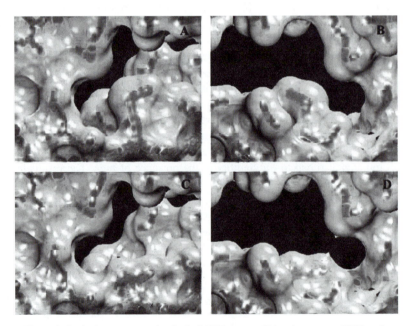

Figure 1. Cavity increase occurring in the M88V mutant. Using the program VMD and a probe radius of 1.4 Å, we analyzed the pdb coordinates of WT bacteriophage and the M88V mutant for the existence of internal cavities in this region. Shown is the region on coat protein in the asymmetric unit of the capsid around residue 88. For the WT phage structure (A), a significant cavity was identified in the neighboring region of the Met88 residue (B). After the replacement of this residue with valine (C), the cavity volume increased to 43 Å^3 (D), reflecting a reduction in the surface area of the residue (59 Å^2) and in the interactions occurring there (from reference 24).

interactions at the viral 5-fold and quasi-6-fold axes. The dlFG mutant was less stable than both the VLPs and the WT virus particles, probably due to the absence of the interdimer protein-protein interactions that normally occur in the capsid. Genetic fusion of the two subunits of the dimer in the single-chain dimer called sc-dlFG stabilized the protein, as did the presence of 34-bp poly(GC) DNA. The use of high pressure permitted us to investigate the different steps involved in phage assembly. The protein-DNA complex formed between dlFG and poly(GC) DNA stabilizes the protein and promotes the complete reversibility of the pressure denaturation process (25). These studies are relevant to the mechanisms by which the interactions in the capsid lattice can be sufficiently stable and specific to ensure assembly, while allowing the structural changes necessary for nucleic acid release during infection.

Using another model of bacteriophage, we had previously demonstrated that the monomeric capsid protein of P22 bacteriophage was very sensitive to pressure and could undergo denaturation at pressures below 2.0×10^8 Pa (32), in contrast to the high resistance to pressure dissociation and denaturation of the assembled T=7 procapsid. P22 procapsid shells dissociated only at high pressures and low temperatures, indicating that they are stabilized by entropy (32). Because the assembly of T=7 double-stranded DNA viruses requires the scaffolding protein, the dissociation was irreversible.

Several single-amino-acid substitution mutants have been utilized to dissect the factors that determine the free-energy stability of the icosahedral lattice of bacteriophage P22 (10, 13). The W48Q mutant shells could be easily dissociated by pressure at room temperature with little dependence on decreasing temperature, suggesting a smaller entropic contribution. In contrast, the lower stabilities found for the G232D and T294I mutants were associated with defective protein cavities (10, 13).

STRUCTURAL TRANSITION IN INSECT AND MAMMALIAN VIRUSES INDUCED BY PRESSURE

To act as infectious particles, viruses have evolved to exert their biological role in a single sequential cycle. First, particles are assembled inside the cells. They then are released to the environment and attach to new host cells. Subsequently, they are disassembled with the release of the genome. Lastly, replication of the genome and transcription of new viral proteins occur. Of these five steps, disassembly of the capsid and unpacking of the nucleic acid are the least understood (20). To be effective, disassembly has to occur fast and at the correct time after endocytosis. The switch for this process is usually attributed to the acidic pH inside the endocytic vesicles, but in vitro many viruses are not uncoated by low pH, or the uncoating occurs slowly, not consistent with the requirement for rapid replication. High pressure provides a powerful tool to explore the uncoating of animal viruses.

The family *Picornaviridae* includes several viruses of great economic and medical importance (34). These viruses have in common a capsid structure composed of 60 copies of four different proteins, VP1 to VP4, and their three-dimensional structures show similar general features (34). Oliveira et al. (29) have described the differences in stability against high pressure and cold denaturation of these viruses. Both poliovirus and rhinovirus are stable against high pressure at room temperature—pressures up to 240 MPa are not enough to promote viral disassembly and inactivation. Within the same pressure range, foot-and-mouth disease virus particles are drastically affected by pressure, with a loss of infectivity of more than 4 log units. The dissociation of polio- and rhinoviruses can be observed only under high pressure at low temperatures in the presence of low concentrations of urea (1 to 2 M). The pressure and low-temperature data reveal clear differences in stability among the three picornaviruses, foot-and-mouth disease virus being the most sensitive, poliovirus being the most resistant, and rhinovirus having intermediate stability (29).

The changes produced by pressure and low temperature on picornaviruses are schematized in Fig. 2. The most important feature is that after a pressure cycle there is reassociation to a noninfectious particle (named P-particle). A more drastic denaturation treatment, such as with high concentrations of urea, results in irreversible unfolding. The scheme also presents a possible explanation for loss of infection that has previously been proposed by other authors to account for heat treatment of some picornaviruses (33, 34), where the defective particle would lose VP4 and/or small molecules (pocket factors) bound to the canyon. The pressure-inactivated picornavirus may resemble the A-particle detected in poliovirus and rhinovirus upon interaction with the host cell, which also lacks the internal capsid protein VP4 (33, 34). The A-particle is substantially less infectious than natural virions and has been considered an intermediate in uncoating. This state is similar to the fusion intermediate state that has been found in enveloped viruses (4, 5, 39). Upon interaction with host cells, the conformations of the coat proteins and envelope glycoproteins have to change, which on one hand leads to noninfectious particles and on the other may lead to the exposure of previously occult epitopes, important for vaccine development. These irreversible conformational changes evoked by high pressure that resemble the changes that occur in vivo are discussed below for most of the viruses we have studied.

We have also used high pressure to evaluate the role of maturation cleavage on flock house virus (FHV) by comparing WT and cleavage-defective mutant (D75N) FHV VLPs (28). FHV is an RNA insect virus; it is nonenveloped and a member of the family *Nodaviridae*. It is composed of a bipartite single-stranded RNA genome packaged in an icosahedral capsid of 180 copies of an identical protein (alpha protein). A fundamental property of many animal viruses is the postassembly maturation required for infectivity. FHV is constructed as a provirion, which matures to an infectious virion by cleavage of alpha protein into beta and gamma subunits.

We found that maturation cleavage leads into a metastable state, in which dissociation of the coat protein is coupled to unfolding. Our results demonstrated that the maturation process targets the particle for an "off-pathway" disassembly, since dissociation is coupled to unfolding. These maturation events are crucial to make the particle infectious. The mature particles were more unstable against pressure than the cleavage-defective particles (Fig. 3), which can be explained by the metastability elicited upon maturation (28). We also have evidence that the gamma subunit is released after a cycle of compression and decompression. A similar situation would occur with picornaviruses, with pressure-induced release of VP4 (Fig. 2). In both cases, pressure induces the formation of a noninfectious particle, apparently a fusion-active state of these nonenveloped viruses.

Figure 3 also shows that the reaction was completely reversible for WT and mutant capsids. The greater sensitivity to pressure of cleaved WT FHV is consistent with the higher dynamics revealed by measurements of fluorescence polarization and nuclear magnetic resonance spectroscopy (28). It is well established that isothermal compressibility is directly related to protein flexibility (due to volume fluctuations) (6).

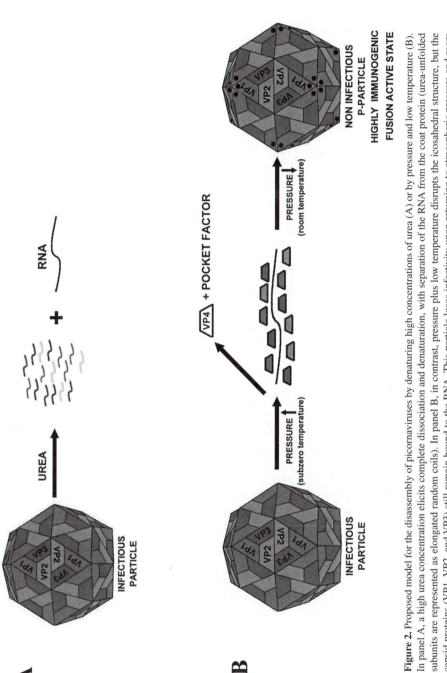

Figure 2. Proposed model for the disassembly of picornaviruses by denaturing high concentrations of urea (A) or by pressure and low temperature (B). In panel A, a high urea concentration elicits complete dissociation and denaturation, with separation of the RNA from the coat protein (urea-unfolded subunits are represented as elongated random coils). In panel B, in contrast, pressure plus low temperature disrupts the icosahedral structure, but the capsid proteins (VP1, VP2, and VP3) still remain bound to the RNA. This particle loses infectivity upon returning to atmospheric pressure and room temperature. This infectivity loss may be due to release of VP4 and the pocket factor (from reference 29).

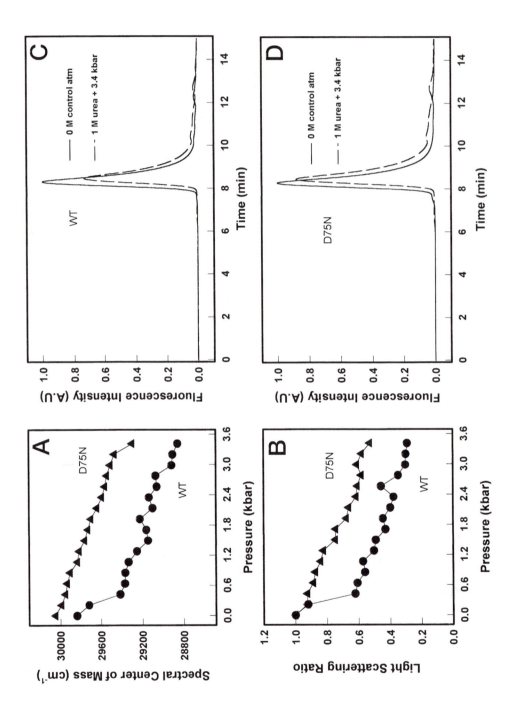

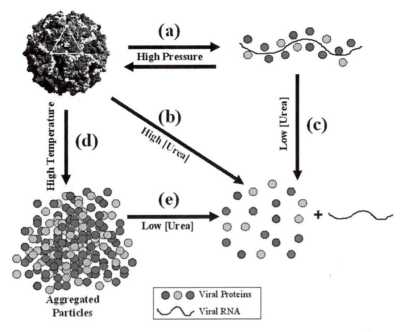

Figure 4. Disassembly/denaturation of FHV by chemical and physical treatments. Shown is the proposed model for the reversible disassembly of FHV by pressure (a) and complete dissociation by high denaturing concentrations of urea (b). (c) Dissociation effect of high pressure and subdenaturing urea concentrations. (d and e) Effect of high temperature without and with subdenaturing (2 M) urea concentration, respectively. The high-temperature treatment induced aggregation of protein subunits. In panels b, c, and e, the treatments led to complete dissociation and denaturation with separation of the RNA from the capsid proteins, whereas high pressure induced disassembly of the particle, with the coat proteins remaining bound to the RNA (from reference 35).

We also investigated the stability of the authentic particles. The data showed that FHV was stable against pressures up to 310 MPa at room temperature. The fluorescence emission and light scattering values showed small changes that were reversible after decompression (Fig. 4). When we combined pressure and subdenaturing urea concentrations (1 M), the changes were more drastic, suggesting dissociation of the capsid. However, these changes were reversible after pressure release. The complete dissociation of FHV could be observed only at high urea concentrations (10 M). There were no significant changes in emission spectra up to 5 M urea. FHV also was stable when we used temperature treatments (high and low) (Fig. 4). We compared the effects of urea and pressure on FHV WT and cleavage-defective

Figure 3. Pressure sensitivity of WT and cleavage-defective FHV VLPs. The pressure stability of FHV WT (circles) and D75N mutant (triangles) VLPs was analyzed by the shift of intrinsic fluorescence spectra (the spectral center of mass [A]) and particle average-size analysis (light scattering [B]). To better appreciate the effect of pressure, the reaction was poised by adding 1 M urea, a subdissociating concentration. Reassembly was evaluated by size exclusion high-performance liquid chromatography for WT FHV (C) and the mutant after decompression (D) (from reference 28). One bar is equal to 10^5 Pa.

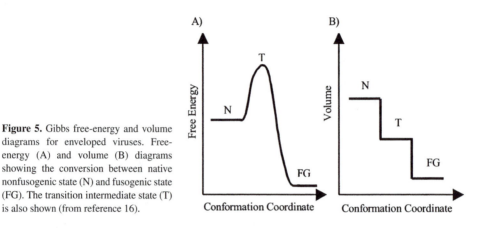

Figure 5. Gibbs free-energy and volume diagrams for enveloped viruses. Free-energy (A) and volume (B) diagrams showing the conversion between native nonfusogenic state (N) and fusogenic state (FG). The transition intermediate state (T) is also shown (from reference 16).

mutant VLPs. The VLPs and authentic particles are distinguishable by protein-RNA interactions, since VLPs pack cellular RNA and native particles contain viral RNA. Our results demonstrated that native particles are more stable than VLPs against physical and chemical treatments. Our data point to the specificity of the interaction between the capsid protein and the viral RNA. This specificity is crucial to the stability of the particle, which makes this interaction an excellent target for drug development (35).

PRESSURE INDUCES THE FUSOGENIC STATE OF MEMBRANE-ENVELOPED VIRUSES

Enveloped animal viruses have to fuse with cellular membranes in order to deliver their genomes into the host cells (4, 5, 39). We have found that high pressure inactivates the enveloped influenza and Sindbis viruses by trapping the particles in the fusion intermediate state (16). In the case of influenza virus and alphaviruses, we demonstrated hydrostatic pressure to trigger a conformational change in the glycoprotein at neutral pHs that is very similar to the change triggered by the low pHs at the endosomes. Our results indicate that pressure rescues the conformation of the metastable, native state (larger volume) into the fusogenic state (smaller volume) by populating a transition-activated state, which although of high energy presents a smaller volume (Fig. 5 and 6). The use of pressure

---

Figure 6. Pressure-induced fusogenic state of Sindbis virus. (A) Schematic representation of pressure-induced membrane fusion activation of Sindbis virus with ghosts. (B) Fusion assay of Sindbis virus with ghosts as measured by means of pyrene excimer/monomer fluorescence ratio intensity. Fusion was measured over a period of 1 h following addition of ghost vesicles to a final concentration of 0.5 mg/ml. □, native virus in the absence of ghosts at pH 7.5; ○, native virus in the presence of ghosts at pH 7.5; ■, pressurized virus in the absence of ghosts at pH 7.5; ▲, pressurized virus in the presence of ghosts at pH 7.5; ●, virus in the presence of ghosts at pH 6.0. (C) Average values ± standard deviations (error bars) after three 1-h experiments for all conditions described for panel (B). Column 1, native virus in the absence of ghosts at pH 7.5; column 2, native virus in the presence of ghosts at pH 7.5; column 3, pressurized virus in the absence of ghosts at pH 7.5; column 4, pressurized virus in the presence of ghosts at pH 7.5; column 5, virus in the presence of ghosts at pH 6.0. The pyrene-labeled lipids in the virus were excited at 330 nm, and the spectra were recorded at wavelengths from 360 to 530 nm. The fluorescence intensities of excimer and monomer emissions were determined at 470 and 389 nm, respectively. The virus concentration was 100 μg/ml (from reference 16).

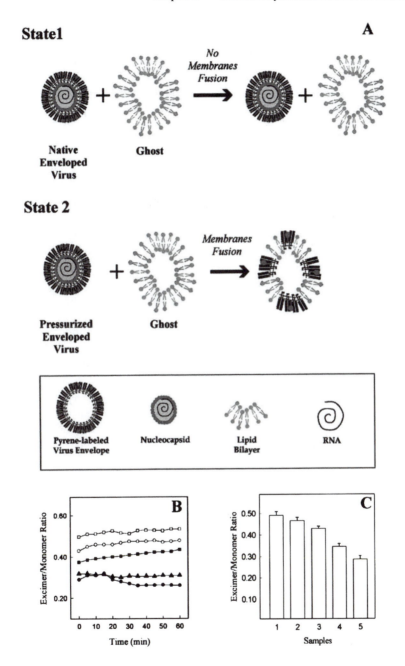

State 1 A

No Membranes Fusion

Native Enveloped Virus   Ghost

State 2

Membranes Fusion

Pressurized Enveloped Virus   Ghost

Pyrene-labeled Virus Envelope   Nucleocapsid   Lipid Bilayer   RNA

B

Excimer/Monomer Ratio

Time (min)

C

Excimer/Monomer Ratio

Samples

to populate fusion-active states can be utilized in the development of new antiviral vaccines and drugs.

The formation of noninfective particles after a cycle of compression and decompression has been demonstrated in many viruses, such as rotaviruses (31), infectious bursal disease virus (IBDV) (40), vesicular stomatitis virus (VSV) (38), simian immunodeficiency virus (21), human immunodeficiency virus (HIV) (27), influenza virus (16), lambda phage (3), alphaviruses (14, 16), and picornaviruses (19, 29). For IBDV, high pressure caused elimination of infectivity, but the virions retained the original immunogenic properties and could elicit high titers of virus neutralizing antibodies (40). In the case of VSV, which is a membrane-enveloped virus, application of a pressure of 250 MPa for sufficient time abolishes infectivity (38). Electron microscopy of the compressed samples shows no detectable dissociation, but the subunits appear displaced from their normal positions by the pressure, as shown by bulges under the virus membrane. With the membrane-enveloped viruses, the membrane glycoprotein undergoes a conformational change induced by pressure similar to the receptor-activated conformational change (16). The finding that influenza virus and HIV envelope glycoproteins undergo pH- or receptor-activated conformational changes indicates that their native states are metastable. For influenza virus, this change is usually referred as the spring-loaded model, in which the fusion-peptide region is thought to insert into the target membrane at an early step of the fusion process (4, 5, 39). For influenza virus, low pH elicits the spring-loaded mechanism.

We find that influenza virus subjected to pressure exposes hydrophobic domains as determined by tryptophan fluorescence and by the binding of bis-8-anilino-1-naphthalenesulfonate, a well-established marker of the fusogenic state in influenza virus (16; A. C. B. Silva, P. Souza-Santos, A. Rosenthal, C. L. A. Silva, M. S. Freitas, D. Foguel, A. M. O. Gomes, A. C. Oliveira, and J. L. Silva, submitted for publication). Pressure also produced an increase in the fusion activity at neutral pH as monitored by fluorescence resonance energy transfer using lipid vesicles labeled with fluorescence probes (16; Silva et al., submitted). These data indicate the change of the labile native state of the envelope complex to a more stable one, mimicking the fusion-active conformation.

In the case of HIV, we also found that tubular structures of the capsid protein were susceptible to pressure dissociation (J. R. Cortines, L. M. T. R. Lima, R. Mohana-Borges, J. K. Lanman, P. E. Prevelige, Jr., and J. L. Silva, submitted for publication). HIV type 1 capsid protein has the property of assembling in vitro into hollow cylinders. Pressure at 250 MPa for 2 h at 25°C induces a significant disassembly of the tubular particles and affects the intra- and intersubunit interactions in the capsid protein structure. Fluorescence spectroscopy and circular dichroism were used to study the protein transition conformations and the stability of the tubular particles. The data demonstrated the existence of an irreversible pressure-induced dissociation process. These studies are very important to demonstrate that in addition to affecting the membrane glycoprotein, pressure provokes a direct perturbation of the protein-protein interactions of the HIV capsid.

PRESSURE INACTIVATION OF VIRUSES AND OTHER MICROORGANISMS—A PASCALIZED VACCINE

Virus inactivation by hydrostatic pressure has been evaluated with a view toward two potential applications: vaccine development and virus sterilization (3, 14, 19, 21, 27, 30, 31, 40; Silva et al., submitted). Concerning viral vaccines, there are basically three types of immunization strategies: use of live (attenuated) particles, use of killed (inactivated) whole-virion particles, and use of subunit vaccines. Immunization is the most efficient way of preventing infectious diseases in animals and humans (1). The antibodies against pressurized virus particles were as effective as those against the intact viruses as measured by their neutralization titer in plaque reduction assay (19, 29, 31, 40). Pressure induced the formation of inactive rotavirus (31), the most important viral agent that causes gastroenteritis, responsible for millions of cases of fatal diarrhea in children from developing countries. The new conformation of the pressurized particle did not result in a loss of immunogenicity. Pressure alters the receptor-binding protein VP4 by triggering changes similar to those produced when the virus interacts with target cells (31). As the changes in VP4 conformation caused by pressure occur prior to virus exposure to target cells, it leads to noninfectious particles and may lead to the exposure of previously occult epitopes, important for vaccine development.

Recently, high pressure has also been used to inactivate leptospires (36), the causative agents of important zoonotic and human diseases. *Leptospira interrogans* serovar Hardjo was subjected to different hydrostatic pressures. Electron microscopy showed dislocation of the outer membrane, partial loss of the helical shape, and extrusion of the axial filament from the cytoplasmic cylinder of the pressurized leptospires. Pressure treatment abolished leptospire infectivity. When the pressure-treated leptospires were inoculated into rabbits, they were highly immunogenic, and they are now being evaluated to formulate a vaccine.

The high titers of the neutralizing antibodies elicited by pressure-inactivated viruses indicate that hydrostatic pressure can be used to prepare whole-virus immunogens. Effective immunization against viruses requires presentation of the whole virus particle to the immune system. The employment of high pressure to prepare antiviral vaccines may have important advantages over other methods such as attenuation, chemical inactivation, or isolated subunit vaccines. Attenuated live viruses can revert after a certain time and may cause the disease that they are intended to prevent, or worsen the real disease. Immunization with isolated subunits has several problems, especially because the immune system recognizes the isolated antigen less effectively than the whole virus. A reliable physical method, such as high pressure, to prepare killed vaccines should not have the same problems. On the other hand, the use of whole virion particles requires a detailed characterization of the kinetics of the inactivation process in order to find the condition under which inactivation is fully obtained (19).

The reason why the pressurized viruses maintain the immunogenic potential probably resides in the fact that the structural changes are very subtle. As we discussed above, pressure treatment seems to mimic the changes that are produced when viruses bind to cellular receptors. We found that pressure-inactivated VSV attaches to the cellular membrane, but it is not internalized by endocytosis (7). The most reasonable explanation is that high pressure leads the envelope protein to the fusion conformation. This would be similar to the inhibition of an enzyme by a transition state analog; the difference here is that a physical tool, pressure, freezes the viral protein in a conformation that cannot bind to the receptor, preventing progression to endocytosis. Therefore, the other steps of the infection cycle are compromised. However, the attachment of inactivated virus, such as VSV, to the host cell (7) in a nonproductive way may be crucial to evoke an immune response, first, because a new epitope might be exposed which leads to efficient production of neutralizing antibodies, and second, because the particle, although not infectious, can be processed by the cell, resulting in a cellular response (CD8$^+$ cytotoxic T lymphocytes).

CONCLUSIONS AND PERSPECTIVES

The reduced stability of virus particles at high pressure and low temperature suggests the potential application to inactivate viruses of medical importance. We have found that several viruses are inactivated by pressure at room temperature. The substantial evidence that high pressure traps viruses in the fusion intermediate states (found with alphaviruses, influenza virus, retroviruses, etc.), not infectious but highly immunogenic, is very promising for vaccine development. Pressure-induced population of fusion-active states can be utilized in the development of new antiviral vaccines and drugs. The viruses stable against pressure at room temperature are inactivated when temperature is decreased under pressure. These studies are important not only as a potential approach to produce antiviral vaccines but also to sterilize biological products, such as blood, plasma, and derivatives.

Acknowledgments. This work was supported by grants from Conselho Nacional de Desenvolvimento Científico e Tecnológico (CNPq), Programa de Núcleos de Excelência (PRONEX), Coordenação de Aperfeiçoamente de Pessoal de Nível Superior (CAPES), Fundação Carlos Chagas Filho de Amparo à Pesquisa no Estado do Rio de Janeiro (FAPERJ), Fundação Universitária José Bonifácio (FUJB), Financiadora de Estudos e Projetos (FINEP), and Instituto Milênio de Biologia Estrutural em Biomedicina e Biotecnologia (IMBEBB).

REFERENCES

1. **Bloom, B. R.** 1996. A perspective on AIDS vaccines. *Science* **272:**1888–1890.
2. **Bonafe, C. F., C. M. Vital, R. C. Telles, M. C. Gonçalves, M. S. Matsuura, F. B. Pessine, D. R. Freitas, and J. Vega.** 1998. Tobacco mosaic virus disassembly by high hydrostatic pressure in combination with urea and low temperature. *Biochemistry* **37:**11097–11105.
3. **Bradley, D. W., R. A. Hess, F. Tao, L. Sciaba-Lentz, A. T. Remaley, J. Laugharn, Jr., and M. Manak.** 2000. Pressure cycling technology: a novel approach to virus inactivation in plasma. *Transfusion* **40:**193–200.
4. **Bullough, P. A., F. M. Hughson, J. J. Skehel, and D. C. Wiley.** 1994. Structure of influenza haemagglutinin at the pH of membrane fusion. *Nature* **371:**37–43.
5. **Carr, C. M., and P. S. Kim.** 1993. A spring-loaded mechanism for the conformational change of influenza hemagglutinin. *Cell* **73:**823–832.
6. **Cooper, A.** 1976. Thermodynamic fluctuations in protein molecules. *Proc. Natl. Acad. Sci. USA* **73:**2740–2741.
7. **Da Poian, A. T., A. M. O. Gomes, R. J. N. Oliveira, and J. L. Silva.** 1996. Migration of vesicular stomatitis virus glycoprotein to the nucleus of infected cells. *Proc. Natl. Acad. Sci. USA* **93:**8268–8273.
8. **Da Poian, A. T., J. E. Johnson, and J. L. Silva.** 1994. Differences in pressure stability of the three components of cowpea mosaic virus: implications for virus assembly and disassembly. *Biochemistry* **33:**8339–8346.
9. **Da Poian, A. T., A. C. Oliveira, and J. L. Silva.** 1995. Cold denaturation of an icosahedral virus. The role of entropy in virus assembly. *Biochemistry* **34:**2672–2677.
10. **De Souza, P. C., Jr., R. Tuma, P. E. Prevelige, J. L. Silva, and D. Foguel.** 1999. Cavity defects in the procapsid of bacteriophage P22 and the mechanism of capsid maturation. *J. Mol. Biol.* **287:**527–538.
11. **Fiers, W.** 1979. Structure and function of RNA bacteriophages, p. 69–204. *In* H. Fraenkel-Conrat and R. R. Wagner (ed.), *Comprehensive Virology.* Plenum, New York, NY.
12. **Foguel, D., and J. L. Silva.** 1994. Cold denaturation of a repressor-operator complex: the role of entropy in protein-DNA recognition. *Proc. Natl. Acad. Sci. USA* **91:**8244–8247.
13. **Foguel, D., C. M. Teschke, P. E. Prevelige, and J. L. Silva.** 1995. Role of entropic interactions in viral capsids: single amino acid substitutions in P22 bacteriophage coat protein resulting in loss of capsid stability. *Biochemistry* **34:**1120–1126.
14. **Freitas, M. S., A. T. Da Poian, O. M. Barth, M. A. Rebello, J. L. Silva, and L. P. Gaspar.** 2006. The fusogenic state of Mayaro virus induced by low pH and by hydrostatic pressure. Hydrostatic pressure induces the fusion-active state of enveloped viruses. *Cell Biochem. Biophys.* **44:**325–335.

15. **Gaspar, L. P., J. E. Johnson, J. L. Silva, and A. T. Da Poian.** 1997. Partially folded states of the capsid protein of cowpea severe mosaic virus in the disassembly pathway. *J. Mol. Biol.* **273:**456–466.

16. **Gaspar, L. P., A. C. B. Silva, A. M. O. Gomes, M. S. Freitas, A. P. Ano Bom, W. D. Schwarcz, J. Mestecky, M. J. Novak, D. Foguel, and J. L. Silva.** 2002. Hydrostatic pressure induces the fusion-active state of enveloped viruses. *J. Biol. Chem.* **277:**8433–8439.

17. **Golmohammadi, R., K. Valegard, K. Fridborg, and L. Liljas.** 1993. The refined structure of bacteriophage MS2 at 2.8 A resolution. *J. Mol. Biol.* **234:**620–639.

18. **Humphrey, W., A. Dalke, and K. Schulten.** 1996. VMD: visual molecular dynamics. *J. Mol. Graph.* **14:**33–38.

19. **Ishimaru, D., D. Sa-Carvalho, and J. L. Silva.** 2004. Pressure-inactivated FMDV: a potential vaccine. *Vaccine* **22:**2334–2339.

20. **Johnson, J. E.** 1996. Functional implications of protein-protein interactions in icosahedral viruses. *Proc. Natl. Acad. Sci. USA* **93:**27–33.

21. **Jurkiewicz, E., M. Villas-Boas, J. L. Silva, G. Weber, G. Hunsmann, and R. M. Clegg.** 1995. Inactivation of simian immunodeficiency virus by hydrostatic pressure. *Proc. Natl. Acad. Sci. USA* **92:**6935–6937.

22. **Lauffer, M. A., and R. B. Dow.** 1941. The denaturation of tobacco mosaic virus at high pressures. *J. Biol. Chem.* **140:**509–518.

23. **Leimkuhler, M., A. Goldbeck, M. D. Lechner, and J. Witz.** 2000. Conformational changes preceding decapsidation of bromegrass mosaic virus under hydrostatic pressure: a small-angle neutron scattering study. *J. Mol. Biol.* **296:**1295–1305.

24. **Lima, S. M. B., D. S. Peabody, J. L. Silva, and A. C. Oliveira.** 2004. Mutations in the hydrophobic core and in the protein-RNA interface affect the packing and stability of icosahedral viruses. *Eur. J. Biochem.* **271:**135–145.

25. **Lima, S. M. B., A. C. Q. Vaz, T. L. F. Souza, D. S. Peabody, J. L. Silva, and A. C. Oliveira.** 2006. Dissecting the role of protein-protein and protein-nucleic acid interactions in MS2 bacteriophage stability. *FEBS J.* **273:**1463–1475.

26. **Mozhaev, V. V., K. Heremans, J. Frank, P. Masson, and C. Balny.** 1996. High pressure effects on protein structure and function. *Proteins* **24:**81–91.

27. **Nakagami, T., H. Ohno, T. Shigehisa, T. Otake, H. Mori, T. Kawahata, M. Morimoto, and N. Ueba.** 1996. Inactivation of human immunodeficiency virus by high hydrostatic pressure. *Transfusion* **36:**475–476.

28. **Oliveira, A. C., A. M. O. Gomes, F. C. L. Almeida, R. Mohana-Borges, A. P. Valente, V. S. Reddy, J. E. Johnson, and J. L. Silva.** 2000. Virus maturation targets the protein capsid to concerted disassembly and unfolding. *J. Biol. Chem.* **275:**16037–16043.

29. **Oliveira, A. C., D. Ishimaru, R. B. Goncalves, T. J. Smith, P. Mason, D. Sa-Carvalho, and J. L. Silva.** 1999. Low temperature and pressure stability of picornaviruses: implications for virus uncoating. *Biophys. J.* **76:**1270–1279.

30. **Oliveira, A. C., A. P. Valente, F. C. L. Almeida, S. M. B. Lima, D. Ishimaru, R. B. Gonçalves, D. Peabody, D. Foguel, and J. L. Silva.** 1999. Hydrostatic pressure as a tool to study virus assembly: inactivated vaccines and antiviral drugs. *NATO ASI Ser. E* **358:**497–513.

31. **Pontes, L., Y. Cordeiro, V. Giongo, M. Villas-Boas, A. Barreto, J. R. Araujo, and J. L. Silva.** 2001. Pressure-induced formation of inactive triple-shelled rotavirus particles is associated with changes in the spike protein Vp4. *J. Mol. Biol.* **307:**1171–1179.

32. **Prevelige, P. E., J. King, and J. L. Silva.** 1994. Pressure denaturation of the bacteriophage P22 coat protein and its entropic stabilization in icosahedral shells. *Biophys. J.* **66:**1631–1641.

33. **Rossmann, M. G.** 1994. The beginnings of structural biology. Recollections, special section in honor of Max Perutz. *Protein Sci.* **3:**1712–1725.

34. **Rueckert, R. R.** 1996. Picornaviridae: the viruses and their replication, p. 609–645. *In* B. N. Fields, D. M. Knipe, and P. M. Howley (ed.), *Fields Virology,* 3rd ed. Lippincott-Raven Publishers, Philadelphia, PA.

35. **Schwarcz, W. D., S. P. C. Barroso, A. M. O. Gomes, J. E. Johnson, A. Schneemann, A. C. Oliveira, and J. L. Silva.** 2004. Virus stability and protein-nucleic acid interaction as studied by high-pressure effects on nodaviruses. *Cell. Mol. Biol.* **50:**419–427.

36. **Silva, C. C., V. Giongo, A. J. Simpson, E. R. Camargos, J. L. Silva, and M. C. Koury.** 2001. Effects of hydrostatic pressure on the *Leptospira interrogans:* high immunogenicity of the pressure-inactivated serovar hardjo. *Vaccine* **19:**1511–1514.

37. **Silva, J. L., D. Foguel, A. T. Da Poian, and P. E. Prevelige.** 1996. The use of hydrostatic pressure as a tool to study viruses and other macromolecular assemblages. *Curr. Opin. Struct. Biol.* **6:**166–175.

38. **Silva, J. L., P. Luan, M. Glaser, E. W. Voss, and G. Weber.** 1992. Effects of hydrostatic pressure on a membrane-enveloped virus: high immunogenicity of the pressure-inactivated virus. *J. Virol.* **66:**2111–2117.

39. **Skehel, J. J., and D. C. Wiley.** 2000. Receptor binding and membrane fusion in virus entry: the influenza hemagglutinin. *Annu. Rev. Biochem.* **69:**531–569.

40. **Tian, S. M., K. C. Ruan, J. F. Qian, G. Q. Shao, and C. Balny.** 2000. Effects of hydrostatic pressure on the structure and biological activity of infectious bursal disease virus. *Eur. J. Biochem.* **267:**4486–4494.

High-Pressure Microbiology
Edited by C. Michiels, D. H. Bartlett, and A. Aertsen
© 2008 ASM Press, Washington, DC

Chapter 3

Effects of High Pressure on Bacterial Spores

Peter Setlow

Bacteria of various *Bacillus* and *Clostridium* species and their close relatives can grow and divide as long as adequate nutrients are available. However, depletion of nutrients in the environment of these organisms initiates the process of sporulation whereby the growing, actively metabolizing cell is converted into a spore (20). The developing spore, often termed the forespore, matures within a surrounding mother cell and is thus termed an endospore. These spores are metabolically dormant, as they catalyze no detectable metabolism of exogenous or endogenous compounds and lack the high-energy compounds such as ATP and reduced pyridine nucleotides found at high levels in growing cells (60). The spores are also extremely resistant to a variety of stress factors, including heat, desiccation, radiation, and chemicals (40, 60, 64). Indeed, spores are one of the most resistant life forms known, and there are reports suggesting that spores can survive for millions of years in their dormant state (11, 31, 70). The growing forms of a variety of sporeformers are found in high levels in soil, as are the spores. Consequently, significant levels of spores and sporeformers are present in many foodstuffs. Growing or stationary-phase cells of a number of sporeformers cause food spoilage as well as food-borne diseases. These diseases include food poisoning caused by *Clostridium perfringens* and *Bacillus cereus,* lethal intoxication caused by the neurotoxin excreted by *Clostridium botulinum,* and anthrax caused by *Bacillus anthracis,* spores of which are a potential bioweapon.

As a consequence of the damage that can be caused by many spore-forming organisms, the food industry has long been concerned with methods to inactivate spores in foods and foodstuffs. High-temperature treatments, whether very high temperatures for short times or high temperatures for long times, are effective in killing spores. However, such treatments almost always have deleterious effects on food quality (see also chapter 10). Consequently, there has been an ongoing search for methods of spore inactivation that will have less damaging effects on foods.

High pressures (50 to 800 MPa) were found to inactivate spores many years ago (26, 40, 56). The killing of spores by pressure is unusual in that very high pressures are often less effective than are lower pressures. This apparent anomaly is because spores are first germinated by high pressures and then killed, and because spore germination is sometimes less

Peter Setlow • Department of Molecular, Microbial and Structural Biology, University of Connecticut Health Center, Farmington, CT 06030-3305.

effective at very high pressures. The initial germination may be essential for efficient spore killing by pressure, since germinated spores are much more sensitive to stress factors, in particular, heat, than are dormant spores (40). Consequently, in order to fully understand the effects of high pressures on spores, it is essential to first understand (i) the structure of dormant spores, (ii) some of the factors that contribute to spore resistance, and (iii) the normal mechanism(s) for germination of bacterial spores, as high-pressure germination uses some components of normal spore germination pathways.

SPORE STRUCTURE, PROPERTIES, AND RESISTANCE

As noted above, spores are metabolically dormant and resistant to a variety of harsh treatments, much more resistant than are growing cells or germinated spores of the same strain. For example, hydrated dormant spores of various strains are generally resistant to ~40°C-higher temperatures than are their corresponding growing cells (71). Consequently, effective spore killing by wet heat requires very high temperatures, especially to ensure the killing of large numbers of spores.

Much of the spores' resistance properties are due to unique aspects of the spore structure. Spores of *Bacillus* and *Clostridium* species have a structure very different from that of growing cells, with a number of layers unique to spores and many spore-specific macromolecules. Most of our detailed knowledge of spore structure and biochemistry has come from studies on spores of *Bacillus* species, in particular, the most well-studied sporeformer, *Bacillus subtilis*. However, generalizations from work with spores of *B. subtilis* and other *Bacillus* species appear to apply to spores of *Clostridium* species.

Exosporium

Starting from the outside in, the initial spore layer is the exosporium (Fig. 1), a structure that is unique to spores. The exosporium is a large balloon-shaped structure on spores of some species, in particular, members of the *B. cereus* group, which includes *B. anthracis, B. cereus,* and *Bacillus thuringiensis* (19). This structure contains a number of spore-specific proteins and glycoproteins (52). However, the function of the exosporium is not

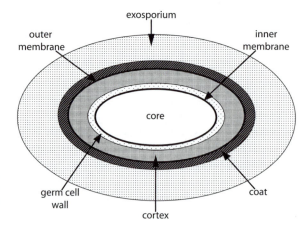

Figure 1. Dormant-spore structure. The various layers of the dormant spore are not drawn to scale. Note that the exosporium is not present in spores of many species.

known, and this structure is not present in spores of many species, including *B. subtilis* (19). The role of the exosporium in spore resistance properties has never been tested explicitly, but this structure seems unlikely to be important in spore resistance.

Coat

The next layer, the coat, is present in all spores and is often composed of a number of layers made up of a large number (>50 in *B. subtilis*) of spore-specific proteins (19, 21). The coat protects spores against lytic enzymes and thus against predation by other organisms (19, 33). The coat also contributes to spore resistance to a number of toxic chemicals (64), and coat-associated proteins play some roles in spore germination (2, 5, 12, 19). However, the coat plays no significant role in spore resistance to heat or high pressure (24, 39, 42). The coat has been suggested to protect the spore from mechanical disruption, but much coat protein can be removed without any loss in spore resistance to disruption (30). During spore germination and outgrowth the coat is at least partially degraded and is eventually shed, but the mechanisms for these events are not known (19).

Outer Membrane

The spore's outer membrane lies underneath the coat. While this membrane is essential in spore formation, it may not be a complete functional membrane in the dormant spore. This membrane, as well as much spore coat protein, can be removed with no effect on spore resistance to high pressure (42).

Cortex

Underlying the outer membrane is the spore cortex, composed almost exclusively of peptidoglycan (PG) (47). Where studied, cortical PG has the repeating disaccharide unit of *N*-acetylglucosamine and muramic acid found in vegetative cell PG. However, a large fraction (~50% in *B. subtilis* spores) of the muramic acid in cortical PG is present as muramic δ-lactam (MAL), a derivative not found in vegetative PG. In addition, ~25% of the cortical muramic acid is attached to only a single alanine residue, again a derivative unique to cortex PG, with similar amounts linked to the usual pentapeptide found in vegetative PG. Invariably, this pentapeptide contains diaminopimelic acid, even if vegetative PG contains lysine. The cortex is synthesized late in sporulation using enzymes present in both the developing spore and surrounding mother cell. This structure appears to be crucial in establishing and maintaining the low water content in the spore core (see below) that is essential for spore dormancy and much spore resistance. The cortex is degraded in the first minutes of spore germination by one of several cortex lytic enzymes (CLEs) that are unique to spores and which target cortex PG exclusively because of the CLEs' specificity for PG containing MAL (48, 63). Cortical PG is also degraded by lysozyme (48).

Germ Cell Wall

The next spore layer is the germ cell wall, again composed of PG, but in this case with a structure likely identical to that of vegetative PG (47). The germ cell wall is not degraded by CLEs, as germ cell wall PG lacks MAL (47, 63). As a consequence, the germ cell wall becomes the cell wall of the germinated and outgrowing spore (63).

Inner Membrane

Underlying the germ cell wall is the spore's inner membrane. This membrane has a phospholipid and fatty acid composition similar to that of the growing cell's plasma membrane (7, 13, 15). However, the inner membrane has a number of unique properties. First, the inner membrane is a major permeability barrier restricting access of small molecules to the spore's central region or core and preventing the loss of small molecules present in the core (15). Molecules as small as methylamine and likely even water cross the inner membrane only slowly, and this membrane's low passive permeability appears to be a significant factor in spore resistance to many DNA-damaging chemicals (15, 72). The reason for this membrane's extremely low permeability is not known. Second, lipid probes in the inner membrane are immobile, as determined by measurements of their fluorescence redistribution after photobleaching (17). However, such probes become mobile when the spore germinates fully and the inner membrane encompassed volume increases ~2-fold. This resultant increase of 1.3- to 1.4-fold in inner membrane surface area takes place without new membrane synthesis or even ATP production (17). Given the importance of the inner membrane in spore germination, our lack of knowledge of the structure of this membrane is a major deficiency.

In addition to its unique properties, the inner membrane has a number of unique proteins. These include the receptors for the nutrient germinants that trigger spore germination, as well as proteins, such as those encoded by the *spoVA* operon, that may be involved in movement of spore core molecules (see below) across this membrane (3, 63, 68, 69). These proteins must function in the inner membrane and are likely immobile in this membrane.

Core

The final spore layer is the core, akin to a growing cell's protoplast and containing the spore's DNA, protein synthetic machinery, and most enzymes. A major feature of the core's environment is a low level of water (24). While growing cells contain ~80% of their wet weight as water, the core's wet weight is only 25 to 55% water depending on the species. The low core water content plays a major role in spore resistance to wet heat, as spores with the lowest core water content are the most wet heat resistant. However, there is no information on the role, if any, of core water content in spore pressure resistance. There is also no knowledge of the levels of free water in the core.

The low core water content is undoubtedly also the major reason for the spore's metabolic dormancy, as enzymes in the core are inactive. Indeed, at least one protein that is mobile in growing cells and fully germinated spores is immobile in the spore core (16). Major contributors to the spore core's low water content include the cortex, which acts as a straitjacket restricting core swelling, and the core's large pool of pyridine-2,6-dicarboxylic acid (dipicolinic acid [DPA]) (Fig. 2). DPA is synthesized in the mother cell during sporulation and taken up into the core of the developing forespore to very high levels—up to 25% of spore core dry weight (24). The DPA taken up replaces core water, and DPA-less spores of *B. subtilis* have ~25% more core water than do DPA-replete spores (41, 43). As a consequence of the reciprocal relationship between DPA and core water content, DPA-less spores are much more heat sensitive than are their DPA-replete brethren (43).

The DPA in spores is predominantly, if not exclusively, in a 1:1 chelate with divalent metal ions, predominantly Ca^{2+}. As a consequence, the spore core is highly mineralized. It has been suggested that the core is in a glass-like state, but this has not been proven (1, 35, 65). The

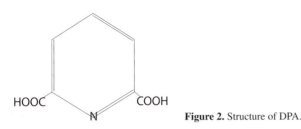

Figure 2. Structure of DPA.

core's depot of DPA is excreted in the first minutes of germination. The means and regulation of DPA excretion, as well as DPA uptake in sporulation, are not known. The SpoVA proteins are likely involved in DPA uptake in sporulation, and perhaps in DPA efflux in germination (24, 66–69). However, the roles of these proteins in DPA movement have not been firmly established, and there is no knowledge of how they might operate or be regulated.

Other unique molecules in the core are a group of small, acid-soluble proteins (SASP) (21, 61). The SASP are synthesized late in sporulation in the developing spore and together comprise ≥10% of total spore protein. Most SASP are of the α/β-type, which binds to and protects spore DNA from damage caused by UV radiation, chemicals, and desiccation. The binding of these proteins to DNA alters DNA structure, and this change as well as the large amount of DPA in the core alter spore DNA photochemistry dramatically (18). The last change is the major reason for spore resistance to UV radiation (62). All SASP are degraded to amino acids following resumption of enzyme activity in the spore core upon completion of spore germination, with this degradation initiated by a novel protease.

In addition to the novel molecules found in the core, the core lacks a number of small molecules present at high levels in the protoplast of growing cells. These include ATP and other nucleoside triphosphates, reduced pyridine nucleotides, and acyl coenzyme A (60). These "high-energy" compounds are lost from the developing forespore late in sporulation, and their absence is a hallmark of the spore's metabolic dormancy. However, the spore contains significant levels of nucleoside monophosphates, oxidized pyridine nucleotides, and coenzyme A, and these are converted to the high-energy forms in the first minutes of spore outgrowth. Spores also contain high levels of 3-phosphoglyceric acid, a compound present in only low levels in growing cells. This 3-phosphoglyceric acid depot is utilized in the first minutes of outgrowth to generate ATP, NADH, and acetyl coenzyme A.

SPORE GERMINATION

Although capable of surviving for many years in a dormant, resistant state, if given the proper stimulus, spores can "return to life" within minutes through the process of spore germination followed by outgrowth (25, 39, 45, 63). Spore germination has been defined in various ways. However, in this review germination is defined as processes taking place after addition of a germination trigger that do not require immediate generation of metabolic energy. Subsequent processes that require metabolic energy are termed spore outgrowth, and this process converts the germinated spore to a growing vegetative cell.

Spore germination can be initiated by a variety of agents, including nutrients, Ca-DPA, salts, cationic surfactants, lysozyme, mechanical abrasion, and high pressures, as discussed below. The overall process has been divided into two stages based on work with *B. subtilis.*

While germination can take only a few minutes for individual spores and outgrowth can take only 20 to 40 min depending on the medium and strain, there is great heterogeneity in the times to complete these processes in a spore population. This heterogeneity is primarily in the time to initiate the germination process, with some spores in a population only germinating after days or longer. These spores are often called superdormant, although there is no understanding of the reasons for superdormancy. In the laboratory, spores of some species can be made to initiate germination more synchronously via the process of activation, most often a sublethal heat treatment (32). While activation is often a reversible process, the mechanism of spore activation is not known.

Nutrient Germination—Stage I

Spore germination can be initiated by a number of different stimuli as noted above, although "natural" germination is almost certainly the process triggered by nutrients (Fig. 3). The nutrients that trigger germination vary in a species- and strain-specific manner, but common ones are L-amino acids, D-sugars, and purine nucleosides (25). Catabolism of nutrients is not required for germination, nor is catabolism of endogenous energy reserves (39, 45, 63). Indeed, the entire process of germination does not require metabolic energy, as noted above. While the nutrient germinant triggers are not metabolized, they are recognized in a stereospecific manner—for example, L-alanine is a good trigger for *B. subtilis* spore germination, while D-alanine is an inhibitor of L-alanine germination. Nutrient germinants are recognized by germinant receptors that are located in the spore's inner membrane, presumably with the nutrient binding site located on the outside of the membrane.

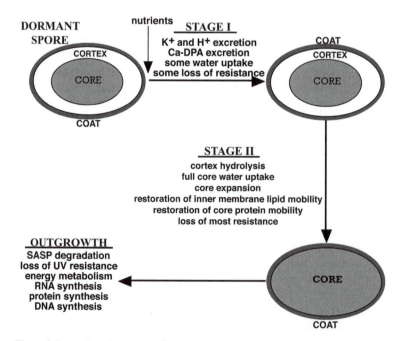

Figure 3. Events in various stages of spore germination. This figure is adapted from Fig. 3 in reference 63.

Spores that lack the germinant receptors germinate extremely poorly with nutrients (44). The germinant receptors are encoded by tricistronic operons, termed *gerA* operon homologs, that are transcribed only late in sporulation and only within the developing spore. All three proteins encoded by a particular *gerA* operon homolog almost certainly associate to form the functional receptor, and all three proteins are essential for receptor function (28, 39, 45). There are almost always multiple *gerA* operon homologs in sporeformers, and each of the encoded receptors has a different specificity for nutrient germinants. While some of these receptors operate alone, in some cases different receptors cooperate to recognize combinations of nutrients (4, 39, 63). *B. subtilis* spores have three functional germinant receptors. The GerA receptor responds to L-alanine, while the GerB and GerK receptors together trigger germination with a mixture of L-asparagine (or L-alanine), D-glucose, D-fructose, and K^+ ions. In addition to the germinant receptors, initiation of germination requires the GerD protein. Little is known about this protein other than that it is synthesized in the forespore in parallel with the germinant receptors.

The A and B proteins of the germinant receptors are almost certainly integral membrane proteins, while the C proteins have a signal sequence and a consensus sequence for covalent addition of diacylglycerol. Addition of diacylglycerol to the appropriate C protein is essential for the function of both the GerA and GerB receptors in *B. subtilis* (27). The sequence of the various proteins of the germinant receptor does not give a clear indication of the function of these proteins. Some of the A proteins show weak sequence homology to a group of basic amino acid transporters found in bacteria (39), but there is no evidence that these receptors actually function as transporters.

The binding of a nutrient germinant to its cognate germinant receptor appears to commit the spore to germinate, and in at least some cases germination proceeds even when the germinant is removed (45). However, the mechanism of this commitment is not known. The first known events in nutrient germination are the release of monovalent ions (e.g., K^+ and H^+), as well as DPA and its chelated divalent cations. Unfortunately, the mechanism of this ion efflux and its triggering by nutrient binding to germinant receptors are not clear. The makeup of the channels or pores through which ions and DPA leave the spore is also not clear, although the SpoVA proteins that are likely located in the spore's inner membrane have been implicated in DPA efflux (23, 67–69). However, there is no idea of how such channels or pores might be "gated."

DPA and other ions that are released early in germination comprise ~25% of the spore core's dry weight. Release of this large amount of dry weight is accompanied by uptake of significant water into the spore core. With *B. subtilis* spores this change can raise the percentage of core wet weight as water from 35 to 45%, resulting in a large decrease in the spore's wet heat resistance, although the stage I germinated spore is much more wet heat resistant than are vegetative cells (58). The stage I germinated spore also retains almost all of the dormant spore's resistance to UV radiation and many chemicals (58). The mechanism for water movement into the spore core is unknown, and *Bacillus* species lack obvious genes for the aquaporins, which facilitate water movement across membranes in other bacteria. Excretion of H^+ early in germination also increases the core pH from ~6.5 to ~7.8. However, the stage I germinated spore remains metabolically dormant, as core water levels are still too low to allow significant protein movement and enzyme action (16, 58). Excretion of DPA and other ions and the associated water uptake complete stage I of spore germination. While there is no striking change in spore morphology during this period, the spore becomes slightly less bright in the phase-contrast microscope, since the ratio of solids to water in the core decreases.

Nutrient Germination—Stage II

The events in stage I of germination trigger subsequent events that comprise stage II. Foremost among stage II events is the hydrolysis of the cortical PG and the release of its fragments into the medium (39, 44, 45, 63). With the loss of this PG layer it is as if a restraining straitjacket has been removed, and thus cortex hydrolysis allows the core to expand by uptake of water to bring the core water content to 80% of the wet weight.

Cortex hydrolysis is normally initiated by one of several redundant CLEs that are specific for cortical PG because of their recognition only of PG containing MAL (12, 36, 41, 47, 58). The CLEs are present in the dormant spore predominantly at the outer surface of the cortex (5, 12), and at least in *Bacillus* species are present in a form that should or could be active but do not act in the dormant spore. In *B. subtilis* one of the two redundant CLEs, termed CwlJ, is activated by the Ca-DPA released in stage I of germination (41). However, the mechanism for activation of the other *B. subtilis* CLE, SleB, is not known, although it may be some physical change in the structure of the cortical PG due to events in stage I of germination. SleB is a lytic glycosylase, but the bond specificity of CwlJ is not known. Spores of *Clostridium* species appear to lack a CwlJ homolog but contain a SleB-like enzyme (36). Another possible CLE in spores of *C. perfringens* is present as a zymogen that is activated by proteolysis in the first minutes of germination (36). As noted above, cortex hydrolysis is accompanied by significant water uptake into the core, but the mechanism of this water uptake is not clear. The germ cell wall must also remodel in some fashion to expand around the swollen core, whose volume has increased ~2-fold—all in the absence of energy metabolism.

With the completion of stage II of germination, the dormant spore with its relatively dehydrated core has been converted to a germinated spore with a core whose hydration level is the same as that of a growing cell. As a consequence, the fully germinated spore has lost its resistance to wet heat and some chemicals, lipids become mobile in the spore's inner membrane, and protein mobility and enzyme action begin in the core (16, 17, 63). One enzyme that becomes active upon completion of germination is a protease, GPR, which initiates the degradation of the DNA-protective α/β-type SASP. With their degradation, spore resistance to DNA-damaging chemicals and UV radiation returns to that of growing cells (45).

Nonnutrient Germination

While the scenario given above for spore germination in response to nutrients is likely followed in the environment, there are a number of other triggers for germination in the test tube, including Ca-DPA, cationic surfactants, inorganic salts, exogenous lysozyme, mechanical abrasion, and high pressure.

Ca-DPA

Exogenous Ca-DPA is a good germinant of spores of most species, in *B. subtilis* spores by activating CwlJ, and bypasses the need for germinant receptors completely (25, 41). Presumably this action of exogenous Ca-DPA mimics the activation of CwlJ by Ca-DPA released from the spore core in stage I of nutrient germination. How activation of this CLE by exogenous Ca-DPA leads to the release of DPA and other ions from the core is not known. While it is possible that Ca-DPA released by some spores in a population triggers the germination of others, this seems unlikely under natural circumstances, as the concentrations of exogenous Ca-DPA needed to stimulate germination are 20 to 60 mM.

Cationic Surfactants

Another group of molecules that can trigger spore germination is cationic surfactants, of which dodecylamine is the best-studied compound (53, 55, 57). Like Ca-DPA, dodecylamine triggers germination by a mechanism that bypasses the germinant receptors. However, dodecylamine does not directly activate a CLE, as *B. subtilis* spores lacking both CwlJ and SleB undergo stage I of germination when exposed to dodecylamine.

The rate of *B. subtilis* spore germination by dodecylamine as measured by DPA release increases with increasing temperatures only up to 60°C (69). Although this temperature optimum is ~20°C higher than that for nutrient and Ca-DPA germination, the presence of a temperature optimum is consistent with dodecylamine triggering spore germination by activating a proteinaceous inner membrane channel allowing Ca-DPA release (69). This Ca-DPA channel may well contain SpoVA proteins (69), although the mechanism whereby dodecylamine triggers Ca-DPA release is not clear.

Inorganic Salts

Germination of spores of some strains, in particular, some *B. megaterium* strains, is triggered by inorganic salts such as KBr (45). Germination by salts appears to use the nutrient germination pathway including the germinant receptors, as inhibitors of germinant receptor action block salt germination (14). Nutrient germination also often requires or is stimulated by inorganic salts (25, 39). However, it is not known if these two phenomena are related.

Lysozyme

Exogenous lytic enzymes such as hen egg white lysozyme have also been used to trigger germination. This is not effective with intact spores, since the spore coat restricts lysozyme's access to the spore cortex. However, if spores are first decoated or contain a mutation in spore coat assembly, exogenous lysozyme can trigger germination by hydrolysis of the spore cortex PG (47). However, unlike CwlJ and SleB, lysozyme will degrade both cortical and germ cell wall PG. Consequently, a decoated spore undergoes osmotic rupture following lysozyme treatment unless treatment is in a hypertonic medium. Again, how cortex degradation from the outside leads to release of Ca-DPA and other ions is not clear, but the germinant receptors are not involved.

Mechanical Abrasion

Spore germination can be triggered by mechanical abrasion, in a process that does not require the germinant receptors (30, 54). Abrasion leads to activation of CLEs, and in *B. subtilis* spores activation of either of the two redundant CLEs by abrasion can trigger spore germination. The mechanism of activation of CLEs by abrasion is not known but may involve physical damage to the spore cortex.

SPORE GERMINATION BY PRESSURE

Pressures of 50 to 800 MPa also trigger spore germination. Although germinated spores are generally much less resistant to harsh treatments than are dormant spores, high pressures alone are not particularly effective in killing germinated spores, and much of the spore killing by high pressure appears to be due to the elevated temperature of pressure treatments (26, 34, 38, 49, 56). It would be informative in this regard to compare growing or stationary-phase cell killing by high pressure with that of fully germinated spores. However, as far as I am aware, no such comparison has been done. In any event, because high pressure alone is not especially effective in killing germinated spores, high-pressure processes designed to kill spores are most often carried out at high temperatures (e.g., 70 to 90°C). Such high temperatures alone will generally kill fully germinated spores but not dormant spores. However, it is not clear that completion of spore germination is necessary for spore killing at high pressures and such high temperatures (9).

Use of elevated temperatures for high-pressure processing generally restricts the pressures used for treatment to high ones (400 to 800 MPa), as at lower pressures spores are not germinated well at elevated temperatures. The reason for the latter phenomenon is that spore germination is triggered by different mechanisms at moderately high pressures (mHP) (50 to 300 MPa) than at ultrahigh pressures (uHP) (300 to 800 MPa). mHP triggers spore germination by activating the spore's germinant receptors (8, 41, 74). This leads to rapid release of DPA and ions followed by cortex hydrolysis, as in nutrient germination. Given the involvement of spore proteins, in particular, the germinant receptors, in mHP germination, it is not surprising that this process shows a temperature optimum of ~40°C with *B. subtilis* spores, when germination is measured by DPA release (Fig. 4). In contrast, uHP germination does not require the germinant receptors (41, 74), and DPA release due to uHP germination of *B. subtilis* spores increases with increasing temperature up to at least 60°C (Fig. 4). While mHP and uHP germination proceeds by different mechanisms, there is some overlap in the pressures that lead to germination by these two mechanisms. However, for the lower limit of pressures giving mHP germination and the upper limit of pressures giving uHP germination, only one mechanism predominates. While there is a significant amount of data on pressures giving germination of spores of many different species, most data on the mechanisms of these effects have been obtained with *B. subtilis* spores because of the molecular genetic tools that can be applied to this organism. Consequently, in the discussion of mHP and uHP germination below, the data discussed have been obtained with *B. subtilis* spores.

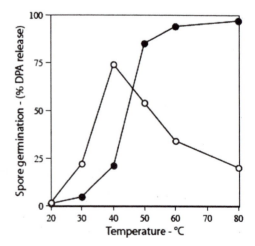

Figure 4. Rates of DPA release from *B. subtilis* spores upon treatment with mHP and uHP at various temperatures. Spores of *B. subtilis* PS533, a derivative of strain 168 carrying plasmid pUB110 (59), were incubated at an optical density at 600 nm of 2 in 50 mM Tris-HCl (pH 7.5) and treated for various times and at various temperatures with either 150 MPa (○) or 500 MPa (●). Spore germination was assessed in duplicate samples by monitoring DPA release by measuring the optical density at 270 nm of the supernatant fluid from 1-ml samples centrifuged in a microcentrifuge. The release of DPA in this experiment was linear with time (data not shown). The variation in values for the duplicate samples was ±7% or less. The data for this experiment are from reference 9.

mHP Germination

There is a variety of evidence that mHP germination is mediated through the spore's germinant receptors (8, 41, 46, 74) (Fig. 5). (i) Deletion of all functional germinant receptors almost eliminates mHP germination of *B. subtilis* spores. (ii) Covalent addition of diacylglycerol to the C protein of most germinant receptors is essential for their function in nutrient germination as noted above, and is also essential for mHP germination. (iii) The *gerD* gene, whose product is required for nutrient germination, is also required for mHP germination. (iv) Elevated levels of various germinant receptors result in higher rates of mHP germination. (v) Core demineralization decreases mHP germination of spores and decreases nutrient germination (24, 29, 45). Unfortunately, while it is clear that germinant receptors mediate mHP germination, it is not clear how mHP activates the germinant receptors. This could be through effects of mHP on the germinant receptors themselves or on the spore's inner membrane, in which the germinant receptors reside, as high pressures can affect both membranes and proteins (6, 10).

While there is yet no definitive evidence for either of the two possibilities noted above, there is evidence that the inner membrane plays a major role in the responsiveness of germinant receptors to mHP. As noted above, the spore's inner membrane has extremely low permeability and lipid probes are immobile in this membrane. This suggests that lipids themselves are relatively immobile in this membrane and thus that the membrane has very low

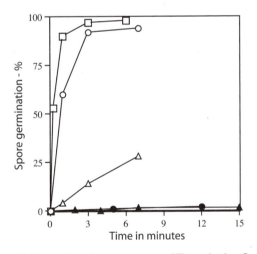

Figure 5. Effects of changes in nutrient receptors on mHP germination. Spores of various isogenic strains of *B. subtilis* at an optical density at 600 nm of 1 were treated with 150 MPa of pressure at 37°C in 50 mM Tris-HCl (pH 7.5). The germination of duplicate samples of the treated spores was assessed by flow cytometry after staining the spores with the nucleic acid stain Syto 16 (Molecular Probes, Eugene, OR) as described previously (8, 67). Dormant-spore nucleic acids are not stained by Syto 16, which only stains fully germinated spores. Symbols represent the various strains of spores used as follows: ○, PS533 (wild type [59]); ●, FB72 (44) (lacks all three functional germinant receptors); △, PS3301 (lacks the only *B. subtilis* lipoprotein diacylglycerol transferase [27]); ▲, FB68 (lacks GerD [46]); and □, PS3476 (8) (contains 20-fold-higher levels of the GerA germinant receptor, the major germinant receptor responding to mHP). The variation in values for the duplicate samples was ±5% or less. Data are from references 8 and 46.

fluidity. One parameter that affects the fluidity of membranes in growing bacteria is growth temperature. Membrane fluidity generally rises at lower growth temperatures, largely through changes in fatty acid composition, and membranes from *B. subtilis* cells grown at lower temperatures have higher levels of branched and unsaturated fatty acids (13, 15). Spore inner membrane permeability also changes as a function of sporulation temperature. When the spore's inner membrane's permeability to methylamine is measured at one temperature, spores made at lower temperatures exhibit much higher rates of methylamine permeation than do spores made at higher temperatures (13, 15). The inner membrane from spores prepared at low temperatures also has a higher level of branched and unsaturated fatty acids. Consequently, the slower germination of spores made at lower temperatures with mHP (8, 50, 51) (Fig. 6) may be due to the higher fluidity of the low-temperature spore's inner membrane. However, this has not been established. Spores with various levels of inner membrane unsaturated fatty acids made at the same temperature also germinate equally well with mHP.

Subsequent to activation of the spore's germinant receptors, mHP germination appears to proceed through the same pathway as nutrient germination—including DPA release followed by cortex hydrolysis (73). Either CwlJ or SleB is needed for *B. subtilis* spore cortex hydrolysis triggered by mHP, as *cwlJ sleB* spores do not complete mHP germination (42). However, mHP does cause DPA release from *cwlJ sleB* spores, as do nutrients, although *cwlJ sleB* spores cannot complete nutrient or mHP germination, since they cannot degrade their cortex PG (42, 58). DPA-less spores of strain FB122 (41), which lacks the *spoVF* operon encoding DPA synthetase as well as *sleB,* do not complete mHP germination, since CwlJ cannot be activated (8). This is also the case for DPA-less FB122 spores incubated with nutrients (41).

A screen for *B. subtilis* mutants blocked in pressure germination resulted in isolation of a strain with a mutation in the *ykvU* gene (2), a gene with no known function. While YkvU has no obvious sequence similarity to proteins of known function, it is a sporulation-specific membrane protein made in the mother cell compartment and is located in the spore's outer layers (22). Mutation of *ykvU* decreases mHP and uHP germination and nutrient and dodecylamine germination, but it has no effect on germination by exogenous Ca-DPA. Detailed analysis of YkvU function not only in pressure germination but also in all germination processes may be informative.

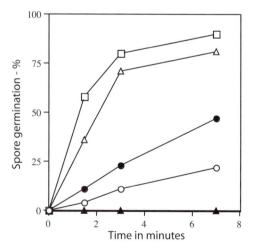

Figure 6. Effects of sporulation temperature and loss of DPA on mHP germination. Spores of strain PS533 (wild type [59]) were prepared at various temperatures, and spores of strain FB122 (DPA-less [41]) were prepared at 37°C. Spores were treated with 150 MPa of pressure at 37°C, and spore germination in duplicate samples was assessed as described in the legend to Fig. 5. The symbols for the spores of the strains used are as follows: ○, PS533 spores prepared at 23°C; ●, PS533 spores prepared at 30°C; △, PS533 spores prepared at 37°C; □, PS533 spores prepared at 44°C; ▲, FB122 (DPA-less) (41) spores prepared at 37°C. The variation in values for the duplicate samples was ±8% or less. Data are from reference 8.

uHP Germination

uHP germination bypasses the spore's germinant receptors (9, 42, 46). In contrast to mHP germination, uHP germination is not abolished by germinant receptor loss, is not blocked by a *gerD* mutation or loss of the lipoprotein diacylglycerol transferase, and is not inhibited by most inhibitors of germinant receptor function (9, 42, 46) (Fig. 7). Completion of uHP germination requires at least one CLE, as *cwlJ sleB* spores of *B. subtilis* cannot complete uHP germination. However, uHP does not trigger germination by direct activation of a CLE as does Ca-DPA, since uHP treatment of *cwlJ sleB* spores results in DPA release. uHP treatment at elevated temperatures, where it is unlikely that CLEs can function, also leads to rapid DPA release (9) (Fig. 4). However, spores of several species incubated at uHP and temperatures of $\geq 110°C$ are actually more stable than are spores incubated at these high temperatures but at ambient pressure (37).

In addition to not requiring the spore's germinant receptors, uHP germination also differs from nutrient and mHP germination in other respects (73). In contrast to nutrient-germinated spores, uHP-treated spores that have released their Ca-DPA undergo little SASP degradation, generate little ATP, and retain their UV resistance as well as some other resistance properties. These observations suggest that uHP triggers spore germination by directly causing Ca-DPA release in some fashion, and that Ca-DPA release then triggers subsequent events in germination. However, the uHP-treated spores appear to remain in stage I of germination for much longer than nutrient- or mHP-germinated spores. Presumably some stage I events needed for progression into stage II of germination take place only slowly in uHP-treated spores. Alternative explanations include that action of CwlJ or SleB is blocked during uHP-triggered Ca-DPA release or that uHP inactivates SleB and/or CwlJ to a significant degree. Whatever the reason for the "pause" in stage I of the uHP-treated spores, they eventually progress through stage II and then on into outgrowth.

As expected if the Ca-DPA release caused by uHP treatment leads to subsequent events in germination, DPA-less spores of a strain lacking DPA synthetase and SleB do not complete uHP germination (9) (Fig. 7). It thus appears that uHP triggers germination by creating pores for Ca-DPA in the spore's inner membrane or by activating preexisting Ca-DPA

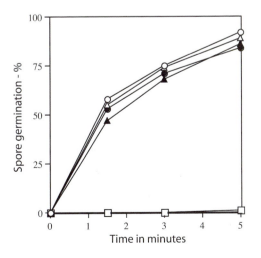

Figure 7. Effect of loss of germinant receptor function or DPA on uHP germination. Spores of various *B. subtilis* strains at an optical density at 600 nm of 1 were treated with 500 MPa of pressure at 50°C, and spore germination in duplicate samples was assessed as described in the legend to Fig. 5. The symbols for the spores of the strains used are as follows: ○, PS533 (wild type [56]); ●, FB72 (lacks all germinant receptors [44]); △, PS3301 (lacks the lipoprotein diacylglycerol transferase [27]); ▲, FB68 (lacks GerD [42]); and □, FB122 (DPA-less [41]). The variation in values for the duplicate samples was ±7% or less. Data are from reference 9.

channels. However, while available evidence suggests that uHP makes the spore's inner membrane permeable to Ca-DPA (38), the mechanism of this effect is unknown and certainly deserves further study.

If uHP acts only to stimulate Ca-DPA movement across the spore's inner membrane, then the physical properties of this membrane could affect uHP germination. In particular, changes in membrane fluidity and/or fatty acid composition might alter uHP germination. Large changes in the fatty acid composition of inner membrane lipids from spores made at a single temperature do not alter spores' responses to uHP germination (Fig. 8). However, spores made at lower temperatures are more responsive to uHP germination than are spores made at higher temperatures (Fig. 8). While it is tempting to ascribe this effect to differences in inner membrane fluidity as described above, it is possible that there are other changes in spores made at different temperatures that cause these differences.

In some respects the germination of spores by uHP resembles germination by dodecylamine, in that nutrient receptors, receptor diacylglycerylation, and GerD are not required. In addition, dodecylamine triggers spore germination by causing Ca-DPA release, and extended dodecylamine treatment leads to loss of other spore small molecules as well. Spores made at lower temperatures also germinate faster with dodecylamine than do spores made at higher temperatures (13), as has also been observed with uHP germination (9). However, spores treated with oxidizing agents germinate faster with dodecylamine, while such treatment has no effect on spore germination with uHP (9, 13). Surprisingly, dodecylamine

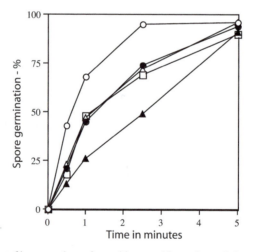

Figure 8. Effect of inner membrane fatty acid composition and sporulation temperature on uHP spore germination. Spores of various *B. subtilis* strains prepared at different temperatures were treated at an optical density at 600 nm of 1 with 500 MPa of pressure at 50°C, and spore germination in duplicate samples was assessed as described in the legend to Fig. 5. The symbols for the spores of the strains used are as follows: ○, PS533 (wild-type [59]) sporulated at 23°C; ●, PS533 sporulated at 37°C; ▲, PS533 sporulated at 44°C; △, PS3628 (lacking fatty acid desaturase and containing no unsaturated fatty acids in the spore's inner membrane [13]) prepared at 37°C; and □, PS3624 (has elevated levels of fatty acid desaturase and >10-fold-higher levels of unsaturated fatty acids in the spore's inner membrane compared to PS533 spores [13]) prepared at 37°C. The variation in values for the duplicate samples was ±7% or less. Data are from reference 9.

strongly inhibits uHP germination of spores (9), although the reason for this effect is not known.

CONCLUSIONS AND THE FUTURE

The major response of bacterial spores to high pressure is the initiation of germination. This response is of major interest to the food industry, since germinated spores are more readily killed than dormant spores, especially by high temperatures. The major stumbling block to more widespread utilization of high pressure for food processing is that high-pressure germination is not complete, and even multiple cycles of high-pressure treatment can leave some superdormant spores untouched. This is also true for nutrient germination. Consequently, a major area for future work is on the nature of these superdormant spores. Are these simply those few spores that are low in some crucial factor that responds to high pressure or other germination stimuli? Is the inner membrane of these superdormant spores significantly different from that of the general population? At present, next to nothing is known about what makes a spore superdormant.

While mHP and uHP can trigger spore germination, uHP appears to be of greater utility in food-processing applications, as it can be combined with elevated temperatures to kill the germinated spores, while mHP germination is retarded at elevated temperatures. It appears likely that uHP triggers spore germination by altering the permeability of the spore's inner membrane in some fashion. Unfortunately, the structure of the spore's inner membrane is not known, and thus it is difficult to imagine how uHP treatment could alter the properties of this membrane. Since membrane-active agents might alter effects of uHP on the spore's inner membrane, the effects of such agents on uHP processing could be a fruitful area for investigation.

The activation of germinant receptors by mHP also likely tells us something important about how these receptors function in the spore's inner membrane. Unfortunately, we do not yet know how these receptor proteins function, nor do we know their structure in the environment of the spore's inner membrane. At some point, perhaps, knowledge of the latter may allow us to understand the effectiveness of mHP in activating these proteins.

Acknowledgments. Work in my laboratory cited in this chapter has been supported by grants from the National Institutes of Health (GM19598) and the Army Research Office to me and from the U.S. Department of Agriculture to me and Dallas G. Hoover.

REFERENCES

1. **Ablett, S., A. H. Darke, P. J. Lillford, and D. R. Martin.** 1999. Glass formation and dormancy in bacterial spores. *Int. J. Food Sci. Technol.* **34:**59–69.
2. **Aertsen, A., I. Van Opstal, S. C. Vanmuysen, E. Y. Wuytack, and C. W. Michiels.** 2005. Screening for *Bacillus subtilis* mutants defective in pressure induced spore germination: identification of *ykvU* as a novel germination gene. *FEMS Microbiol. Lett.* **243:**385–391.
3. **Alberto, F., L. Botella, F. Carlin, C. Nguyen-the, and V. Brouselle.** 2005. The *Clostridium botulinum* GerBA germination protein is located in the inner membrane of spores. *FEMS Microbiol. Lett.* **253:**231–235.
4. **Atluri, S., K. Ragkousi, D. E. Cortezzo, and P. Setlow.** 2006. Cooperativity between different nutrient receptors in germination of spores of *Bacillus subtilis* and reduction of this cooperativity by alterations in the GerB receptor. *J. Bacteriol.* **188:**28–36.
5. **Bagyan, I., and P. Setlow.** 2002. Localization of the cortex lytic enzyme CwlJ in spores of *Bacillus subtilis*. *J. Bacteriol.* **184:**1289–1294.

6. **Bartlett, D. H.** 2002. Pressure effects on *in vivo* microbial processes. *Biochim. Biophys. Acta* **1595:**367–381.
7. **Bertsch, L. L., P. P. Bonsen, and A. Kornberg.** 1969. Biochemical studies of bacterial sporulation and germination. XIV. Phospholipids in *Bacillus megaterium. J. Bacteriol.* **98:**75–81.
8. **Black, E. P., K. Koziol-Dube, D. Guan, J. Wei, B. Setlow, D. E. Cortezzo, D. G. Hoover, and P. Setlow.** 2005. Factors influencing the germination of *Bacillus subtilis* spores via the activation of nutrient germinant receptors. *Appl. Environ. Microbiol.* **71:**5879–5887.
9. **Black, E. P., J. Wei, S. Atluri, D. E. Cortezzo, K. Koziol-Dube, D. G. Hoover, and P. Setlow.** 2007. Analysis of factors influencing the rate of germination of spores of *Bacillus subtilis* by very high pressure. *J. Appl. Microbiol.* **102:**65–76.
10. **Braganza, L. F., and D. L. Worcester.** 1986. Structural changes in lipid bilayers and biological membranes caused by hydrostatic pressure. *Biochemistry* **25:**7484–7488.
11. **Cano, R. J., and M. Borucki.** 1995. Revival and identification of bacterial spores in 25- to 40-million-year-old Dominican amber. *Science* **268:**1060–1064.
12. **Chirakkal, H., M. O'Rourke, A. Atrih, S. J. Foster, and A. Moir.** 2002. Analysis of spore cortex lytic enzymes and related proteins in *Bacillus subtilis* endospore germination. *Microbiology* **148:**2383–2392.
13. **Cortezzo, D. E., K. Koziol-Dube, B. Setlow, and P. Setlow.** 2004. Treatment with oxidizing agents damages the inner membrane of spores of *Bacillus subtilis* and sensitizes the spores to subsequent stress. *J. Appl. Microbiol.* **97:**838–852.
14. **Cortezzo, D. E., B. Setlow, and P. Setlow.** 2004. Analysis of the action of compounds that inhibit the germination of spores of *Bacillus* species. *J. Appl. Microbiol.* **96:**725–741.
15. **Cortezzo, D. E., and P. Setlow.** 2005. Analysis of factors influencing the sensitivity of spores of *Bacillus subtilis* to DNA damaging chemicals. *J. Appl. Microbiol.* **98:**606–617.
16. **Cowan, A. E., D. E. Koppel, B. Setlow, and P. Setlow.** 2003. A soluble protein is immobile in dormant spores of *Bacillus subtilis* but is mobile in germinated spores: implications for spore dormancy. *Proc. Natl. Acad. Sci. USA* **100:**4209–4214.
17. **Cowan, A. E., E. M. Olivastro, D. E. Koppel, C. A. Loshon, B. Setlow, and P. Setlow.** 2004. Lipids in the inner membrane of dormant spores of *Bacillus* species are immobile. *Proc. Natl. Acad. Sci. USA* **101:**7733–7738.
18. **Douki, T., B. Setlow, and P. Setlow.** 2005. Photosensitization of DNA by dipicolinic acid, a major component of spores of *Bacillus* species. *Photochem. Photobiol. Sci.* **4:**591–597.
19. **Driks, A.** 1999. The *Bacillus subtilis* spore coat. *Microbiol. Mol. Biol. Rev.* **63:**1–20.
20. **Driks, A.** 2002. Overview: development in bacteria: spore formation in *Bacillus subtilis. Cell. Mol. Life Sci.* **59:**389–391.
21. **Driks, A.** 2001. Proteins of the spore core and coat, p. 527–535. *In* A. L. Sonenshein, J. A. Hoch, and R. Losick (ed.), Bacillus subtilis *and Its Closest Relatives: from Genes to Cells.* ASM Press, Washington, DC.
22. **Eichenberger, P., S. T. Jensen, E. M. Conlon, C. van Ooij, J. Silvaggi, J. E. Gonzalez-Pastor, M. Fujita, S. Ben-Yehuda, P. Stragier, J. S. Liu, and R. Losick.** 2003. The σ^E regulon and the identification of additional sporulation genes in *Bacillus subtilis. J. Mol. Biol.* **327:**945–972.
23. **Errington, J.** 1993. *Bacillus subtilis* sporulation: regulation of gene expression and control of morphogenesis. *Microbiol. Rev.* **57:**1–33.
24. **Gerhardt, P., and R. E. Marquis.** 1989. Spore thermoresistance mechanisms, p. 43–63. *In* I. Smith, R. A. Slepecky, and P. Setlow (ed.), *Regulation of Prokaryotic Development.* ASM Press, Washington, DC.
25. **Gould, G. W.** 1969. Germination, p. 397–444. *In* G. W. Gould and A. Hurst (ed.), *The Bacterial Spore.* Academic Press, London, England.
26. **Gould, G. W., and A. J. H. Sale.** 1969. Initiation of germination of bacterial spores by hydrostatic pressure. *J. Gen. Microbiol.* **60:**335–346.
27. **Igarashi, T., B. Setlow, M. Paidhungat, and P. Setlow.** 2004. Analysis of the effects of a *gerF (lgt)* mutation on the germination of spores of *Bacillus subtilis. J. Bacteriol.* **186:**2984–2991.
28. **Igarashi, T., and P. Setlow.** 2005. Interaction between individual protein components of the GerA and GerB nutrient receptors that trigger germination of *Bacillus subtilis* spores. *J. Bacteriol.* **187:**2513–2518.
29. **Igura, N., Y. Kamimura, M. S. Islam, M. Shimoda, and I. Hayakawa.** 2003. Effects of minerals on resistance of *Bacillus subtilis* spores to heat and hydrostatic pressure. *Appl. Environ. Microbiol.* **69:**6307–6310.
30. **Jones, C. A., N. L. Padula, and P. Setlow.** 2005. Effect of mechanical abrasion on the viability, disruption and germination of spores of *Bacillus subtilis. J. Appl. Microbiol.* **99:**1484–1494.

31. **Kennedy, M. J., S. L. Reader, and L. M. Swierczynski.** 1994. Preservation records of micro-organisms: evidence of the tenacity of life. *Microbiology* **140:**2513–2529.

32. **Keynan, A., and Z. Evenchik.** 1969. Activation, p. 359–396. *In* G. W. Gould and A. Hurst (ed.), *The Bacterial Spore.* Academic Press, London, England.

33. **Klobutcher, L. A., K. Ragkousi, and P. Setlow.** 2006. The *Bacillus subtilis* spore coat provides "eat resistance" during phagosomal predation by the protozoan *Tetrahymena thermophila. Proc. Natl. Acad. Sci. USA* **103:**165–170.

34. **Knorr, D.** 1999. Novel approaches in food-processing technology: new technologies for preserving foods and modifying function. *Curr. Opin. Biotechnol.* **10:**485–491.

35. **Leuschner, R. G. K., and P. J. Lillford.** 2003. Thermal properties of bacterial spores and biopolymers. *Int. J. Food Microbiol.* **80:**131–143.

36. **Makino, S., and R. Moriyama.** 2002. Hydrolysis of cortex peptidoglycan during bacterial spore germination. *Med. Sci. Monit.* **8:**RA119–RA127.

37. **Margosch, D., M. A. Ehrmann, R. Buckow, V. Heinz, R. F. Vogel, and M. G. Gänzle.** 2006. High-pressure-mediated survival of *Clostridium botulinum* and *Bacillus amyloliquefaciens* endospores at high temperature. *Appl. Environ. Microbiol.* **72:**3476–3481.

38. **Margosch, D., M. G. Gänzle, M. A. Ehrmann, and R. F. Vogel.** 2004. Pressure inactivation of *Bacillus* endospores. *Appl. Environ. Microbiol.* **70:**7321–7328.

39. **Moir, A., B. M. Corfe, and J. Behravan.** 2002. Spore germination. *Cell. Mol. Life Sci.* **59:**403–409.

40. **Nicholson, W. L., N. Munakata, G. Horneck, H. J. Melosh, and P. Setlow.** 2000. Resistance of *Bacillus* endospores to extreme terrestrial and extraterrestrial environments. *Microbiol. Mol. Biol. Rev.* **64:**548–572.

41. **Paidhungat, M., K. Ragkousi, and P. Setlow.** 2001. Genetic requirements for induction of germination of spores of *Bacillus subtilis* by Ca^{2+}-dipicolinate. *J. Bacteriol.* **183:**4886–4893.

42. **Paidhungat, M., B. Setlow, W. B. Daniels, D. Hoover, E. Papafragkou, and P. Setlow.** 2002. Mechanisms of initiation of germination of spores of *Bacillus subtilis* by pressure. *Appl. Environ. Microbiol.* **68:**3172–3175.

43. **Paidhungat, M., B. Setlow, A. Driks, and P. Setlow.** 2000. Characterization of spores of *Bacillus subtilis* which lack dipicolinic acid. *J. Bacteriol.* **182:**5505–5512.

44. **Paidhungat, M., and P. Setlow.** 2000. Role of Ger proteins in nutrient and nonnutrient triggering of spore germination in *Bacillus subtilis. J. Bacteriol.* **182:**2513–2519.

45. **Paidhungat, M., and P. Setlow.** 2001. Spore germination and outgrowth, p. 537–548. *In* A. L. Sonenshein, J. A. Hoch, and R. Losick (ed.), Bacillus subtilis *and Its Closest Relatives: from Genes to Cells.* ASM Press, Washington, DC.

46. **Pelczar, P. L., T. Igarashi, B. Setlow, and P. Setlow.** 2007. Role of GerD in germination of *Bacillus subtilis* spores. *J. Bacteriol.* **189:**1090–1098.

47. **Popham, D. L.** 2002. Specialized peptidoglycan of the bacterial endospore: the inner wall of the lockbox. *Cell. Mol. Life Sci.* **59:**426–433.

48. **Popham, D. L., J. Helin, C. E. Costello, and P. Setlow.** 1996. Muramic lactam in peptidoglycan of *Bacillus subtilis* spores is required for spore outgrowth but not for spore dehydration or heat resistance. *Proc. Natl. Acad. Sci. USA* **93:**15405–15410.

49. **Raso, J., and G. Barbosa-Canovas.** 2003. Nonthermal preservation of foods using combined processing techniques. *Crit. Rev. Food Sci. Nutr.* **43:**265–285.

50. **Raso, J., G. Barbosa-Canovas, and B. G. Swanson.** 1998. Sporulation temperature affects initiation of germination and inactivation by high hydrostatic pressure of *Bacillus cereus. J. Appl. Microbiol.* **85:**17–24.

51. **Raso, J., M. M. Gongora-Nieto, G. V. Barbosa-Canovas, and B. G. Swanson.** 1998. Influence of several environmental factors on the initiation of germination and inactivation of *Bacillus cereus* by high hydrostatic pressure. *Int. J. Food Microbiol.* **44:**125–132.

52. **Redmond, C., L. W. Baillie, S. Hibbs, A. J. Moir, and A. Moir.** 2004. Identification of proteins in the exosporium of *Bacillus anthracis. Microbiology* **150:**355–363.

53. **Rode, L. J., and J. W. Foster.** 1960. The action of surfactants on bacterial spores. *Arch. Mikrobiol.* **36:**67–94.

54. **Rode, L. J., and J. W. Foster.** 1960. Mechanical germination of bacterial spores. *Proc. Natl. Acad. Sci. USA* **46:**118–128.

55. **Rode, L. J., and J. W. Foster.** 1961. Germination of bacterial spores with alkyl primary amines. *J. Bacteriol.* **81:**768–779.

56. **Sale, A. J. H., G. W. Gould, and W. A. Hamilton.** 1970. Inactivation of bacterial spores by hydrostatic pressure. *J. Gen. Microbiol.* **60:**323–334.
57. **Setlow, B., A. E. Cowan, and P. Setlow.** 2003. Germination of spores of *Bacillus subtilis* with dodecylamine. *J. Appl. Microbiol.* **95:**637–648.
58. **Setlow, B., E. Melly, and P. Setlow.** 2001. Properties of spores of *Bacillus subtilis* blocked at an intermediate stage in spore germination. *J. Bacteriol.* **183:**4894–4899.
59. **Setlow, B., and P. Setlow.** 1996. Role of DNA repair in *Bacillus subtilis* spore resistance. *J. Bacteriol.* **178:** 3486–3495.
60. **Setlow, P.** 1994. Mechanisms which contribute to the long-term survival of spores of *Bacillus* species. *J. Appl. Bacteriol.* **76:**49S–60S.
61. **Setlow, P.** 1995. Mechanisms for the prevention of damage to the DNA in spores of *Bacillus* species. *Annu. Rev. Microbiol.* **49:**29–54.
62. **Setlow, P.** 2001. Resistance of spores of *Bacillus* species to ultraviolet light. *Environ. Mol. Mutagen.* **38:** 97–104.
63. **Setlow, P.** 2003. Spore germination. *Curr. Opin. Microbiol.* **6:**550–556.
64. **Setlow, P.** 2006. Spores of *Bacillus subtilis:* their resistance to radiation, heat and chemicals. *J. Appl. Microbiol.* **101:**514–535.
65. **Stecchini, M. L., M. Del Torre, E. Venir, A. Morettin, P. Furlan, and E. Maltini.** 2006. Glassy state in *Bacillus subtilis* spores analyzed by differential scanning calorimetry. *Int. J. Food Microbiol.* **106:**286–290.
66. **Tovar-Rojo, F., M. Chander, B. Setlow, and P. Setlow.** 2002. The products of the *spoVA* operon are involved in dipicolinic acid uptake into developing spores of *Bacillus subtilis. J. Bacteriol.* **184:**584–587.
67. **Vepachedu, V. R., and P. Setlow.** 2004. Analysis of the germination of spores of *Bacillus subtilis* with temperature sensitive *spo* mutations in the *spoVA* operon. *FEMS Microbiol. Lett.* **239:**71–77.
68. **Vepachedu, V. R., and P. Setlow.** 2005. Localization of SpoVAD to the inner membrane of spores of *Bacillus subtilis. J. Bacteriol.* **187:**5677–5682.
69. **Vepachedu, V. R., and P. Setlow.** 2007. Role of SpoVA proteins in release of dipicolinic acid during germination of *Bacillus subtilis* spores triggered by dodecylamine or lysozyme. *J. Bacteriol.* **189:**1565–1572.
70. **Vreeland, R. H., W. D. Rosenzweig, and D. W. Powers.** 2000. Isolation of a 250 million-year-old halotolerant bacterium from a primary salt crystal. *Nature* **407:**897–900.
71. **Warth, A. D.** 1980. Heat stability of *Bacillus cereus* enzymes within spores and in extracts. *J. Bacteriol.* **143:**27–34.
72. **Westphal, A. J., B. P. Price, T. J. Leighton, and K. E. Wheeler.** 2003. Kinetics of size changes of individual *Bacillus thuringiensis* spores in response to changes in relative humidity. *Proc. Natl. Acad. Sci. USA* **100:**3461–3466.
73. **Wuytack, E. Y., S. Boven, and C. W. Michiels.** 1998. Comparative study of pressure-induced germination of *Bacillus subtilis* spores at low and high pressure. *Appl. Environ. Microbiol.* **64:**3220–3224.
74. **Wuytack, E. Y., J. Soons, F. Poschet, and C. W. Michiels.** 2000. Comparative study of pressure- and nutrient-induced germination of *Bacillus subtilis* spores. *Appl. Environ. Microbiol.* **66:**257–261.

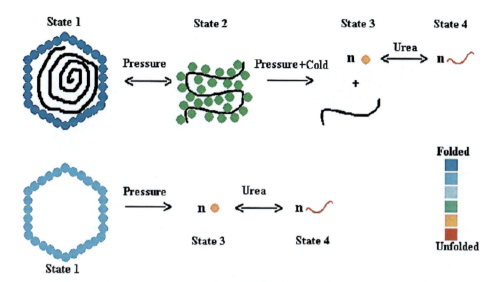

Color Plate 1 (chapter 2). Schematic representation of pressure and cold disassembly of icosahedral viruses. The degree of unfolding of coat proteins is represented by a color code (scale at right). The results indicate that the coat proteins can exist in at least four states: (i) the native conformation in the virus capsid; (ii) bound to RNA when the virus is dissociated by pressure at room temperature, assuming a conformation that retains the information for reassembly; (iii) free subunits in a molten-globule conformation when the virus is dissociated by low temperature under pressure; and (iv) free subunits completely unfolded by high concentrations of urea. Empty capsids of single-stranded RNA viruses dissociate into partially unfolded coat proteins with characteristics of a molten globule, different from the completely unfolded state obtained at high concentrations of urea (bottom panel). Particles containing RNA disassemble reversibly into a ribonucleoprotein complex. At subzero temperatures under pressure, coat proteins release the RNA and also achieve a molten globule-like conformation (top panel) (from reference 15).

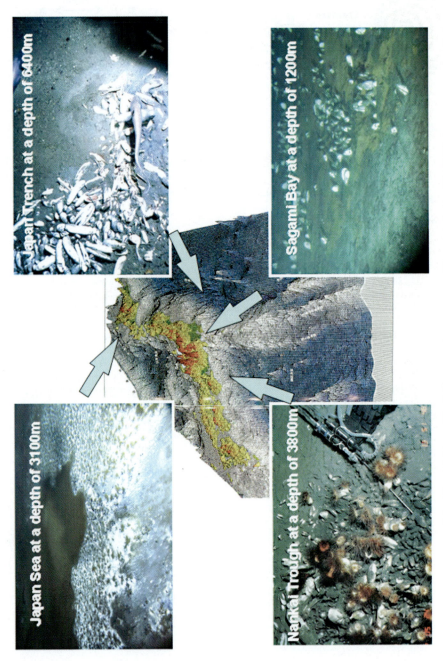

Color Plate 2 (chapter 13). Cold-seep environments around Japan, studied by JAMSTEC.

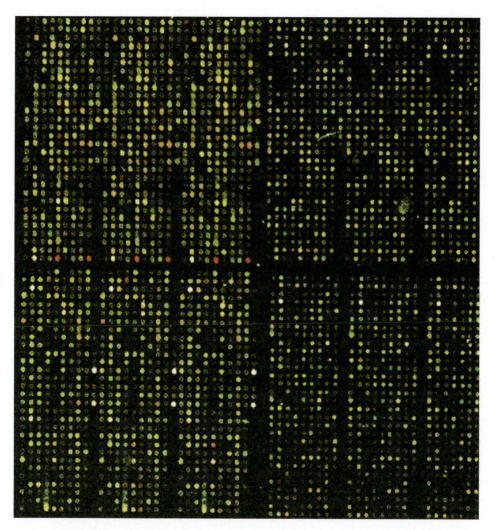

Color Plate 3 (chapter 18). Fluorescent image of an SS9 microarray prepared and examined as previously described (23).

High-Pressure Microbiology
Edited by C. Michiels, D. H. Bartlett, and A. Aertsen
© 2008 ASM Press, Washington, DC

Chapter 4

Inactivation of *Escherichia coli* by High Pressure

Bernard M. Mackey and Pilar Mañas

Following the renewed interest in high hydrostatic pressure as a food preservation technology in Japan in the late 1980s, there has been much effort to characterize the factors that determine microbial resistance to pressure and to define pressure treatments that will allow safe and palatable food products to be produced using this technology (see chapter 10). *Escherichia coli* is of special interest in high-pressure studies because of its importance as an agent of food-borne disease and also because it is an ideal model for studying the mechanisms of inactivation by very high pressures.

High-pressure processing is essentially a pasteurization process when applied at ambient temperature, and it is therefore essential for safety reasons to ensure that the treatments used will eliminate the most pressure-resistant vegetative pathogens from the treated foods. The species *Escherichia coli* includes several enterohemorrhagic serotypes that can cause severe illness, have a low infectious dose, and have been implicated in several large outbreaks of food-borne illness. *Escherichia coli* O157:H7 is the organism of most concern in the United States and Europe, but other enterohemorrhagic serovars, including O111, O26, O145, O103, and O145, have also caused food-borne illness and may be more common than O157 in some countries. Comparisons of pressure resistance among vegetative food-borne pathogens revealed that some strains of *E. coli* O157:H7 were the most resistant so far encountered (8, 16, 75). When defining pressure processes to be used in food preservation, it is therefore important to ensure that the pressures and exposure times used are capable of reducing numbers of the most resistant *E. coli* O157 strains by an adequate margin of safety.

The first experiments on the inactivation of *E. coli* by high pressure appear to be those conducted by Roger in 1895 (cited by Zobell [98]), who found that exposure of cultures to 290 MPa for 10 min at 25 to 30°C killed most of the cell population but did not completely sterilize the culture. Subsequently, Larson et al. in 1918 found that *E. coli* was killed by exposure to 600 MPa for 14 h but could be recovered after exposure to 300 MPa for a similar time (53). These early studies gave the first indications of the times and pressures needed to inactivate *E. coli* but did not provide quantitative information on how rapidly cells die. This was investigated in 1946 by Johnson and Lewin, who showed that the kinetics of inactivation

Bernard M. Mackey • Department of Food Biosciences, The University of Reading, Whiteknights, Reading RG6 6AP, United Kingdom.   ***Pilar Mañas*** • Departamento de Producción Animal y Ciencia de los Alimentos, Facultad de Veterinaria, Miguel Servet 177, 50013, Zaragoza, Spain.

were approximately first order (42). At temperatures between 22.5 and 37°C the rate of inactivation increased with increasing pressure up to 42 MPa, but at 46.9°C, a temperature which was somewhat lethal to cells, elevated pressure decreased the inactivation rate compared with the rate at ambient pressure. The authors noted that this behavior is similar to that seen with protein denaturation. In recent years there have been many studies examining the inactivation of *E. coli* strains by pressure treatment in different types of food (see chapter 10).

Escherichia coli is a convenient experimental model to use in studying the mechanisms of inactivation by high pressure because its biochemical and physiological properties have been extensively documented and the annotated genomes of several strains are now available, including those of the standard laboratory strain *E. coli* K-12 and *E. coli* O157:H7. Studies on the effects of pressure on *E. coli* physiology began in the late 1940s and early 1950s and were aimed largely at identifying cellular processes that are pressure sensitive and therefore limiting for growth at moderately high pressures (11, 12, 98). Cell division, DNA replication, and synthesis of RNA and protein are each inhibited at slightly different pressures between about 35 and 90 MPa, but these effects are reversible; following decompression, growth and cell division can resume once more. In this chapter we examine the effects of lethal pressures above 100 MPa on cellular structure and function of *E. coli* and attempt to distinguish between events that may be associated with loss of viability and those that are not. This is summarized in a provisional model of the mechanisms of inactivation in *E. coli* based on our current understanding.

STRATEGIES FOR STUDYING THE MECHANISMS OF INACTIVATION BY HIGH PRESSURE

What Do We Mean by a Viable Cell?

The living microbial cell is a complex homeostatic unit comprising many different interlinked and interdependent cellular structures and metabolic processes. Identifying a particular structure or function whose impairment by pressure leads to death is therefore a difficult task. Before attempting this analysis, it is necessary first to define what we mean by a viable cell. Living things may be defined as autopoietic entities that can multiply and evolve by natural selection (35). The term autopoietic, meaning "able to self-construct," implies a capacity for using free energy and raw materials to create an ordered structure that is able to maintain homeostasis under varying external conditions. Based on this, our working definition is that a viable *E. coli* cell is one that has the *potential* to replicate indefinitely under suitable conditions. In practice, viability must be determined by demonstrating reproductive ability, i.e., the ability to form colonies on agar, turbidity in broth, or the ability to undergo cell division under the microscope. It follows that so-called viability indicators based on fluorescent probes of membrane integrity and/or metabolic activity do not strictly measure viability per se, although they do measure important attributes of living cells.

Critical Cell Components

In principle it should be possible to identify key cellular components which if destroyed or made nonfunctional would lead inevitably to cell death. These vital components have two essential properties: they must (i) be indispensable for autopoiesis and replication and (ii) be irreplaceable if rendered nonfunctional. Recently, Miles (63) defined such components as "critical components," and we have adopted this terminology. Critical components thus described may not necessarily correspond to the products of essential genes, because although these comply with requirement i, they do not necessarily comply with requirement ii. For example, an enzyme might perform an indispensable role in cell replication, but loss of the enzyme following exposure to an adverse agent may not necessarily be lethal if the enzyme can be resynthesized once the stress is removed.

Examples of components that are both indispensable and irreplaceable if rendered nonfunctional include elements of the transcription and translation apparatus such as RNA polymerase or ribosomes. Because these items are required for their own synthesis, they would be irreplaceable if lost. Physical loss of cytoplasmic membrane integrity will usually be lethal because the consequent loss of homeostasis will preclude repair of or resynthesis of the damaged membrane. However, under some circumstances even cells with damaged cytoplasmic membranes can recover if placed in a suitable medium that allows metabolism to continue. For example, cells of *E. coli* whose membranes were made permeable by treatment with the pore-forming colicin K could multiply when placed in a neutral medium with high concentrations of potassium and magnesium (50). Clearly, irreparable loss of the information encoded by an essential gene would be lethal, as would complete loss of the ability to generate energy. Around 620 of the 4,291 protein-encoding genes in *E. coli* were found to be indispensable for aerobic growth on a rich medium (32) (http://www.genome.wisc.edu/resources/essential.htm), although the number of genes eventually designated as essential may prove to be less than this. At this stage we do not know whether they all encode critical components. Starting from the idea that cell death ensues when critical cell components are reduced to below a level critical for life, Miles (63) derived a general equation relating loss of viability to the number of critical components per cell and the probability of a component sustaining critical damage.

Experimental Approaches to Studying the Mechanisms of Inactivation by High Pressure

Three separate but complementary approaches can be recognized in studies of cellular inactivation by lethal agents such as pressure.

(i) The first approach is examining changes in cell physiology or morphology that occur in pressure-treated cells and correlating these changes with loss of viability. This can be done by comparing the pressures at which the respective events occur or by comparing the rates of change of different processes at a constant pressure.

(ii) The second approach is isolation of pressure-sensitive or pressure-resistant mutants and identification of the genes and gene products responsible for the altered resistance.

(iii) The third approach is identification of genes up- or down-regulated under sublethal pressures and subsequent examination of the effect on pressure resistance of null mutations in, or overexpression of, the corresponding genes.

The first approach is illustrated in Fig. 1. Loss of viability is measured as loss of colony-forming ability, while events labeled A to E represent independent measurements of changes in the structure or state of cell components such as the membrane, cellular proteins, ribosomes, or the nucleoid, or changes in physiological cellular functions such as respiratory activity, maintenance of membrane potential, intracellular pH, or uptake or efflux systems. Events that happen before loss of colony-forming ability (Fig. 1A and B) or

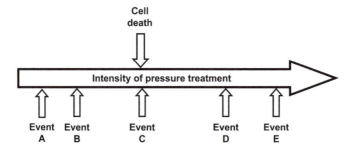

Figure 1. Scheme of experimental approach. Events labeled A to E represent independent measurements of changes in the cell structures or functions caused by pressure. Events that happen before loss of colony-forming ability (A and B) or after it (D and E) are most unlikely to be the cause of cell death, whereas events occurring coincidentally with loss of colony-forming ability (C) may be responsible for death.

after it (Fig. 1D and E) are most unlikely to be the cause of cell death, whereas events occurring coincidentally with loss of colony-forming ability (Fig. 1C) may be responsible for death. Ideally, comparisons should be made between the pressures at which the rates of change of the various processes are maximal, but for practical reasons it may only be possible to identify the onset pressures for some processes or the pressure range (window) over which the event occurs. An example of this general approach is the use of differential scanning calorimetry (DSC) to compare thermal denaturation of cell components with loss of viability in *E. coli* (58, 59, 64), which revealed that cell death coincided with denaturation of the 30S ribosomal subunit. Further investigations are needed to confirm such associations, because coincidence does not prove a causal relationship. The true lethal event might, for example, occur at the same time as the one that was measured which was nonlethal but happened to coincide with loss of viability. Despite the limitations of this approach, it does permit one to eliminate some irrelevant events and focus attention on ones that may be critical to loss of viability. Confirmatory tests following identification of a candidate event for the cause of cell death would be to examine the correlation in cells whose pressure resistance has been manipulated by growth under different conditions or in mutants with increased or decreased resistance to pressure. It should be noted, however, that the event leading to loss of viability might change depending on physiological state, e.g., with different phases of growth.

FACTORS AFFECTING THE RESISTANCE OF *ESCHERICHIA COLI* TO HIGH PRESSURE

The confirmatory tests for identifying events as being lethal may be carried out with cells which show different degrees of resistance to pressure, e.g., in different physiological states, treated in particular media, or subjected to different pressure intensities. It is therefore of key importance to understand the factors that affect piezotolerance and to choose the experimental treatment conditions accordingly. For general information on factors affecting bacterial resistance to pressure, the reader is referred to chapter 10. In this section, a broad overview is given of the relative importance of the diverse factors modifying resistance of *E. coli* to pressure.

Physiological Factors

Among the various physiological factors which affect bacterial pressure resistance, the strain, growth phase, and growth temperature have been the most extensively studied in *E. coli*. The relative resistance of natural isolates of *E. coli* to high-pressure treatments varies very widely (8, 16, 75, 78). Some strains of the enterohemorrhagic serotype O157:H7 have remarkably high resistance compared with other vegetative bacteria (75). For instance, a treatment of 8 min at 500 MPa in phosphate-buffered saline (PBS) achieved less than a 10-fold reduction of *E. coli* O157:H7 strain C9490 (72). Differences in survival among strains of *Escherichia coli* after the application of similar pressure treatments can exceed 6 to 8 log cycles (16).

The reason for this intraspecies variation in piezoresistance is not fully known, but available evidence indicates that it could be largely, although not completely, related to RpoS activity (78). RpoS is an alternative sigma factor, also referred to as σ^s, which regulates the general stress response in gram-negative bacteria. This response is triggered as cells enter stationary phase or are exposed to acid, osmotic, or other stresses and brings about an increase in resistance to a broad range of hostile conditions (93). Differences in pressure resistance among natural isolates of *E. coli* were related to heterogeneity among the *rpoS* alleles that resulted in differences in RpoS activity during growth (78). Interestingly, cells in the exponential phase of growth show very similar degrees of pressure resistance, and though the subsequent increase in resistance as cells enter stationary phase was related to RpoS activity, there was some increase in resistance even in strains with truncated RpoS molecules (16, 72, 78). This reinforces the view that the RpoS sigma factor plays an essential role in piezoresistance but also implies that other stationary-phase functions are involved. Genome-wide expression profiling in *E. coli* has identified 481 genes that are up-regulated under the influence of RpoS, and of these, 11% are involved in coping with the detrimental effects of stress (93). It is not yet clear which particular components of the response are most important in conferring resistance to lethal pressures. The limited information available indicates that strains with different sensitivities to pressure, and correspondingly different RpoS activities, display some differences in their susceptibility to cytoplasmic membrane permeabilization (16).

Growth temperature is an important factor modifying bacterial resistance to heat, cold shock, and many other damaging agents, and modification of the fluidity of the membrane through changes in the fatty acid composition (homeoviscous adaptation [67]) is considered to be one of the contributory mechanisms. Cells grown at lower temperatures generally have a higher percentage of unsaturated fatty acids in the membrane and thus more fluid membranes (79). The influence of growth temperature on pressure resistance of *E. coli* varies depending on the growth phase of the culture (19). In exponential-phase cells, pressure resistance is greatest in cells grown at 10°C and then decreases progressively with increasing growth temperature, but in stationary-phase cells, resistance is lowest for cells grown at 10°C, increases with increasing growth temperature up to a maximum at 30 to 37°C, and then decreases at 45°C. In other words, for stationary-phase cells there is a peak for maximum pressure resistance at temperatures around the optimum for growth (Fig. 2) (19). The specific role of membrane fluidity in pressure resistance of *E. coli* is discussed later in this chapter. Cell concentration also modifies bacterial resistance to pressure. Furukawa et al. (27) demonstrated that the inactivation rate of *E. coli* in PBS at 100 MPa and

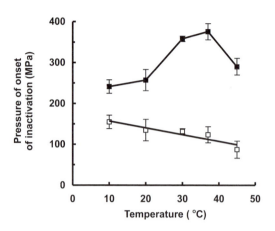

Figure 2. Effect of growth temperature on pressure resistance of exponential-phase (□) or stationary-phase (■) cells of *E. coli* NCTC 8164. Pressure resistance is expressed as the pressure of onset of cell death (± standard error). Reproduced from *Applied and Environmental Microbiology* (19) with permission.

45°C increased as the initial cell concentration decreased. This effect was attributed to the formation of clumps during pressurization such that cells within the clumps were protected against inactivation by pressure, but this was not demonstrated conclusively and the proposed mechanism remains to be confirmed.

Environmental Factors

The medium in which microbes are pressurized can have a very large effect on pressure resistance. Complex media such as foods tend to be protective, whereas acid conditions or the presence of antimicrobial substances decreases resistance. Many studies on the pressure resistance of *Escherichia coli* have been carried out in laboratory media such as broths or buffer solutions. In general broths exert a greater protective effect than buffers. For instance, a treatment of 500 MPa for 5 min achieved a 7-log inactivation of *E. coli* O157:H45 in PBS and only 3.5-log reduction in tryptone soya broth (TSB)-yeast extract plus 0.5% NaCl (85). Phosphate buffer undergoes a pressure-induced change in pH of about 0.4 pH unit per 100 MPa (40). This may contribute to the lower recovery in PBS than in TSB, but since TSB also contains 2.5 g of potassium phosphate per liter, there may be undefined protective constituents in TSB that account for the higher recovery. Foods such as oysters, meat, liquid egg, and milk further protect *E. coli* against inactivation by high pressure (75, 85). The reason for the protective effect of foods is not known, and it is probably due to the combination of specific piezoprotective effects of food components and/or more general protection by physical-chemical parameters.

Milk has proven to be particularly protective for *E. coli,* and this has been partly attributed to its high calcium content (31, 38, 75). The presence of divalent cations seems to play an important role in bacterial piezoresistance. *E. coli* cells pressurized in a medium containing EDTA are sensitized to high pressure, and conversely, cells pressurized in a medium containing calcium ions become more piezotolerant (38). These facts suggest the existence of pressure targets in *E. coli* cells whose stability is highly dependent on calcium binding. However, it is difficult to propose any particular target because many bacterial structures and functions need divalent cations, including the outer membrane, ribosomes, and DNA-binding proteins, among others.

The addition of other salts and sugars also increases piezotolerance of *E. coli* (85, 91). The presence of sucrose exerts considerable piezoprotection even at concentrations as low as 10%, which induces a very small decrease of the water activity (91). This would indicate that the protective effect is linked to the nature of the solute added and not simply to the water activity. In fact, studies carried out with gram-positive bacteria have shown that a similar piezoprotective effect is attained by media containing either 3.0 M sodium chloride or 0.5 M sucrose, whose water activities were 0.917 and 0.985, respectively (66). The nature of the piezoprotective effect of these compounds is not clear, but a decrease of the transition-phase temperature of cell membranes and stabilization of intracellular targets by compatible solutes accumulated in the cytoplasm have been suggested (66, 91).

The pH of the pressurizing medium has been one of the most-studied factors in *E. coli,* mainly because of its practical significance, since juices are regarded as ideal foods to be preserved by high-pressure treatments. The resistance of *E. coli* to high pressure decreases as the pH of the suspending buffer is lowered (7, 30). For instance, treatment at 300 MPa for 15 min at 20°C reduced viable numbers of a pressure-resistant mutant of *E. coli* by only 0.4 log cycle at pH 7.0 but by 2.9 log cycles at pH 3.0 (30). In fruit juices, *E. coli* survival is partly related to the pH, i.e., in more acidic juices the inactivation attained is generally greater (13, 30, 55). An interesting finding is that cells that survive the initial pressure treatment die off during subsequent storage under acid conditions (30, 43, 71). This phenomenon should be taken into account when establishing appropriate treatment intensities for acid products, since it would allow milder treatments, leading to reduced sensory and nutritional losses.

Process Factors

In most *Escherichia coli* strains studied, pressures below 100 MPa when applied at ambient temperature (ca. 20°C) do not exert a lethal effect within typical process times of 2 to 20 min (19). At higher pressures lethality increases with increasing pressure up to 700 MPa, which typically is the maximum used in practice. Bacterial cells are more piezosensitive when pressurized at low temperatures (below 25 to 30°C) (19). Above that peak value, an increase in temperature leads to faster bacterial inactivation (7, 74). The effect of treatment temperature seems to be related to the effects of temperature on membrane fluidity (see below) and also on protein stability (86).

The effect of pressurization time on the degree of inactivation achieved depends on the treatment conditions. For instance, Van Opstal et al. (92) reported that the inactivation curves of *E. coli* MG1655 in carrot juice (log fraction of survivors versus treatment time) were linear, whereas in buffer concave upward curves were observed. In many cases, the first 4- to 5-log reductions can be described by first-order kinetics, and inactivation rates can be described by using conventional decimal reduction times (the time needed at a given pressure to reduce viable numbers by a factor of 10). However, inactivation curves that go beyond this level show concave upward profiles, and in many cases tailing phenomena are observed. The occurrence of tails is usually attributed to the presence of a more resistant subpopulation of cells. In principle, this could arise from physiological differences between cells within an isogenic population or from a genetically distinct resistant subpopulation. Although it is possible to isolate resistant mutants by several cycles of pressure treatment and outgrowth, attempts to demonstrate a resistant fraction by subculturing once from the resistant tail have generally failed to demonstrate a stable resistant phenotype. Recently, however, this procedure did reveal a genetically distinct resistant fraction in cell populations of a clinical isolate of *E. coli* O157 (70). This suggests that the mutation rate to pressure resistance may be quite high, as also evidenced by the ease of isolating resistant mutants using relatively few cycles of pressure treatment and outgrowth (37).

The inactivation rates of *E. coli* KUEN 1504 in broth, milk, peach juice, and orange juice obtained by Erkmen and Dogan (26) were interpreted as first order, and although this is convenient for calculating inactivation at different pressures, it was challenged by Davey and Phua (24), who showed that the inactivation rates actually decreased with time. Estimation of treatment lethality using simple first-order kinetics could overestimate treatment lethality and pose a risk to both safety and stability of pressure-processed foods; consequently, the development of mathematical models which appropriately describe the non-log linear kinetics of inactivation of microbes by high pressure is needed. Various approaches have been tested, including the Arrhenius, linear Arrhenius, log logistic, square root (or Belehrádek), and polynomial forms (20, 24, 92), but there is currently no consensus on the best models to use. Factors such as the pressure come-up time and the increase in temperature due to adiabatic heating add further complication to predicting the lethality of treatments. Models based on resistance distributions, such as the Van Boekel equation, which assumes a Weibull distribution of resistance among the bacterial population, are gaining attention due to their flexibility, which allows description even of survival curves with shoulders, which have occasionally been reported (92). However, a difficulty that is sometimes encountered with these approaches is that the parameters of the primary models do not always vary monotonically with pressure, and this limits their ability to predict behavior under changing conditions. Recently two other approaches to modeling non-log linear kinetics have been introduced. In the first of these, inactivation was assumed to be a first-order process in which the specific rate varies inversely with time (49). In the second approach (52), the inactivation curves were assumed to be biphasic, consisting of an exponential region and a tail. Biphasic inactivation was modeled by using an inverse or "mirror image" version of the Baranyi growth model (10). Both approaches allowed the derivation of secondary models that could predict inactivation rates at different pressures.

At present, all inactivation models lack a physiological basis, which prevents the introduction of important factors such as the development of stress resistance responses and cellular repair in damaged cells. The development of predictive models with a biological basis is needed not only for high hydrostatic pressure but also for such traditional technologies for bacterial inactivation as heat.

EFFECTS OF HIGH PRESSURE ON *ESCHERICHIA COLI* CELLS

Morphological Changes Caused by High Pressure

The relationship between loss of viability and pressure-induced morphological changes was examined in *E. coli* J1, a commensal strain that displayed large differences in resistance between exponential- and stationary-phase cells (61) (Fig. 3). This difference in resistance is very similar to that seen in exponential- and stationary-phase cells of *E. coli* O157 strain C9490, which is also notably pressure resistant (16, 72). The cytoplasm of healthy untreated exponential- or stationary-phase cells appeared dark and homogeneous when viewed under phase-contrast optics (Fig. 4A and B), but after treatment at pressures of 200 MPa or above the cytoplasm appeared more granular in both cell types (Fig. 4C and D), and small rounded areas of high refractility were visible in stationary-phase cells. The

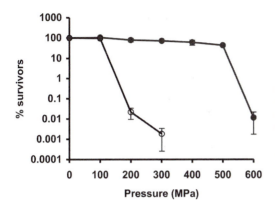

Figure 3. Pressure resistance of *E. coli* J1 in exponential (○) and stationary (●) phases: percentage of survivors after 8 min of treatment at different pressures. Each point corresponds to the mean of at least three independent experiments. The initial inoculum levels were 5×10^9 cells/ml for stationary-phase cells and 10^8 cells/ml for exponential-phase cells. Reproduced from *Applied and Environmental Microbiology* (61) with permission.

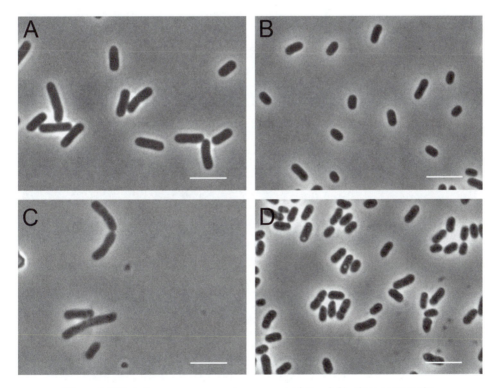

Figure 4. Effect of pressure on general appearance of *E. coli* J1: phase-contrast photomicrographs of exponential- and stationary-phase cells. (A) Exponential-phase cells, untreated (100% viable cells); (B) stationary-phase cells, untreated (100% viable cells); (C) exponential-phase cells treated with 300 MPa for 8 min (0.002% viable cells); (D) stationary-phase cells treated with 600 MPa for 8 min (0.01% viable cells). Bar marker, 2 μm. Reproduced from *Applied and Environmental Microbiology* (61) with permission.

cytoplasm of exponential-phase cells also became lighter after pressure treatment, implying loss of material from the cell (Fig. 4C). The composition of the refractile bodies is not known, but staining with fluorescein isothiocyanate (FITC) (Fig. 5D) suggests that they contain protein, possibly including denatured ribosomal protein (see below).

One of the most obvious changes in both exponential- and stationary-phase cells is condensation of the nucleoid, which can be seen in cells stained with the fluorescent DNA-binding probe 4′,6′-diamidino-2-phenylindole (DAPI) (Fig. 5A to C). Nucleoid condensation has also been observed in *E. coli* cells growing at 34 MPa, a pressure which inhibits cell division and causes filamentation (11). However, the appearance of the nucleoid in the two conditions is somewhat different: at sublethal pressures the condensed nucleoids are generally rather symmetrical and central to the cell axis, similar to those seen in chloramphenicol-treated cells, whereas in cells treated at 200 MPa or above the nucleoids were often skewed across the cell, or pulled to one side next to the membrane. The nucleoid condensation in sublethal pressures is presumably an active process, whereas at lethal

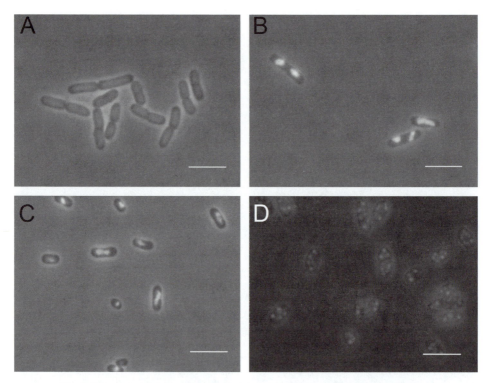

Figure 5. Pressure-induced condensation of the nucleoid and aggregation of cellular protein in *E. coli* J1. Phase-contrast/fluorescence photomicrographs of exponential-phase cells, left untreated (100% viable) and stained with DAPI (A); exponential-phase cells treated with 400 MPa for 8 min (<0.0001% viable cells) and stained with DAPI (B); stationary-phase cells treated with 400 MPa for 8 min (61% viable cells) and stained with DAPI (C); and stationary-phase cells treated with 300 MPa for 8 min (73% viable) and stained with FITC (D). Bar marker, 2 μm. Reproduced from *Applied and Environmental Microbiology* (61) with permission.

pressures the effect may be the result of physical changes in the cell, e.g., denaturation of DNA-binding proteins, separation of aggregated proteins from nucleic acids or disruption of connections between nascent proteins and the membrane. These morphological changes can also be seen in electron micrographs, which clearly show the unusual conformation of the nucleoid and its fibrillar DNA and the appearance of dark condensed regions that are presumably protein (P. Chilton and B. M. Mackey, unpublished results) (Fig. 6). Electron micrographs of pressure-treated cells of *Salmonella enterica* serovar Thompson are very similar in appearance to those of *E. coli* (56), and pressure-induced condensation of the nucleoid has also been seen in the gram-positive organisms *Listeria monocytogenes, Lactobacillus plantarum,* and *Lactobacillus viridescens* (73, 96).

A pressure value between 100 and 200 MPa seemed to be the threshold for the initiation of nucleoid changes and protein aggregation in both exponential- and stationary-phase cells, despite the enormous difference in pressure sensitivity between them. These events

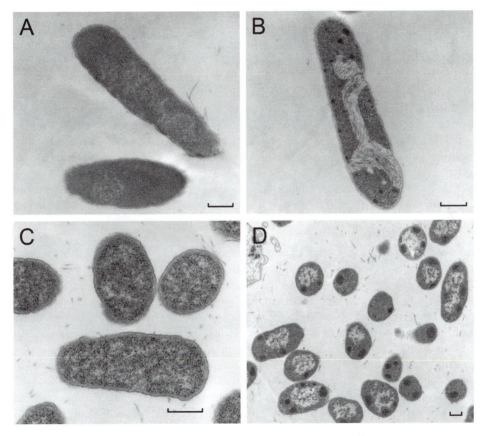

Figure 6. Morphological changes in *E. coli* caused by high pressure. Shown are transmission electron micrographs of thin sections of untreated exponential-phase cells (A), exponential-phase cells pressure treated at 200 MPa for 2 min (30% survival) (B), untreated stationary-phase cells (C), and stationary-phase cells pressure treated at 400 MPa for 4 min (<1% survival) (D). Bar marker, 0.25 μm.

are more or less coincident with the onset of cell death in exponential-phase cells but not in stationary-phase cells and are thus very unlikely to be the cause of death in the latter.

Effects of Pressure on Membrane Structure and Function

High pressure compresses the acyl chains of membrane phospholipids and promotes a phase change from liquid crystalline to gel states which may affect membrane permeability and the functioning of membrane-bound enzymes. Leakage of cell contents occurs in pressure-treated cells (83), and the leaked material may include large molecules such as proteins and RNA (61). The relationship between cytoplasmic membrane damage and loss of viability following pressure treatment was examined in *E. coli* strains C9490, H1071, and NCTC 8003, which showed high, medium, and low degrees of resistance to pressure, respectively, in stationary phase but similar degrees of pressure resistance in exponential phase (72). Loss of membrane integrity was assessed as loss of osmotic responsiveness or as increased uptake of the fluorescent dye propidium iodide (PI). In exponential-phase cells, a loss of viability occurred at pressures between 100 and 200 MPa and coincided with the loss of membrane integrity in all three strains. In stationary-phase cells of the pressure-resistant strain C9490, very little loss of viability occurred below 500 MPa, whereas in strain NCTC 8003, the onset of cell death began at 100 MPa. Very little uptake of PI was observed at any pressure in either of these strains. Thus, in exponential-phase cells, death coincides with permanent loss of membrane integrity, but in stationary-phase cells there is no simple relationship between membrane damage and loss of viability. Strains NCTC 8003 and H1071 both have mutations in *rpoS* (78), and their greater pressure sensitivity than that of strain C9490 in stationary phase may therefore be related to cell functions other than or additional to membrane resilience to pressure.

Further investigations of membrane integrity were undertaken with *E. coli* strain J1 using a microscopic method that allowed the responses of individual cells to be observed (61). Figure 7 shows the osmotic responses of untreated and pressurized cells when placed on agar containing 0.75 M NaCl. Untreated cells, which possess an intact cytoplasmic membrane, showed a very refractile cytoplasm caused by loss of water and consequent concentration of the cytoplasm under hypertonic conditions (Fig. 7A and C). Exponential-phase cells appeared uniformly refractile and slightly shrunken in appearance (Fig. 7A), whereas in stationary-phase cells the refractile cytoplasm had apparently become separated from the poles of the cell (Fig. 7C). In cells with a nonfunctional membrane, water and salt can pass freely through the membrane, and there is no osmotic response and no condensation of the cytoplasm, which therefore appears dark grey. Exponential-phase cells pressurized for 8 min at 100 MPa remained viable and retained their ability to respond to an osmotic challenge, but those pressurized at 200 MPa, where more than 99.99% of the population was dead, had completely lost the osmotic response (Fig. 7B). Stationary-phase cells pressurized for 8 min at 200 and 500 MPa, where the percentages of viable cells were 79 and 43%, respectively, maintained their ability to plasmolyze (Fig. 7D and E). Only those cells pressurized at 600 MPa showed a loss of osmotic responsiveness, as can be seen in Fig. 7F, in which some of the cells do not show a refractile cytoplasm. However, although more than 99.99% of the cells pressurized at 600 MPa were dead, approximately 60% were still able to plasmolyze. This confirms the relationship between membrane damage and loss

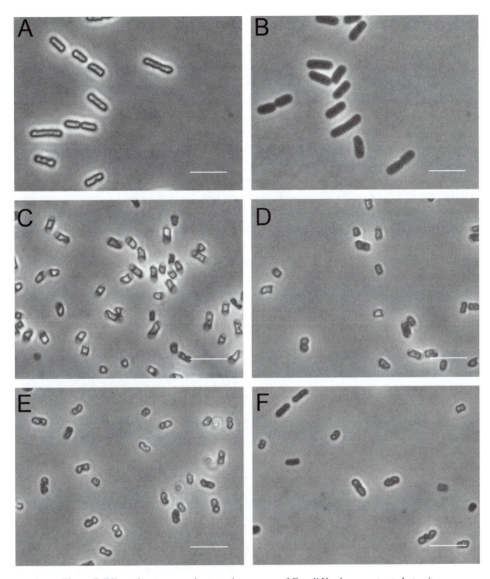

Figure 7. Effect of pressure on the osmotic response of *E. coli* J1: phase-contrast photomicrographs of exponential- and stationary-phase cells placed on agar containing 0.75 M NaCl. (A) Exponential-phase cells, untreated (100% viable cells); (B) exponential-phase cells treated with 200 MPa for 8 min (0.02% viable cells); (C) stationary-phase cells, untreated (100% viable cells); (D) stationary-phase cells treated with 200 MPa for 8 min (79% viable cells); (E) stationary-phase cells treated with 500 MPa for 8 min (43% viable cells); (F) stationary-phase cells treated with 600 MPa for 8 min (0.01% viable cells). Bar marker, 2 μm. Reproduced from *Applied and Environmental Microbiology* (61) with permission.

of viability in exponential-phase cells but also showed that some stationary-phase cells die with apparently intact membranes.

Structural changes in the membrane were examined using the lipophilic dye FM 4-64, which binds preferentially to the cytoplasmic membrane (Fig. 8). Pressurization at 300 MPa caused visible disruption of the envelopes of exponential cells, including the formation of vesicles or buds of lipid material coming out from the cells, and the formation of local structures resembling invaginations protruding into the cytoplasm. Gross loss of membrane material in this way would presumably result in permanent loss of membrane integrity. No structural changes in the membrane could be detected in stationary-phase cells at pressures up to 500 MPa. The formation of membrane vesicles has been observed during growth of *E. coli* at high pressure and following exposure to heat or osmotic stress (11, 14, 45, 82). Surface indentations rather than vesicles were seen in scanning electron micrographs of stationary-phase cells of *Salmonella enterica* serovar Typhimurium pressurized for 10 min at 325 MPa (89). The mechanisms responsible for these perturbations are not known, though Beney et al. (15) showed that the formation of buds on liposomes was a consequence of differences in compressibility between water and phospholipid. During compression the vesicle decreases in size and water passes out of the vesicle. During decompression the expanding phospholipid layer becomes too large for the water it contains, so buds form to increase the surface/volume ratio. This is a possible mechanism, but we do not know yet whether it would apply to bacterial cell membranes in their native state.

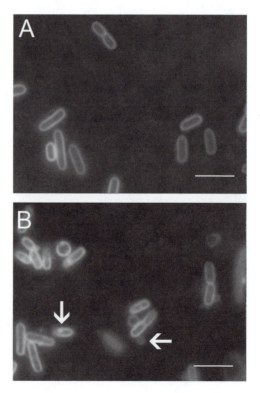

Figure 8. Effect of pressure on membrane integrity of exponential-phase cells of *E. coli* J1: fluorescence photomicrographs of cells stained with lipophilic membrane probe FM 4-64. (A) Untreated cells (100% viable cells); (B) cells pressurized at 300 MPa for 8 min (0.002% viable cells). Bar marker, 2 µm. Arrows show internal and external vesicles. Reproduced from *Applied and Environmental Microbiology* (61) with permission.

Permeabilization of the outer membrane during compression of *E. coli* cells has been demonstrated by the release of periplasmic enzymes (39) and the increased uptake of the fluorescent probe 1-*N*-phenylnaphthylamine (NPN) (28). This permeabilization explains the sensitization of *E. coli* cells to compounds added to the pressurization medium, such as lysozyme, which otherwise would not enter the cell. Increased outer membrane permeability occurs before the onset of sublethal injury or cell death, thus eliminating this structural change as being responsible for pressure inactivation (28). In *Salmonella enterica* serovar Typhimurium, pressure treatment caused a considerable loss of protein from the outer membrane (76). In some cases, outer membrane proteins seem to disappear totally, apart from the major proteins OmpA and LamB. This would have drastic effects on membrane properties, and it would be interesting to identify the extruded proteins and characterize the functional properties of the residual membranes.

Apart from loss of membrane physical integrity, pressure might conceivably cause other more subtle perturbations of membrane function that could lead to cell death. Smelt et al. (84) suggested that inactivation of membrane-bound F_oF_1 proton-translocating ATPase could lead to acidification of the cytoplasm and contribute to death at higher pressures. Evidence to support this idea was provided by Wouters et al. (96), who showed that ATPase activity, proton-extruding capacity, and ability to maintain a transmembrane proton gradient were all impaired in *Lactobacillus plantarum* after treatment at 250 MPa. The membrane-bound F_oF_1 ATPase of *E. coli* is also sensitive to pressure (62) but is not essential for pH homeostasis in this organism. Mutational loss of F_oF_1 ATPase is not lethal in *E. coli* provided that a source of fermentable carbohydrates is available. Indeed, the inability to generate a proton motive force is also not necessarily lethal (36). ATPase is therefore not a priori a critical component in *E. coli*, although its inactivation could conceivably contribute to loss of viability by restricting the amount of ATP available to repair other damaged structures. Certain enzymes of the *E. coli* tricarboxylic acid cycle are also somewhat sensitive to pressure (98) but are unlikely to be a cause of death for the same sort of reasons, particularly since glycolytic enzymes are apparently more pressure resistant.

Pressure-treated *E. coli* cells are nevertheless impaired in their ability to maintain internal pH homeostasis under acid conditions (48, 71), and cells of *Salmonella enterica* serovar Typhimurium are affected similarly (89). The precise mechanisms are not known, though loss of potassium from the cell (89) suggests that the potassium transporters and other components of the pH homeostatic systems may be nonfunctional in pressure-treated cells. Kilimann et al. (48) noted that pressure-induced loss of a transmembrane ΔpH correlated with cell death in *E. coli* TMW 2.479 and showed that operation of the glutamate and arginine decarboxylase pathways of acid protection did not affect survival or internal pH during pressure treatment, but improved the ability of cells to maintain a high internal pH during recovery from pressure damage. Loss of pH homeostasis was not believed to account for loss of viability in the pressure-resistant *E. coli* O157 strain C9490, because at an external pH of 3.5 the internal pH values of native and pressure-treated cells were similar at about 4.9 but only pressure-treated cells died at this pH within a 1-h period of incubation (71). Loss of pH homeostasis is an indication of membrane damage and compromises cell survival under acid conditions, but whether this loss of function constitutes a general mechanism of cell death is not yet clear.

Evidence of a different type of membrane perturbation in stationary-phase cells of *E. coli* came from the observation that cells become leaky under pressure but can apparently reseal after decompression. Thus, when PI is present in the suspending medium during pressurization, it is taken up at pressures above 100 MPa, but when added after decompression, there is very little uptake of the dye (72). The degree of staining during pressurization is generally much greater in pressure-sensitive strains than in pressure-resistant ones, indicating a greater degree of transient membrane leakiness in the former. Pressure-sensitive strains may also show some loss of osmotic responsiveness after resealing, which also suggests a greater degree of membrane damage under pressure in these strains (72). The decrease in viable numbers proved to be related to the intensity of staining during pressurization (Fig. 9) (P. Mañas and B. M. Mackey, unpublished data), suggesting that cell death might be related to the extent of transient membrane leakiness. This suggests a different cause of death, possibly distinct from permanent loss of membrane integrity, but the putative mechanism(s) is not yet known.

Effects of Membrane Fluidity on Pressure Resistance

As mentioned above, the pressure resistance of exponential-phase cells decreases with increasing growth temperature, whereas the pressure resistance of stationary-phase cells shows a broad optimum, around 30 to 37°C. However, in both exponential- and stationary-phase cells membrane fluidity increased continuously with increasing growth temperature, as reflected in the phase transition temperature measured by DSC and the ratio of unsaturated to saturated fatty acids in the phospholipids (19). By altering the temperature of pressurization as opposed to growth temperature, it was shown that increasing membrane fluidity increased pressure resistance of both exponential- and stationary-phase cells. However, in stationary-phase cells, the effect of membrane fluidity is superimposed on stationary-phase functions (possibly regulated by RpoS), which have their maximum effect at intermediate temperatures. An involvement of factors other than membrane fluidity

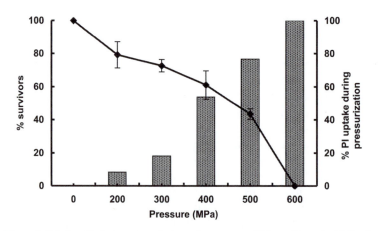

Figure 9. Relationship between transient membrane permeability and loss of viability in stationary-phase *E. coli* J1. The relative extent of transient membrane permeability is expressed as a percentage of the maximum fluorescence reading of cells pressurized in the presence of PI (bars). Viability was determined by viable counts of cells given the same treatment (◆).

in pressure resistance is indicated by the absence of changes in membrane fatty acids in pressure-resistant mutants of *E. coli* MG1665 (37).

In a different experimental approach, the pressure resistance of *E. coli* was examined in an unsaturated fatty acid auxotroph supplemented with oleic ($C_{18:1}$), linoleic ($C_{18:2}$), or elaidic (*trans* $C_{18:1}$) acid (18). Although elaidic acid is an unsaturated fatty acid, its *trans* configuration allows it to pack more closely and thus behave like a saturated fatty acid. Exponential-phase cells grown on linoleic acid were somewhat more pressure resistant than those grown on oleic acid, but the differences between them were small in comparison with cells grown on elaidic acid. These cells, which would have more rigid membranes, were much more pressure sensitive (Fig. 10A). In stationary-phase cells the inactivation curves were more complex; during the initial phase of inactivation, cells grown on linoleic and oleic acids were somewhat more resistant than those grown on elaidic acid, but in the latter phase of inactivation, elaidic acid-grown cells showed a slightly more resistant tail (Fig. 10B). Nevertheless, the overall differences between stationary-phase cells grown on

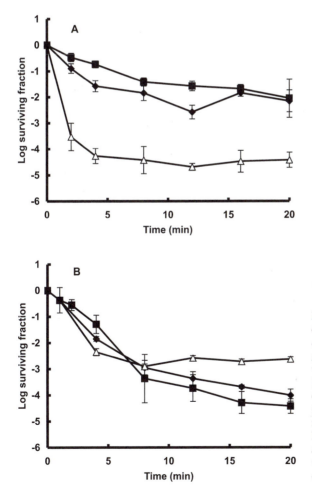

Figure 10. Effect of membrane fatty acid composition on pressure resistance. *E. coli* K1060, a fatty acid auxotroph, was grown at 37°C in medium supplemented with linoleic ($C_{18:2}$ [■]), oleic ($C_{18:1}$ [♦]), or elaidic (*trans* $C_{18:1}$ [△]) acid to exponential phase (A) or stationary phase (B), and pressure resistance was determined at 200 or 300 MPa, respectively. Reproduced from *Advances in High Pressure Bioscience and Biotechnology* (18) with the kind permission of Springer Science and Business Media.

the different supplements were quite small. These experiments thus confirm the conclusion that cells with more fluid membranes are more pressure resistant and that membrane fluidity per se has a much smaller influence in stationary-phase than in exponential-phase cells. The conclusion that pressure resistance of *E. coli* increases with increasing membrane fluidity is in accord with observations in *Lactobacillus plantarum* (79, 84, 87).

The reason why more fluid membranes are less susceptible to pressure damage is not immediately obvious. Pressure causes closer packing of the hydrocarbon chains of phospholipids and in this sense has a similar effect to cooling. Rapid cooling is known to prevent the lateral phase separation of phospholipid and protein domains, leading to the formation of packing faults in the gelled membrane and leakage of cell components (54). Whether pressure-induced phase transitions cause leakage in the same way is not known, though it is interesting that cells grown at lower temperatures are also more resistant to cold shock. Alternative possibilities are that fatty acid composition could affect the ability of the membranes to recover their original physical state after decompression or may affect phospholipid compressibility and the tendency to lose membrane material in the form of vesicles.

Effects of Pressure on Ribosomes and the Transcription and Translation Apparatus

Ribosomes undergo a large decrease in volume on dissociation into subunits of 240 ml/mol or greater depending on buffer composition, pressure, and conformational state (34, 81). During protein synthesis, the elongation factor Tu (EF-Tu)-dependent binding of aminoacyl-tRNA to the ribosomal A site followed by the peptidyl transfer reaction converts tight ribosomal couples to loose couples, while the EF-G-dependent translocation step and subsequent release of EF-G convert loose couples to tight couples once more. In a buffer system containing polyamines and 10 mM Mg^{2+}, functional ribosomal complexes were stable up to 100 MPa and tight couples were stable at pressures up to 60 MPa (34). However, when the magnesium concentration was decreased to 4 mM, considered to be closer to physiological conditions, there was significant destabilization of charged ribosomal complexes (34). Analysis of ribosomal particles in different conformations at 4 mM Mg^{2+} showed the posttranslocational complex to be the most pressure-sensitive step in protein synthesis, with a midpoint dissociation at about 70 MPa, similar to the pressure of 67 MPa, which blocks protein synthesis in vivo (11, 12).

Ribosomal subunit dissociation is fully reversible below 100 MPa and is not therefore a likely cause of cell death. Irreversible denaturation of ribosomes at higher pressures has been studied in *E. coli* NCTC 8164 using DSC (68). Previous cell fractionation experiments with the same strain had established that the main endotherm peak observable in DSC traces of whole cells corresponded with denaturation of the ribosomes (59). Measurement of the area of the main peak thus provides a measure of the quantity of intact ribosomes present in whole cells. When cells of *E. coli* 8164 were exposed to pressures of between 50 and 250 MPa for 20 min, the loss of ribosome-associated enthalpy was linearly related to the decrease in viability. This could be interpreted as showing that death is a direct effect of pressure-induced denaturation of the ribosome, but this is not necessarily true. If death of an individual cell occurred when the number of functional ribosomes fell below a certain critical threshold, the viability-versus-enthalpy curve would be sigmoidal, assuming that damage was evenly distributed throughout the ribosome population but individual cells contained different amounts of ribosomes (see Discussion in reference 68). The actual curve was more or less linear within experimental error, suggesting that this is not the case. A linear response would be consistent with an all-or-nothing model in which all of the ribosomes of killed cells were destroyed, while those of surviving cells remained largely unaffected. This model suggests that a catastrophic event affecting individual cells triggered the destruction of a majority of ribosomes, which would inevitably result in cell death. An all-or-nothing response could occur, for example, if all the ribosomes in a particular cell were destabilized as a result of membrane damage and loss of magnesium from the cell. The destruction of the 30S ribosomal subunit in sublethally heat-injured cells of *Salmonella enterica* serovar Typhimurium and *Staphylococcus aureus* is an indirect consequence of loss of intracellular magnesium and subsequent attack on the destabilized ribosome by ribonucleases (90). A similar process is conceivable in pressure-treated cells, or, since magnesium also stabilizes the ribosome at moderately high pressures (34, 81), its loss from a cell may predispose ribosomes in that cell to denaturation. Ribosome denaturation under pressure has also been demonstrated by using DSC in *Staphylococcus aureus, E. coli* O157, and *Leuconostoc mesenteroides* (9, 44). Because ribosomes qualify as critical components, and ribosome denaturation occurs over the same pressure range as loss of viability, their loss must be considered as a possible cause of cell death, but further work is needed to confirm this.

The pressure stability of RNA polymerase has been examined in *E. coli* and in the piezophile *Shewanella violacea* (46). The *E. coli* holoenzyme was dissociated at 140 MPa, whereas the *S. violacea* enzyme was not affected at this pressure. Both enzymes were inactivated at 150 MPa, but pressure sensitivity is greatly influenced by the conformation of the enzyme such that transcribing complexes are unusually resistant to pressure and are not fully dissociated even at 200 MPa (25). RNA polymerase is a critical component and must be considered as a cause of pressure inactivation in *E. coli,* but more needs to be known about its inactivation in relation to cell death and other events such as membrane damage.

Effects of Pressure on DNA

DNA synthesis and chromosome replication in *E. coli* are both very sensitive to pressure, with chromosome replication, the more sensitive of these processes, being halted at about 50 MPa (12). Inhibition of DNA synthesis at these moderate pressures is reversible, but there appears to be little information about the irreversible effects of higher pressures on the DNA replication machinery. DNA gyrase undergoes a progressive and irreversible inactivation at pressures between 55 and 83 MPa in vitro (21). This enzyme is essential for growth in *E. coli,* but we have no information on its possible role in death of pressure-treated cells.

The DNA molecule is stabilized by hydrogen bonding between complementary base pairs and hydrophobic interactions between the neighboring stacks of base pairs. Hydrogen bonds are slightly stabilized by pressure, and pressures up to 1,000 MPa appear to have little effect on the native structure of DNA. However, a transition from B to Z conformation occurs in poly(GC) nucleotides at pressures of 600 MPa, and various changes in plasmid properties have been reported, including an increase in supercoiling, similar to that caused by a decrease in temperature or an increase in ionic strength, and an increase in transforming ability. However, no direct effects of pressure on DNA primary structure have been reported. In view of this, the preliminary finding of Chilton et al. (22) that some DNA degradation occurred after pressure treatment and that mutations in *recA* (but not *uvrA*) sensitized *E. coli* AB1157 to pressure were unexpected. It was hypothesized that pressure might activate endogenous endonucleases or bring them into contact with DNA. Degradation was reduced in a strain lacking endonuclease A, but pressure resistance was not affected (22; Chilton and Mackey, unpublished). Some elegant studies recently completed by Aertsen et al. (3, 4) have uncovered a novel pressure-induced SOS response in *E. coli* in which double strand breaks in DNA were the inducing signal. In this case the strand breaks were a cellular response to sublethal pressure brought about by activation of the Mrr restriction endonuclease. Inactivation of Mrr slightly improved survival of *E. coli* after high-pressure treatment, whereas inactivation of LexA and RecA caused slight sensitization. These studies show that some DNA damage may arise indirectly as a result of pressure, but because DNA damage occurs by an indirect mechanism, its role in cell survival is likely to be highly dependent on cell physiology and pressurization conditions. From available evidence it would appear to play a relatively minor role in cell death compared with other forms of damage.

Cellular Stress Responses and Pressure Resistance

The RpoS-dependent general stress response plays a major role in determining pressure resistance, but since so many genes are affected it is difficult to know which particular components of the response are most important. More specific information on the likely determinants of pressure resistance in *E. coli* has come from studying responses to pressure itself (see chapter 5). Exposure of exponential-phase cells to a pressure that allowed cell elongation but not multiplication resulted in a transient increase in synthesis of 55 pressure-induced proteins, 11 of which were heat shock proteins and 4 of which were cold shock proteins (94). The heat shock proteins included the chaperones DnaK, GrpE, GroES, and GroEL and the ClpB, ClpP, and Lon proteases. Heat shock proteins DnaK, ClpP, ClpX, and Lon were also synthesized in exponential-phase *E. coli* cells during recovery from 15 min of exposure to pressures of 150 MPa (5). Exposure to the lower pressure of 20 MPa resulted in synthesis of GroES but not GroEL, DnaK, or GrpE (33). In thermally stressed cells, the DnaK system is the most effective chaperone system for coping with misfolded and aggregated protein in vivo (65). Large proteins are most vulnerable to thermal unfolding and aggregation, and DnaK and ClpB act in concert to prevent aggregation taking place in the first place and also to bring about solubilization of any aggregates that do form (65). Substrates of the DnaK chaperone include key proteins of the transcription and translation apparatus that qualify as critical components. The idea that heat shock proteins might be important in resisting lethal pressures is strengthened by the observation that heat shock causes an increase in pressure resistance (5). However, we do not yet know which proteins are most susceptible to pressure and, of these, which ones are critical for survival. There appear to be subtle differences in the role of heat shock proteins in microbial responses to heat and pressure, because pressure shock failed to elicit an increase in heat or pressure resistance in *E. coli* (5). It is also worth noting that an increase in pressure resistance of *E. coli* strain H1071 following heat shock was associated with an increase in resilience of the cytoplasmic membrane to pressure-induced permeabilization (72).

Two major studies have examined the response of *E. coli* to pressure using DNA microarray procedures. Ishii et al. (41) compared gene expression in early- and late-exponential-phase cells grown at 0.1 MPa and 30 or 50 MPa. In early exponential phase, expression of 469 and 313 genes was altered at 30 and 50 MPa, respectively, while in late exponential phase, expression of 698 and 662 genes was affected. Heat and cold stress responses were induced simultaneously by the elevated pressure. The responses were complex, and genes up-regulated in early exponential phase were not necessarily the same as those up-regulated in late exponential phase. Gene expression was affected throughout all functional classes in both growth phases, with genes involved in energy metabolism, transporters, DNA-binding proteins, and translation being especially affected in late-exponential-phase cells. An *E. coli* mutant with a deletion in *hns*, encoding a DNA-binding regulatory protein, exhibited great pressure sensitivity, suggesting that the H-NS protein was a possible transcriptional regulator for adaptation to high-pressure stress. Malone et al. (60) investigated the response of stationary-phase cells of a pressure-resistant strain of *E. coli* O157:H7, EC-88, to a sublethal pressure of 100 MPa for 15 min. More than 100 genes were up-regulated or down-regulated following pressure treatment, but in only 36 of these was the change regarded as significant (60). The major functional categories affected were (i) stress responses, (ii) thiol-disulfide redox system, (iii) Fe-S cluster assembly, (iv) spontaneous mutation, and (v) several miscellaneous genes (60). An important conclusion from this study was that high pressure adversely affects the cell's redox homeostasis. This is consistent with earlier studies by Aertsen et al. (1) showing that pressure induces oxidative stress in *E. coli*.

Pressure-Resistant and Pressure-Sensitive Mutants

Pressure-resistant mutants of *E. coli* selected by pressure cycling show elevated basal levels of many heat shock proteins (5), providing further support for the vulnerability of proteins to pressure damage and the role of heat shock proteins in pressure resistance. Interestingly, these mutants were also more resistant to oxidative stress (1), implying that pressure damage is likely to be multifactorial in nature. The Lon protease is a component of the heat shock response which serves to break down denatured cellular protein. Mutations in *lon* sensitized *E. coli* to pressures between 100 and 200 MPa, but not 250 MPa (2). In this case the pressure sensitivity of *lon* mutants was associated with SulA-dependent inhibition of cell division as a result of the SOS response rather than disposal of denatured protein (2). A separate set of pressure-resistant mutants of *E. coli* isolated by Gao et al. (29) had elevated levels of three proteins, one of which showed high identity with a known outer membrane protein. This is an intriguing observation because the outer membrane is not believed to be a critical target (28).

Malone et al. (60) investigated the effect on pressure resistance of several of the genes identified in their transcriptional analysis by comparing survival in otherwise isogenic pairs after treatment at 400 MPa for 5 min. Genes identified as having the greatest effect on resistance are listed in Tables 1 and 2, together with results from previous studies. Genes that were up- or down-regulated by mild pressure shock but which had no effect on resistance to lethal pressure are listed in Table 3. The *rpoS* gene was again confirmed as being an important determinant of pressure resistance, while the *otsA* gene, which is regulated by RpoS, also had a slight effect on resistance, suggesting a role for trehalose in pressure tolerance. Both *rpoE*, encoding the periplasmic stress response regulator, and *nlpI*, encoding a membrane lipoprotein, were up-regulated following sublethal pressure. A mutation in either of these sensitized cells to pressure, and hence denaturation of periplasmic polypeptides,

Table 1. Mutations causing a decrease in the resistance of *Escherichia coli* to lethal pressures

Gene	Product function	Reference(s)
rpoS	Sigma S; regulator of general stress response	1, 60, 78
rpoE	Sigma E; regulator of periplasmic stress response	60
dps	Stress response DNA-binding protein	60
trxA	Thioredoxin 1, redox factor, carrier protein	60
trxB	Thioredoxin reductase, FAD/NAD(P) binding	60
hns	Transcription regulator, DNA-binding protein HLP-II	41
hns stpA[a]	The *stpA* product is a DNA-binding protein with chaperone activity	60
nlpIA	NlpI lipoprotein	60
nlpIB	NlpI lipoprotein	60
otsA	Trehalose-6-phosphate phosphatase; osmoregulation	60
katE	HPII hydroperoxidase	1
oxyR	Peroxide activated transcription factor	1
sodAB	Superoxide dismutase	1
soxS	Regulator of superoxide response regulon	1

[a]Double mutant.

may contribute to pressure inactivation. Neither *hns,* encoding a transcriptional regulator, nor *stpA,* which encodes a similar protein, showed significant changes in expression, and a mutation in either gene separately caused no change in pressure resistance. However, a double mutant was pressure sensitive (Table 1), which, taken with the results of Ishii et al. (41), points to a critical role of the regulatory protein H-NS in pressure resistance. The *ibpAB* mutant showed a small but not statistically significant decrease in pressure resistance, implying that the small heat shock chaperone may be less important in pressure resistance than the major heat shock proteins mentioned above.

Aertsen et al. (1) showed, using leaderless alkaline phosphatase as a probe, that pressure treatment induces endogenous oxidative stress in *E. coli* and that the lethal effect of pressure was increased by mutations in *oxyR, sodAB,* or *soxS.* The authors proposed that under some circumstances inactivation of *E. coli* by pressure could occur as a consequence of a suicide mechanism involving an oxidative burst. The importance of oxidative stress as a mechanism of cell death was supported by the work of Malone et al. (60), who examined pressure resistance of mutants defective in thioreductase activity or in assembly of iron-sulfur complexes. *Escherichia coli* contains two cytoplasmic thioreductases, encoded by *trxA* and *trxB,* which act as disulfide-reducing proteins. Strains with mutations in *trxA* or *trxB* were significantly more pressure sensitive than the parent strains (60). It was suggested that pressure denatures proteins in a way that exposes $-SH$ and S-S to catalytic agents that cause protein denaturation and impairment of redox balance (60). The thiol-disulfide redox system may thus protect cells by facilitating proper protein folding and maintaining redox homeostasis.

Table 2. Mutations causing an increase in resistance of *Escherichia coli* to lethal pressures

Gene	Product function	Reference
sufABCDSE	Assembly of Fe-S clusters	60
iscU	FeS cluster template protein	60
hscA	Chaperone involved in assembly of Fe-S clusters	60
fdx	Ferredoxin, electron carrier protein, involved in assembly of Fe-S clusters	60
fnr	Transcriptional regulator of aerobic and anaerobic growth and osmotic response	60

Table 3. Genes whose transcription is affected by sublethal pressure but which cause no significant change in the pressure resistance of *Escherichia coli*

Gene	Product function	Reference
ybdQ	Universal stress protein, flavoprotein (UP12)	60
cspA	Major cold shock protein	60
ibpAB	Small heat shock protein	60
iscR	Repressor involved in assembly of Fe-S clusters	60
yafN-yafP	Conserved proteins with possible roles in spontaneous mutation	60
hns	Transcription regulator, DNA-binding protein HLP-II	60
stpA	The *stpA* product is a DNA-binding protein with chaperone activity	60
rbsD	Membrane-associated component of D-ribose uptake system	60
yfiD	Putative formate acetyltransferase	60
eno	Enolase	60

Iron-sulfur clusters are prosthetic groups of proteins that engage in redox reactions, an example being FNR, a transcriptional regulator of genes involved in anaerobic metabolism. In *E. coli* there are two operons that are involved in the assembly of Fe-S clusters, the main one being *iscSUA-hscB-fdx,* while *sufABCDSE* operates under conditions of iron limitation. Mutations in genes of either of these operons brought about an *increase* in pressure resistance, as did a mutation in *fnr* itself (60). Malone et al. (60) proposed that pressure releases iron from Fe-S clusters, leading to the formation of reactive oxygen species within the cell via the Fenton reaction (60). A lowering of the cellular levels of Fe-S clusters would thus be expected to diminish the oxidative damage caused by this mechanism. The Dps protein in *E. coli* is a DNA-binding protein that also binds iron and protects DNA against oxidative damage mediated by hydrogen peroxide. A *dps* mutant was more pressure sensitive than the parent strain, in line with the proposed involvement of cellular iron in mediating oxidative stress in pressure-treated cells (60). In addition to their role in protecting essential proteins from denaturation or aggregation, chaperones such as DnaK could possibly prevent the formation of reactive oxygen species by reducing denaturation of Fe-S-containing proteins.

Cold shock shares some common features with pressure in that it decreases membrane fluidity and inhibits translation. The cold shock response is an adaptive mechanism which enables ribosomes to continue translation at low temperature and thus enables cells to adjust their physiology to the low-temperature conditions. An increase in pressure resistance upon cold shock was shown in *Listeria monocytogenes* (95), *Staphylococcus aureus* (69), and *Lactobacillus sanfranciscensis* (80). An apparent increase in pressure resistance in *E. coli* during centrifugation turned out to be due to chilling, so it seems that cold shock enhances pressure resistance in *E. coli* as well (17). The cold shock protein CspA of *E. coli* is an RNA chaperone essential for translation at low temperature. A *cspA* mutant was not significantly sensitive to pressure (60), though since there is considerable redundancy among cold shock proteins, it will be necessary to examine strains with multiple deletions before the role of these proteins in pressure resistance is known.

CELL RECOVERY AFTER HIGH-PRESSURE TREATMENT

Microorganisms that survive the action of stressing agents are likely to have sustained damage to many cell components. These sublethally injured cells are more fastidious in their growth requirements but can repair the damage and outgrow if the environmental conditions are suitable (57, 97). Studying the occurrence and repair of cellular injury reveals useful information about the mechanisms involved in bacterial inactivation and resistance. It is now well established that permeabilization of the outer membrane is either transient (28, 39) or quickly repaired, since resistance to lysozyme or hydrophobic compounds such as bile salts is regained shortly after decompression. Chilton et al. (23) studied the biosynthetic requirements for the repair of the outer and cytoplasmic membrane functions of *E. coli* K-12 strain AB1157 after a pressure treatment of 400 MPa for 2 min. Just after decompression, more than 99% of the surviving population was sensitive to the presence of bile salts in the recovery agar, but resistance was completely regained within 1 h of incubation in TSB. Resistance to bile salts in *E. coli* is due to restricted diffusion through the porin proteins of the outer membrane combined with active removal from the cell, principally via the AcrAB-TolC multidrug efflux system (88). Since this efflux system is dependent on a proton motive force, loss and recovery of bile salts resistance could be caused by collapse and recovery of the transmembrane proton gradient. However, the tetracycline efflux system is also dependent on proton motive force but high pressure does not sensitize *E. coli* to tetracycline, so it appears that the AcrA-TolC system, which consists of three molecular components, is more sensitive to the direct effects of pressure than the Tet pump, which consists of a single component (47). Recovery of bile salts resistance was not prevented by sodium azide or inhibitors of RNA and protein synthesis (23), so a spontaneously reversible physical disorganization of the outer membrane could also contribute to loss and recovery of resistance to bile salts and other agents.

Enhanced sensitivity to acid conditions or salt concentrations is generally attributed to impairment of cytoplasmic membrane functions (7, 23, 30, 43, 71). Chilton et al. (23), in the same series of experiments as described above, demonstrated that incubation of damaged cells in TSB permitted the cytoplasmic membrane properties to be restored and the tolerance to salt to be regained. However, this process took much longer than the repair of outer membrane damage. The addition of sodium azide delayed the repair process and the addition of rifampin or chloramphenicol completely inhibited it, indicating a requirement for energy production and RNA and protein synthesis, respectively. Repair of cytoplasmic membrane damage in *E. coli* is highly metabolically demanding, contrary to the automatic repair process suggested for the outer membrane. Therefore, pressure causes sublethal damage to both outer and cytoplasmic membranes, but the damage is of a different kind in each, presumably reflecting the different structural organizations of the two envelope components.

Other cellular structures and functions show evidence of damage or disruption by pressure that can be reversed or repaired. This is the case for the nucleoid and cytoplasmic protein. For instance, cytoplasmic protein is extensively aggregated after treatments which cause very little decrease of viability in stationary-phase cells of *E. coli* (Fig. 5). Figure 11 shows stationary-phase cells of *E. coli* J1 pressurized and then maintained in recovery medium for 4 h and stained with FITC for protein visualization (Mañas and Mackey, unpublished). As can be seen in the photograph, protein aggregates had disappeared from some cells, while others that still contained aggregates had started to divide. Experiments with DSC carried out by Niven et al. (68) also suggest the apparent reparable nature of some extreme cellular changes, such as loss of ribosome conformation. Peaks corresponding to ribosomes in the DSC thermograms of pressurized *E. coli* NCTC 8164 cells slowly increased in area during incubation of the cells after decompression. This increase in

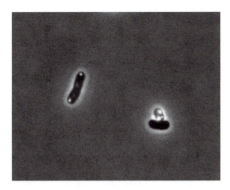

Figure 11. Multiplication of stationary-phase cells containing protein aggregates stained with FITC.

ribosome-associated enthalpy was not related to the eventual recovery of damaged cells, because the cell population continued to lose viability during this period. This may indicate that restoration of ribosome conformation recovery was taking place in dead cells, and therefore cellular repair of some cellular targets was occurring simultaneously with further cell death. However, it is not known whether the increased enthalpy is actually due to ribosomes recovering their native conformation or whether the disrupted ribosome components assume some other conformational state within the cell.

An intriguing recent development in the area of sublethal injury is the observation that apparently dead cells of *E. coli* or *Salmonella* serovar Typhimurium, i.e., those that were unable to form colonies on a rich nonselective agar after being pressure treated at 400 to 550 MPa, could recover culturability when incubated for several days at 20 to 25°C in PBS or phosphate buffer (51, 77). This phenomenon clearly bears similarities to the so-called viable-but-nonculturable state reported for other organisms (57), but its physiological basis is unknown.

The studies by Aertsen et al. (1) mentioned above showed that recovery of pressure-treated *E. coli* cells was enhanced under anaerobic conditions, which prevent cell killing by an indirect mechanism involving oxidative stress. In this sense some survivors may be regarded as sublethally injured and capable of recovery under appropriate conditions. It is important to bear in mind that oxidative stress can arise from a number of sources, including the generation of reactive oxygen-containing species due to metabolic imbalance (6), the release of iron from Fe-S clusters as proposed by Malone et al. (60), and exogenous oxidants such as those present in autoclaved media that have been exposed to sunlight (57). Increased oxidative stress therefore appears to be an important consequence of pressure treatment in *E. coli*, but its contribution to cell killing is likely to be highly dependent on the physiological state of cells and on conditions during pressure treatment and recovery.

MODEL FOR MECHANISMS OF INACTIVATION OF *ESCHERICHIA COLI* BY HIGH PRESSURE

In reviewing the relationship between loss of viability and loss of cell functions, we have identified loss of membrane integrity as a critical event leading to cell death in *E. coli*. In addition, other possible critical events have been identified, and a scheme showing how these may all be interrelated is shown in Fig. 12. The relationship between membrane damage

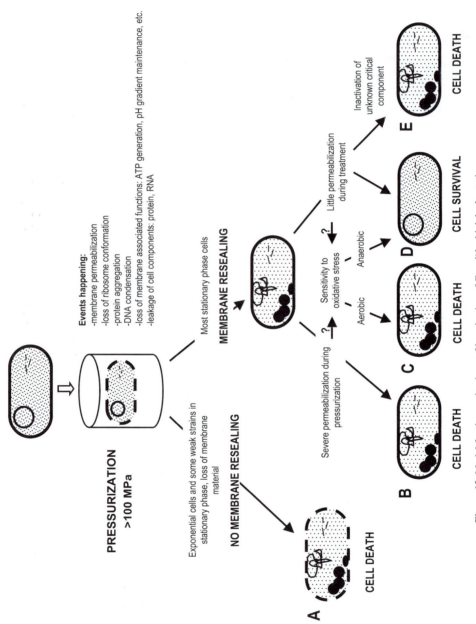

Figure 12. Model for the mechanisms of inactivation of *E. coli* by high hydrostatic pressure.

and cell death is not straightforward, and we have identified a range of possible scenarios. Pressurized cells undergo permeabilization of their outer and cytoplasmic membranes to various degrees and may recover their original structure and function also to various degrees. Accordingly, we can divide cells into two groups: (i) those unable automatically to reseal membranes after decompression and (ii) those in which membranes do reseal. Group i would include exponential-phase cells of all strains so far studied and stationary-phase cells of some weak strains. In these cells (cell A in Fig. 12), membrane damage during compression is very severe and resealing is precluded, possibly because membrane material has been lost as vesicles. In this case we can attribute cell death directly to permanent loss of membrane integrity, because the maintenance of homeostasis and cell metabolism is impossible in cells with leaky membranes. Of course it is possible that other critical targets are affected at similar pressures, but this remains irrelevant in the sense that other functions cannot operate in any case in a cell that cannot control its internal environment.

For cells which can reseal their membranes, i.e., cells in group ii, a more complicated picture emerges. In these cells, the extent to which the membranes are permeabilized during compression and reseal upon decompression is variable, depending on the intrinsic resistance of the strain. In the most resistant organisms such as *E. coli* O157:H7 strain C9490, where the membrane damage is slight and there is little leakage of cell contents, recovery is possible (cell D in Fig. 12). In the more sensitive strains where more extensive disruption occurs, death ensues. We suggest that in these cells transient membrane damage is the initial event leading to loss of viability, but not necessarily the actual cause of death. A relationship between transient permeabilization and cell death has been found (Fig. 9), but we do not yet know the nature of the consequent lethal event. Several possible causes of death may be imagined, including (i) extensive loss of cell contents, including vital components; (ii) irrevocable changes in the intracellular environment that destabilize critical components; or (iii) activation of degradative enzymes such as ribonucleases, proteases, or phospholipases (cell type B in Fig. 12). In this case, a secondary event could be the final cause of death. Alternatively, death could be caused by loss of membrane functionality because although the membranes reseal to the extent of excluding PI, they could still be leaky to other molecules or have lost some other vital property. This appears to be true in some weak strains that show partial loss of osmotic responsiveness after pressure treatment, although cells can exclude PI. Loss of pH homeostasis and loss of salt tolerance are also indications of such a loss of function, though if cells have not been lethally permeabilized during treatment, these properties are recoverable under favorable conditions. As noted above, an oxidative burst is a strong candidate for an indirect cause of death. This may be related to membrane disruption but equally could involve a separate mechanism, independent of transient permeabilization during pressure treatment. In this case recovery is dependent on aerobic or anaerobic recovery conditions (cells C and D in Fig. 12). It should be noted that a proportion of the population of tough strains that can reseal their membranes at moderate pressures may be unable to do so at very high pressures. For example, for *E. coli* strain J1 about 40% of the population had lost the ability to plasmolyze after treatment at 600 MPa, which killed 99.99% of the population (Fig. 7). At these high pressures it seems that some cells of strain J1 die as a consequence of membrane damage, while others in the population, with apparently intact membranes, die from other causes (cell E in Fig. 12).

Increased levels of heat shock chaperones enhance cellular pressure resistance, strongly suggesting that protein denaturation could contribute to loss of viability. It is clear, however, that cells can withstand extensive aggregation of cytoplasmic protein without dying provided that the membrane remains intact, allowing resynthesis or disaggregation of proteins to take place. Loss of particular proteins may be more relevant in this regard than the extent of total protein aggregation. Some essential proteins such as DNA gyrase and RNA polymerase are inactivated at pressures between 100 and 200 MPa in vitro, but whether their loss in vivo is a primary lethal event remains to be established. In *E. coli* NCTC 8164, ribosomes undergo denaturation in the same pressure range as loss of viability, and ribosome loss has been suggested as a cause of death in other organisms (9, 44). This presumptive evidence needs to be tested further in strains having different degrees of resistance to pressure.

Referring back to the scheme shown in Fig. 1, we identify the arrow denoting event C as corresponding to a pressure-induced change in the state of the cytoplasmic membrane that causes either (i) irreversible disruption of membrane structure or its functional integrity or (ii) secondary changes that lead indirectly to cell death. Cellular resistance increases or decreases if this arrow moves along the pressure axis to the right or left, respectively. However, if the arrow passes beyond the pressure at which another critical component is inactivated (event D), then the cause of death will change. For example, if a critical protein were to be denatured before membrane integrity was lost, then death would occur in cells with intact membranes. This may be happening at high pressures in some cells of the more pressure-resistant strains which are able to reseal their membranes without loss of function.

There are clearly many gaps in our knowledge that will need to be addressed in the future. We do not know the structural features of cell envelopes that determine their resilience to pressure, nor do we fully understand the mechanisms by which pressure disrupts membrane structure and function. The role of protein denaturation, including cytoplasmic and membrane proteins, in cell death remains to be determined, as does the role of chaperones in protecting against pressure damage. The interrelationships between the various manifestations of pressure-induced damage, both direct and indirect, and cell death will be a fascinating topic of study for the future. We hope that the model we have presented will provide a useful conceptual framework that can be modified as new information becomes available.

Acknowledgment. We are grateful to P. Chilton for providing unpublished electron micrographs of pressure-treated *E. coli.*

REFERENCES

1. **Aertsen, A., P. De Spiegeleer, K. Vanoirbeek, M. Lavilla, and C. W. Michiels.** 2005. Induction of oxidative stress by high hydrostatic pressure in *Escherichia coli. Appl. Environ. Microbiol.* **71:**2226–2231.
2. **Aertsen, A., and C. W. Michiels.** 2005. SulA-dependent hypersensitivity to high pressure and hyperfilamentation after high-pressure treatment of *Escherichia coli lon* mutants. *Res. Microbiol.* **156:**233–237.
3. **Aertsen, A., and C. W. Michiels.** 2005. Mrr instigates the SOS response after high pressure stress in *Escherichia coli. Mol. Microbiol.* **58:**1381–1391.
4. **Aertsen, A., R. Van Houdt, K. Vanoirbeek, and C. W. Michiels.** 2004. An SOS response induced by high pressure in *Escherichia coli. J. Bacteriol.* **186:**6133–6141.
5. **Aertsen, A., K. Vanoirbeek, P. De Spiegeleer, J. Sermon, K. Hauben, A. Farewell, T. Nystrom, and C. W. Michiels.** 2004. Heat shock protein-mediated resistance to high hydrostatic pressure in *Escherichia coli. Appl. Environ. Microbiol.* **70:**2660–2666.

6. **Aldsworth, T. G., R. L. Sharman, and C. E. Dodd.** 1999. Bacterial suicide through stress. *Cell. Mol. Life Sci.* **56:**378–383.

7. **Alpas, H., N. Kalchayanand, F. Bozoglu, and B. Ray.** 2000. Interactions of high hydrostatic pressure, pressurization temperature and pH on death and injury of pressure-resistant and pressure-sensitive strains of foodborne pathogens. *Int. J. Food Microbiol.* **60:**33–42.

8. **Alpas, H., N. Kalchayanand, F. Bozoglu, A. Sikes, C. P. Dunne, and B. Ray.** 1999. Variation in resistance to hydrostatic pressure among strains of food-borne pathogens. *Appl. Environ. Microbiol.* **65:**4248–4251.

9. **Alpas, H., J. Lee, F. Bozoglu, and G. Kaletunç.** 2003. Evaluation of high hydrostatic pressure sensitivity of *Staphylococcus aureus* and *Escherichia coli* O157:H7 by differential scanning calorimetry. *Int. J. Food Microbiol.* **87:**229–237.

10. **Baranyi, J., and T. A. Roberts.** 1994. A dynamic approach to predicting bacterial growth in food. *Int. J. Food Microbiol.* **23:**277–294.

11. **Bartlett, D. H.** 1992. Microbial life at high pressure. *Sci. Prog. Oxford* **76:**479–496.

12. **Bartlett, D. H.** 2002. Pressure effects on in vivo microbial processes. *Biochim. Biophys. Acta* **1595:**367–381.

13. **Bayindirli, A., H. Alpas, F. Bozoglu, and M. Hizal.** 2006. Efficiency of high pressure treatment on inactivation of pathogenic microorganisms and enzymes in apple, orange, apricot and sour cherry juices. *Food Control* **17:**52–58.

14. **Beney, L., Y. Mille, and P. Gervais.** 2004. Death of *Escherichia coli* during rapid and severe dehydration is related to lipid phase transition. *Appl. Microbiol. Biotechnol.* **65:**457–464.

15. **Beney, L., J. Perrier-Cornet, M. Hayert, and P. Gervais.** 1997. Shape modification of phospholipid vesicles induced by high pressure: influence of bilayer compressibility. *Biophys. J.* **72:**1258–1263.

16. **Benito, A., G. Ventoura, M. Casadei, T. Robinson, and B. Mackey.** 1999. Variation in resistance of natural isolates of *Escherichia coli* O157 to high hydrostatic pressure, mild heat and other stresses. *Appl. Environ. Microbiol.* **65:**1564–1569.

17. **Casadei, M. A., and B. M. Mackey.** 1997. Effect of centrifugation on the pressure resistance of exponential phase cells of *Escherichia coli* 8164. *Lett. Appl. Microbiol.* **25:**397–400.

18. **Casadei, M. A., and B. M. Mackey.** 1999. Use of a fatty acid auxotroph to study the role of membrane fatty acid composition on the pressure resistance of *Escherichia coli,* p. 51–54. *In* H. Ludwig (ed.), *Advances in High Pressure Bioscience and Biotechnology.* Proceedings of the International Conference on High Pressure Bioscience and Technology, Heidelberg, 1998. Springer-Verlag, Berlin, Germany.

19. **Casadei, M. A., P. Mañas, G. W. Niven, E. Needs, and B. M. Mackey.** 2002. Role of membrane fluidity in pressure resistance of *Escherichia coli* NCTC 8164. *Appl. Environ. Microbiol.* **68:**5965–5972.

20. **Chen, H.** 2007. Use of linear, Weibull, and log logistic functions to model pressure inactivation of seven foodborne pathogens in milk. *Food Microbiol.* **24:**197–204.

21. **Chilikuri, L. N., P. A. G. Fortes, and D. H. Bartlett.** 1997. High pressure modulation of *Escherichia coli* DNA gyrase activity. *Biochem. Biophys. Res. Commun.* **239:**552–556.

22. **Chilton, P., N. S. Isaacs, B. M. Mackey, and R. Stenning.** 1997. The effects of high hydrostatic pressure on bacteria, p. 225–228. *In* K. Heremans (ed.), *High Pressure Research in the Biosciences and Biotechnology.* Proceedings of the XXXIVth Meeting of the European High Pressure Research Group, Leuven, Belgium, September 1–5, 1996. Leuven University Press, Leuven, Belgium.

23. **Chilton, P., N. S. Isaacs, P. Mañas, and B. M. Mackey.** 2001. Metabolic requirements for the repair of pressure-induced damage to the outer and cytoplasmic membranes of *Escherichia coli.* *Int. J. Food Microbiol.* **71:**101–104.

24. **Davey, K. R., and S. T. G. Phua.** 2005. Re: Erkmen and Dogan 2004 Food Microbiology 21, 181–185. *Food Microbiol.* **22:**483–487.

25. **Erijman, L. E., and R. M. Clegg.** 1998. Reversible stalling of transcription and elongation complexes by high pressure. *Biophys. J.* **75:**453–462.

26. **Erkmen, O., and C. Dogan.** 2004. Kinetic analysis of *Escherichia coli* inactivation by high hydrostatic pressure in broth and foods. *Food Microbiol.* **21:**181–185.

27. **Furukawa, S., S. Noma, M. Shimoda, and I. Hayakawa.** 2002. Effect of initial concentration of bacterial suspensions on their inactivation by high hydrostatic pressure. *Int. J. Food Sci. Technol.* **37:**573–577.

28. **Gänzle, M. G., and R. F. Vogel.** 2001. On-line fluorescence determination of pressure-mediated outer membrane damage in *Escherichia coli.* *Syst. Appl. Microbiol.* **24:**477–485.

29. **Gao, X., J. Li, and K. C. Ruan.** 2001. Barotolerant *E. coli* induced by high hydrostatic pressure. *Acta Biochim. Biophys. Sin.* **33:**77–81.

30. **Garcia-Graells, C., K. J. A. Hauben, and C. W. Michiels.** 1998. High-pressure inactivation and sublethal injury of pressure-resistant *Escherichia coli* mutants in fruit juices. *Appl. Environ. Microbiol.* **64:**1566–1568.
31. **Garcia-Graells, C., B. Masschalck, and C. W. Michiels.** 1999. Inactivation of *Escherichia coli* in milk by high-hydrostatic pressure treatment in combination with antimicrobial peptides. *J. Food Prot.* **62:**1248–1254.
32. **Gerdes, S. Y., M. D. Scholle, J. W. Campbell, G. Balázsi, E. Ravasz, M. D. Daugherty, A. L. Somera, N. C. Kyrpides, I. Anderson, M. S. Gelfand, A. Bhattacharya, V. Kapatral, M. D'Souza, M. V. Baev, Y. Grechkin, F. Mseeh, M. Y. Fonstein, R. Overbeek, A.-L. Barabási, Z. N. Oltvai, and A. L. Osterman.** 2003. Experimental determination and system level analysis of essential genes in *Escherichia coli* MG1655. *J. Bacteriol.* **185:**5673–5684.
33. **Gross, M., I. J. Kosmowsky, R. Lorenz, H. P. Molitoris, and R. Jaenicke.** 1994. Response of bacteria and fungi to high-pressure stress as investigated by 2-dimensional polyacrylamide-gel electrophoresis. *Electrophoresis* **15:**1559–1565
34. **Gross, M., K. Lehle, R. Jaenicke, and K. H. Nierhaus.** 1993. Pressure-induced dissociation of ribosomes and elongation cycle intermediates. Stabilising conditions and identification of the most sensitive functional state. *Eur. J. Biochem.* **218:**463–468.
35. **Harold, F. M.** 2001. Postscript to Schroedinger: so what is life? *ASM News* **67:**611–616.
36. **Harold, F. M., and P. C. Maloney.** 1996. Energy transduction by ion currents, p. 283–306. *In* F. C. Neidhardt, R. Curtiss III, J. L. Ingraham, E. C. C. Lin, K. B. Low, B. Magasanik, W. S. Reznikoff, M. Riley, M. Schaechter, and H. E. Umbarger (ed.), Escherichia coli *and* Salmonella: *Cellular and Molecular Biology.* ASM Press, Washington, DC.
37. **Hauben, K. J. A., D. H. Bartlett, C. C. F. Soontjens, K. Cornelis, E. Y. Wuytack, and C. W. Michiels.** 1997. *Escherichia coli* mutants resistant to inactivation by high hydrostatic pressure. *Appl. Environ. Microbiol.* **63:** 945–950.
38. **Hauben, K. J. A., K. Bernaerts, and C. W. Michiels.** 1998. Protective effect of calcium on inactivation of *Escherichia coli* by high hydrostatic pressure. *J. Appl. Microbiol.* **85:**678–684.
39. **Hauben, K. J. A., E. Y. Wuytack, C. C. F. Soontjens, and C. W. Michiels.** 1996. High pressure transient sensitization of *Escherichia coli* to lysozyme and nisin by disruption of outer-membrane permeability. *J. Food Prot.* **59:**350–355.
40. **Heremans, K.** 1995. High pressure effects on biomolecules, p. 81–97. *In* D. A. Ledward, D. E. Johnston, R. G. Earnshaw, and A. P. M. Hasting (ed.), *High Pressure Processing of Foods.* Nottingham University Press, Sutton Bonington Leicestershire, United Kingdom.
41. **Ishii, A., T. Oshima, T. Sato, K. Nakasone, H. Mori, and C. Kato.** 2005. Analysis of hydrostatic pressure effects on transcription in *Escherichia coli* by DNA microarray procedure. *Extremophiles* **9:**65–73.
42. **Johnson, F. H., and I. Lewin.** 1946. The disinfection of *E. coli* in relation to temperature, hydrostatic pressure and quinine. *J. Cell. Comp. Physiol.* **28:**23–45.
43. **Jordan, S. L., C. Pascual, E. Bracey, and B. M. Mackey.** 2001. Inactivation and injury of pressure-resistant strains of *Escherichia coli* O157 and *Listeria monocytogenes* in fruit juices. *J. Appl. Microbiol.* **91:**463–469.
44. **Kaletunç, G., J. Lee, H. Alpas, and F. Bozoglu.** 2004. Evaluation of structural changes induced by high hydrostatic pressure in *Leuconostoc mesenteroides.* *Appl. Environ. Microbiol.* **70:**1116–1122.
45. **Katsui, N., T. Tsuchido, R. Hiramatsu, S. Fujikawa, M. Takano, and I. Shibasaki.** 1982. Heat-induced blebbing and vesiculation of the outer membrane of *Escherichia coli.* *J. Bacteriol.* **151:**1523–1531.
46. **Kawano, H., K. Nakasone, M. Matsumoto, Y. Yoshida, R. Usami, C. Kato, and F. Abe.** 2004. Differential pressure resistance of the activity of RNA polymerase isolated from *Shewanella violacea* and *Escherichia coli.* *Extremophiles* **8:**367–375.
47. **Kawari. T., H. Ogihara, S. Furukawa, R. Aono, M. Kishima, Y. Inagi, A. Irie, A. Ida, and M. Yamasaki.** 2005. High hydrostatic pressure treatment impairs AcrAB-TolC pump resulting in differential loss of deoxycholate tolerance in *Escherichia coli.* *J. Biosci. Bioeng.* **100:**613–616.
48. **Kilimann, K. V., C. Hartmann, R. F. Vogel, and M. G. Ganzle.** 2005. Differential inactivation of glucose and glutamate-dependent acid resistance of *Escherichia coli* TMW 2.497 by high-pressure treatments. *Syst. Appl. Microbiol.* **28:**663–671.
49. **Klotz, B., D. L. Pyle, and B. M. Mackey.** 2007. A new mathematical modeling approach for predicting microbial inactivation by high hydrostatic pressure. *Appl. Environ. Microbiol.* **73:**2468–2478.
50. **Kopecky, A. L., D. P. Copeland, and J. E. Lusk.** 1973. Viability of *Escherichia coli* treated with colicin K. *Proc. Natl. Acad. Sci. USA* **72:**4631–4634.

51. **Koseki, S., and K. Yamamoto.** 2006. Recovery of *Escherichia coli* ATCC 25922 in phosphate buffered saline after treatment with high hydrostatic pressure. *Int. J. Food Microbiol.* **110:**108–111.

52. **Koseki, S., and K. Yamamoto.** 2007. A novel approach to predicting microbial inactivation kinetics during high pressure processing. *Int. J. Food Microbiol.* **116:**275–282.

53. **Larson, W. P., T. B. Hartzell, and H. Diehl.** 1918. The effect of high pressures on bacteria. *J. Infect. Dis.* **22:**271–279.

54. **Leder, I. G.** 1972. Interrelated effects of cold shock and osmotic pressure on the permeability of the membrane to permease accumulated substrates. *J. Bacteriol.* **111:**211–219.

55. **Linton, M., J. M. J. McClements, and M. F. Patterson.** 1999. Survival of *Escherichia coli* O157:H7 during storage in pressure-treated orange juice. *J. Food Prot.* **62:**1038–1040.

56. **Mackey, B., K. Forestière, N. S. Isaacs, R. Stenning, and B. Brooker.** 1994. The effect of high hydrostatic pressure on *Salmonella thompson* and *Listeria monocytogenes* examined by electron microscopy. *Lett. Appl. Microbiol.* **19:**429–432.

57. **Mackey, B. M.** 2000. Injured bacteria, p. 315–341. *In* B. M. Lund, A. C. Baird-Parker, and G. W. Gould (ed.), *The Microbiological Safety and Quality of Food.* Aspen Publishers Inc., Gaithersburg, MD.

58. **Mackey, B. M., C. A. Miles, S. E. Parsons, and D. A. Seymour.** 1991. Thermal denaturation of whole cells and cell components of *Escherichia coli* examined by differential scanning calorimetry. *J. Gen. Microbiol.* **137:**2361–2374.

59. **Mackey, B. M., C. A. Miles, D. A. Seymour, and S. E. Parsons.** 1993. Thermal denaturation and loss of viability in *Escherichia coli* and *Bacillus stearothermophilus*. *Lett. Appl. Microbiol.* **16:**56–58.

60. **Malone, A. S., Y.-K. Chung, and A. E. Yousef.** 2006. Genes of *Escherichia coli* O157:H7 that are involved in high-pressure resistance. *Appl. Environ. Microbiol.* **72:**2661–2671.

61. **Mañas, P., and B. M. Mackey.** 2004. Morphological and physiological changes induced by high hydrostatic pressure in exponential and stationary-phase cells of *Escherichia coli*: relationship with cell death. *Appl. Environ. Microbiol.* **70:**1545–1554.

62. **Marquis, R. E., and G. R. Bender.** 1987. Barophysiology of prokaryotes and proton translocating ATPases, p. 65–73. *In* H. W. Jannasch, R. E. Marquis, and A. M. Zimmerman (ed.), *Current Perspectives in High Pressure Biology.* Academic Press, London, United Kingdom.

63. **Miles, C. A.** 2006. Relating cell killing to the inactivation of critical components. *Appl. Environ. Microbiol.* **72:**914–917.

64. **Miles, C. A., B. M. Mackey, and S. E. Parsons.** 1986. Differential scanning calorimetry of bacteria. *J. Gen. Microbiol.* **132:**939–952.

65. **Mogk, A., T. Tomoyasu, P. Goloubinoff, S. Rüdiger, D. Röder, H. Langen, and B. Bukau.** 1999. Identification of thermolabile *Escherichia coli* proteins: prevention and reversion of aggregation by DnaK and ClpB. *EMBO J.* **18:**6934–6949.

66. **Molina-Höppner, A., W. Doster, R. F. Vogel, and M. G. Ganzle.** 2004. Protective effect of sucrose and sodium chloride for *Lactococcus lactis* during sublethal and lethal high-pressure treatments. *Appl. Environ. Microbiol.* **70:**2013–2020.

67. **Morein, S., A. Andersson, L. Rilfors, and G. Lindblom.** 1996. Wild-type *Escherichia coli* cells regulate the membrane lipid composition in a "window" between gel and non-lamellar structures. *J. Biol. Chem.* **271:** 6801–6809.

68. **Niven, G. W., C. A. Miles, and B. M. Mackey.** 1999. The effects of hydrostatic pressure on ribosome conformation in *Escherichia coli*: an *in vivo* study using differential scanning calorimetry. *Microbiology* **145:** 419–425.

69. **Noma, S., and I. Hayakawa.** 2002. Barotolerance of *Staphylococcus aureus* is increased by incubation at below 0°C prior to hydrostatic pressure treatment. *Int. J. Food Microbiol.* **80:**261–264.

70. **Noma, S., D. Kajiyama, N. Igura, M. Shimoda, and I. Hayakawa.** 2006. Mechanisms behind tailing in the pressure inactivation curve of a clinical isolate of *Escherichia coli*. *Int. J. Food Microbiol.* **109:**103–108.

71. **Pagán, R., S. Jordan, A. Benito, and B. M. Mackey.** 2001. Enhanced acid sensitivity of pressure-damaged *Escherichia coli* O157 cells. *Appl. Environ. Microbiol.* **67:**1983–1985.

72. **Pagán, R., and B. M. Mackey.** 2000. Relationship between membrane damage and cell death in pressure-treated *Escherichia coli* cells: differences between exponential and stationary phase cells and variation among strains. *Appl. Environ. Microbiol.* **66:**2829–2834.

73. **Park, S. W., K. H. Sohn, J. H. Shin, and H. J. Lee.** 2001. High hydrostatic pressure inactivation of *Lactobacillus viridescens* and its effect on ultrastructure of cells. *Int. J. Food Sci. Technol.* **36:**775–781.

74. **Patterson, M. F., and D. J. Kilpatrick.** 1998. The combined effect of high hydrostatic pressure and mild heat on inactivation of pathogens in milk and poultry. *J. Food Prot.* **61:**432–436.

75. **Patterson, M. F., M. Quinn, R. Simpson, and A. Gilmour.** 1995. Sensitivity of vegetative pathogens to high hydrostatic pressure treatment in phosphate-buffered saline and foods. *J. Food Prot.* **58:**524–529.

76. **Ritz, M., M. Feulet, N. Orange, and M. Federighi.** 2000. Effects of high hydrostatic pressure on membrane proteins of *Salmonella typhimurium. Int. J. Food Microbiol.* **55:**115–119.

77. **Ritz, M., M. F. Pilet, F. Jugiau, F. Rama, and M. Federighi.** 2006. Inactivation of *Salmonella* Typhimurium and *Listeria monocytogenes* using high-pressure treatments: destruction or sublethal stress? *Lett. Appl. Microbiol.* **42:**357–362.

78. **Robey, M., A. Benito, R. H. Hutson, C. Pascual, S. F. Park, and B. M. Mackey.** 2001. Variation in resistance to high hydrostatic pressure and *rpoS* heterogeneity in natural isolates of *Escherichia coli* O157:H7. *Appl. Environ. Microbiol.* **67:**4901–4907.

79. **Russell, N. J., R. I. Evans, P. F. ter Steeg, J. Hellemons, A. Verhuel, and T. Abee.** 1995. Membranes as a target for stress adaptation. *Int. J. Food Microbiol.* **28:**255–261.

80. **Scheyhing, C. H., S. Hormann, M. A. Ehrmann, and R. F. Vogel.** 2004. Barotolerance is inducible by preincubation under hydrostatic pressure, cold-, osmotic, and acid-stress conditions in *Lactobacillus sanfranciscensis* DSM 20451(T). *Lett. Appl. Microbiol.* **39:**284–289.

81. **Schulz, E., H.-D. Ludeman, and R. Jaenicke.** 1976. High pressure equilibrium studies on the dissociation-association of *E. coli* ribosomes. *FEBS Lett.* **64:**40–43.

82. **Schwarz, H., and A. L. Koch.** 1995. Phase and electron microscopic observations of osmotically induced wrinkling and the role of endocytotic vesicles in the plasmolysis of the Gram negative cell wall. *Microbiology* **141:**3161–3170.

83. **Shigehisa, T., T. Ohmori, A. Saito, S. Taji, and R. Hayashi.** 1991. Effects of high hydrostatic pressure on characteristics of pork slurries and inactivation of microorganisms associated with meat products. *Int. J. Food Microbiol.* **12:**207–216.

84. **Smelt, J. P. P. M., A. G. F. Rijke, and A. Hayhurst.** 1994. Possible mechanism of high-pressure inactivation of microorganisms. *High Pressure Res.* **12:**199–203.

85. **Smiddy, M., L. O'Gorman, R. D. Sleator, J. P. Kerry, M. F. Patterson, A. L. Kelly, and C. Hill.** 2005. Greater high-pressure resistance of bacteria in oysters than in buffer. *Innovative Food Sci. Emerg. Technol.* **6:**83–90.

86. **Sonoike, K., T. Setoyama, Y. Kuma, and S. Kobayashi.** 1992. Effect of pressure and temperature on the death rate of *Lactobacillus casei* and *Escherichia coli. Colloq. INSERM* **224:**297–301.

87. **ter Steeg, P. F., J. C. Hellemons, and A. E. Kok.** 1999. Synergistic actions of nisin, sublethal high pressure, and reduced temperature on bacteria and yeast. *Appl. Environ. Microbiol.* **65:**4148–4154.

88. **Thanassi, D. G., L. W. Cheng, and H. Nikaido.** 1997. Active efflux of bile salts by *Escherichia coli. J. Bacteriol.* **179:**2512–2518.

89. **Tholozan, J. L., M. Ritz, F. Jugiau, M. Federighi, and J. P. Tissier.** 2000. Physiological effects of high hydrostatic pressure treatments on *Listeria monocytogenes* and *Salmonella typhimurium. J. Appl. Microbiol.* **88:**202–212.

90. **Tomlins, R. I., and Z. J. Ordal.** 1976. Thermal injury in bacteria, p. 153–190. *In* F. A. Skinner and W. B. Hugo (ed.), *Inhibition and Inactivation of Vegetative Microbes.* Academic Press, London, United Kingdom.

91. **Van Opstal, I., S. C. M. Vanmuysen, and C. W. Michiels.** 2003. High sucrose concentration protects *E. coli* against high pressure inactivation but not against high pressure sensitisation to the lactoperoxidase system. *Int. J. Food Microbiol.* **88:**1–9.

92. **Van Opstal, I., S. C. M. Vanmuysen, E. Y. Wuytack, B. Masschalck, and C. W. Michiels.** 2005. Inactivation of *E. coli* by high hydrostatic pressure at different temperatures in buffer and carrot juice. *Int. J. Food Microbiol.* **98:**179–191.

93. **Weber, H., T. Polen, J. Heuveling, V. F. Wendisch, and R. Hengge.** 2005. Genome-wide analysis of the general stress response network in *Escherichia coli:* σ^s-dependent genes, promoters, and sigma factor selectivity. *J. Bacteriol.* **187:**1591–1603.

94. **Welch, T. J., A. F. Farewell, F. C. Neidhardt, and D. H. Bartlett.** 1993. Stress response of *Escherichia coli* to elevated hydrostatic pressure. *J. Bacteriol.* **175:**7170–7177.

95. **Wemekamp-Kamphuis, H. H., A. K. Karatzas, J. A. Wouters, and T. Abee.** 2002. Enhanced levels of cold shock proteins in *Listeria monocytogenes* LO28 upon exposure to low temperature and high hydrostatic pressure. *Appl. Environ. Microbiol.* **68:**456–463.

96. **Wouters, P. C., E. Glaasker, and J. P. P. M. Smelt.** 1998. Effects of high pressure on inactivation kinetics and events related to proton efflux in *Lactobacillus plantarum. Appl. Environ. Microbiol.* **64:**509–514.
97. **Wuytack, E. Y., L. D. T. Phuong, A. Aertsen, K. M. F. Reyns, D. Marquenie, B. De Ketelaere, B. Masschalck, B. M. I. Van Opstal, A. M. J. Diels, and C. W. Michiels.** 2003. Comparison of sublethal injury induced in *Salmonella enterica* serovar Typhimurium by heat and by different nonthermal treatments. *J. Food Prot.* **66:**31–37.
98. **Zobell, C. E.** 1970. Pressure effects on morphology and life processes of bacteria, p. 85–130. *In* A. M. Zimmerman (ed.), *High Pressure Effects on Cellular Processes.* Academic Press, New York, NY.

Chapter 5

Cellular Impact of Sublethal Pressures on *Escherichia coli*

Abram Aertsen and Chris W. Michiels

Like temperature, pressure is an elementary thermodynamic parameter that is inevitably present throughout the biosphere. While surface-adapted microorganisms dwell at atmospheric pressure (0.1 MPa), piezophilic (pressure-loving) or pressure-tolerant bacteria in deep-sea niches are adapted to growth at pressures up to 100 MPa (9, 24, 55). Nevertheless, in the food industry also, mesophiles are faced with high pressure (HP) when they are subjected to pascalization, an emerging preservation process treating foods with HP (100 to 1,000 MPa) for short times (32, 40, 43).

However, the cellular impact of HP on bacteria has not been examined to the same degree as that of heat. Although the extrapolation of thermodynamic principles and the in vitro effects on biomolecules have indicated that HP in general causes protein denaturation and affects membrane fluidity (8, 22, 35) (see chapter 1), it can be expected that due to intrinsic differences not all cellular processes will suffer equally from HP stress. Moreover, living systems might be able to better withstand or counteract some of the resulting deleterious effects over others. Therefore, the need remains to comprehend the actual perception of HP stress and the physiological response it evokes in living bacteria. Overall, this bacterial piezophysiology has been examined by two different approaches. While a first valuable source of insight originates from the comparison of atmospheric to HP-adapted bacteria to identify the specific molecular alterations that enable the latter to live at HP conditions, a second source stems from the characterization of cellular changes that accompany a shift from low (atmospheric) pressure to HP.

Such studies not only have been undertaken for dedicated piezophiles but also have proved worthwhile in mesophiles, such as *Escherichia coli*. The fact that in its evolution *E. coli* has probably never experienced significant pressure changes makes this bacterium less biased, in a sense, to study the impact of HP. Indeed, while piezophiles during evolution are best accustomed to HP, the same cellular adaptations that aid them to withstand or cope with the deleterious influences of HP will tend also to mask their perception of HP stress.

Abram Aertsen and Chris W. Michiels • Centre for Food and Microbial Technology, Department of Microbial and Molecular Systems (M2S), Faculty of Bioscience Engineering, K.U.Leuven, Kasteelpark Arenberg 22, B-3001 Leuven, Belgium.

Moreover, evolution has designed these extremophiles to live at deep-sea niches, where HP typically coincides with extreme thermal and/or chemical conditions, preventing their particular adaptations or response to be attributed to HP only.

In this chapter we focus on the molecular effects and genetic consequences of sublethal pressures on *E. coli,* while chapter 4 of this book discusses the impact of lethal HP treatment.

EFFECT OF HP ON ESSENTIAL MICROBIAL PROCESSES

Early key experiments examining the behavior of bacteria under HP probed their ability to still perform basic cellular functions such as DNA replication, transcription, and translation. These activities were originally assayed in vivo by, respectively, looking at the incorporation of radioactive labeled thymine, uracil, and leucine by *E. coli* cultures under HP (56) and were later studied in more detail to pinpoint the exact molecular defect.

When studying DNA replication, it appeared that between 25 and 45 MPa the kinetics of thymine incorporation by *E. coli* reflected a synchronous replication process, while at higher pressures (50 to 80 MPa) only an initial period of thymine incorporation could be sustained, after which further incorporation was abolished. Finally, incubation at pressures of >95 MPa resulted in the immediate cessation of thymine incorporation. These observations suggest that at around 50 MPa existing replication forks can finish their round, although the initiation of new rounds of DNA replication is inhibited. In line with this hypothesis, a sudden burst in thymine incorporation was observed directly after the release of a culture maintained at 55 MPa, indicating that at this pressure most of the replication machinery is intact and remains poised for the resumption of activity (56). At 90 MPa it is likely that the machinery itself becomes inactive in *E. coli,* although it should be noted that DNA replication slowly recovers after pressure release.

Transcription, in turn, starts to be affected at 20 MPa and is completely inhibited at 80 MPa (56). However, this inhibition is reversible and transcription readily resumes upon the release of HP. Interestingly, detailed in vitro studies with the *E. coli* RNA polymerase (RNAP) have pointed out that inactivation of an RNAP molecule actually depends on its action at the time HP is applied (15, 16). It seems that starting at 50 MPa, free RNAP molecules gradually become susceptible to an irreversible dissociation, since the HP-dissociated subunits undergo conformational changes rendering them unable to reform the active enzyme (15). On the other hand, and consistent with the in vivo observations (56), RNAP molecules that are actively participating in a stable ternary transcribing complex at the time of HP exposure appear to be only reversibly inhibited, being able to rapidly resume transcription at the normal rate after decompression from up to 180 MPa (16).

Finally, the in vivo radiolabeling experiments of Yayanos and Pollard (56) showed that protein synthesis is totally inhibited at between 60 and 70 MPa, although up to these pressures inhibition is readily reversible. A more detailed in vivo study found that the block in protein synthesis did not occur at the level of amino acid permeability, aminoacyl-tRNA formation, or maintenance of polysomal integrity (46). However, despite this polysomal integrity and the stability of nascent proteins, no further incorporation of new amino acids could be observed at 67 MPa, indicating that the inhibition did occur during elongation of the growing peptide chain (23, 46). Subsequent in vitro experiments therefore focused on the ribosomes to further delineate the site of HP inhibition. Here it was demonstrated that at 67 MPa ribosomes still retained their ability to form a peptide bond per se, suggesting that HP inhibition resulted either from the inability of aminoacyl-tRNA to bind the ribosome-mRNA complex or from compromising the translocational mechanism (47).

From these early observations it readily becomes evident that cellular HP effects can progress from very specific targets. In addition, it seems that after pressures up to ~100 MPa, most essential processes can readily resume or at least recover, indicating that no lethal damage is incurred to cell. The absence of obvious lethality allows the response and behavior of *E. coli* during HP growth or after HP shock to be studied. In the following sections the effects of such sublethal HP exposure on *E. coli* physiology are discussed.

PLEIOTROPIC EFFECTS OF GROWTH AT HP

One of the first clear observations related to HP microbiology was the fact that *E. coli* and a number of other terrestrial bacteria were intrinsically able to grow, albeit slower, at pressures up to ~50 MPa, 500 times greater than the pressure they are naturally accustomed to (59). Furthermore, atmospherically grown cultures of *E. coli* can even withstand pressures between 50 and 100 MPa provided they are exposed for only short periods (<1 h). At pressures exceeding 100 MPa, however, *E. coli* increasingly loses viability within minutes (57).

Interestingly, early observations of *E. coli* under growth-permitting pressures readily revealed two of the most visible phenotypes of HP stress in *E. coli* to date, namely, that the bacteria lost motility (37) and grew as filaments (58). While loss of motility under HP has not been documented further, the phenomenon of filamentous growth has attracted more attention. Filamentation is readily observed with a number of other stresses and reveals that production of biomass (i.e., actual growth) proceeds in the absence of cell division. Since the processes of DNA replication, chromosome segregation, and cell division are to some extent coupled in the cell (51), several groups have tried to elucidate which of these processes is in fact HP sensitive and therefore presents the weakest link in the prevention of filamentous growth under HP.

Soon after the discovery of filamentous growth under HP, it was found that *E. coli* filaments (grown at 45 MPa) did contain roughly the same amount of DNA as normal, atmospherically grown cells, although they were on average about 2.5 times larger and heavier (58). The subsequent suggestion that DNA replication was affected by HP was later supported by the data of Yayanos and Pollard (56) discussed above. Although it might be reasonable to assume that HP inhibition of DNA replication, in turn, also leads to abolition of cell division, the molecular mechanism of this connection remains unclear (see also "Is There Also DNA Damage during HP Growth?" below).

In addition to these early microscopic observations, more molecular approaches have revealed profound alterations in the abundance of cell surface proteins of *E. coli* cells grown at HP. A distinct observation in this context was that during HP growth (30 MPa), lambda phage infection was prevented because synthesis of the LamB porin, being the physical attachment site of the phage particle, was repressed together with the entire maltose regulon wherein *lamB* is embedded (44). Furthermore, reduced expression of OmpC and OmpF was also observed during growth at sublethal pressures. Interestingly, this repression was shown to be independent of the EnvZ-OmpR signal transduction cascade that is normally in charge of the regulation of these outer membrane proteins (17).

Another case of alternative regulation imposed by HP was illustrated with the *E. coli*
lac promoter. Under natural, atmospheric conditions this promoter is repressed by the LacI
protein, and repression can be relieved with the addition of lactose. In the cell the lactose is
converted to allolactose, which physically binds to LacI and releases the repressor from its
binding site on the *lac* promoter (54). It was shown that *E. coli* cells harboring a plasmid
with a reporter gene downstream of the *lac* promoter exhibited an 80- to 90-fold-increased
promoter activity during growth at 30 MPa compared to that at atmospheric pressure (30).
Interestingly, it was found that the plasmid copy number also increased during HP growth.
However, the copy number increased only 2- to 3-fold and was therefore unable to account
for the total increase in *lac* promoter activity under HP. It is still unknown whether activation of this *lac* promoter is due to direct HP effects on the DNA-binding capacity of the
LacI repressor protein or on plasmid topology (e.g., supercoiling), or, rather, stems from an
indirect effect of HP on the cell (10).

These studies underscore the peculiar and pleiotropic effects of HP on cellular physiology. The following sections of this chapter elaborate on the molecular dissection of piezophysiology in *E. coli* and focus on the cellular response that is evoked by HP.

HP INDUCTION OF THE HEAT AND COLD SHOCK RESPONSES

Heat and Cold Shock Induction during HP Growth

The first insight in the actual stress response of *E. coli* cells to HP was provided by
Welch et al. (53), who used two-dimensional sodium dodecyl sulfate-polyacrylamide gel
electrophoresis to demonstrate the synthesis of a specific set of proteins during anaerobic
HP growth for 60 to 90 min at 55 MPa. Many of these pressure-induced proteins were up-
regulated transiently, and among them 11 heat shock proteins (HSPs, such as ClpB, ClpP,
Lon, RpoH, DnaK, GroEL, GroES, and GrpE) and 4 cold shock proteins (CSPs, such as
RecA and H-NS) were identified. The simultaneous expression of HSPs and CSPs that
emerged during this proteomic approach was later confirmed by genome-wide transcription profiling of *E. coli* growing at 30 and 50 MPa (27).

At first sight, induction of both HSPs and CSPs seems contradictory, since they combat
stresses that are mutually exclusive. Moreover, so far no other stress has been shown to com-
bine both sets of these stress proteins. However, the different recovery strategies employed
by HSPs and CSPs can coexist. Indeed, HSPs deal with the refolding or destruction of heat-
denatured proteins, while CSPs restore membrane fluidity and accurate protein translation at
low temperatures (7, 49). As discussed earlier, protein denaturation, decreased membrane
fluidity, and translation inhibition also belong to the pleiotropic cellular effects associated
with HP exposure and might therefore permit the gathering of both HSPs and CSPs.

Although the exact nature of this unique coinduction by HP growth remains to be elucidated, a common denominator in the heat and cold shock responses could stem from the
specific HP effects on the ribosomes. Indeed, by using different ribosome-targeting antibiotics, VanBogelen and Neidhardt (50) demonstrated that some of these compounds induced HSPs, while others induced CSPs. This observation led them to suggest that ribosomes could function as cellular thermosensors and, depending on the received signal,
could trigger either HSP or CSP production. Perhaps HP has a unique effect on *E. coli* ribosomes, triggering both their capacities to direct HSP and CSP production.

Heat Shock Induction after HP Shock

HSP induction was also observed in *E. coli* after a short sublethal HP shock (150 MPa, 15 min) (6), further suggesting that coping with protein denaturation remains an important activity in pressurized cells, even for short times at higher pressures. The necessity of HSPs could be confirmed by two subsequent observations. First, it was demonstrated that a sublethal heat shock prior to a lethal HP shock conferred high levels of pressure resistance on *E. coli* cells, indicating that loading the cells with HSPs provides significant protection against HP-induced damage (6). Second, it was discovered that mutants of *E. coli* exhibiting an extreme resistance to HP shock (up to 800 MPa for 15 min) (20) possess constitutively increased levels of HSPs (6), most likely explaining their resistance. Moreover, increased levels of HSPs were also observed for piezoresistant mutants of *Listeria monocytogenes* (29).

These observations point to a caveat for the use of HP in food preservation. The presence of HSPs, either induced by HP or constitutively present in emerging HP-resistant mutants, could give rise to cross-resistance to other stress factors. Indeed, it was recently found that HP-resistant mutants show increased resistance to oxidative stress (1). Moreover, when in food processing inspired by hurdle technology (34) a heat treatment precedes a HP treatment, the heat shock-mediated HP resistance could actually compromise the synergy of this treatment order.

HP INDUCTION OF THE SOS RESPONSE

SOS Induction after HP Shock

Although HP induction of heat and cold shock responses conform to the extrapolation of thermodynamic HP effects on protein denaturation and membrane fluidity, respectively, induction of a third stress regulon was quite surprising. Apparently, a sublethal HP treatment is able to trigger the SOS response in *E. coli* (5) (Fig. 1). The SOS regulon is mounted in response to DNA damage and encompasses the induction of DNA repair proteins. At the molecular level, when *E. coli* suffers DNA damage, single-strand DNA (ssDNA) is exposed that is sensed by the RecA protein. When RecA binds to ssDNA, it becomes activated and stimulates the autocleavage of the LexA repressor protein, in turn triggering the expression of more than 40 genes collectively marked as the SOS response (18, 33). Since an intact LexA protein is needed to repress the promoters of the genes belonging to the SOS regulon, it was first assumed that LexA itself might suffer from direct HP-induced denaturation, thereby failing to repress SOS genes. However, it was demonstrated that HP-mediated alleviation of LexA repression entirely depended on the presence of both a functional RecA and a cleavable LexA protein (5), indicating in fact a bona fide SOS response.

This HP-induced SOS response has some interesting side effects. One of them involves the action of SulA, a member of the SOS regulon that physically binds to the FtsZ protein (12, 25). FtsZ is a key protein in the cell division process that initiates septum formation by the gathering and polymerization of FtsZ monomers at the division plane. Interference by SulA with FtsZ ring formation prevents cell division (26, 39) and is part of the mechanism that aborts cell division for the time necessary to repair the damaged DNA. Phenotypically this results in elongation of the cell after HP treatment, an effect that is even more

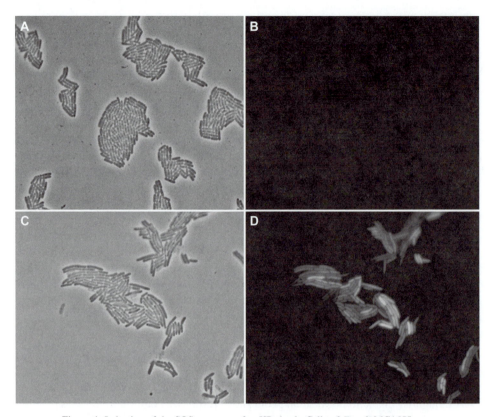

Figure 1. Induction of the SOS response after HP shock. Cells of *E. coli* MG1655 containing a reporter plasmid with the promoter of *sulA* fused to the green fluorescent protein gene *(gfp)* were either left untreated (A and B) or shocked by HP (100 MPa, 15 min, 20°C) (C and D). After treatment, cultures were incubated for 3 h at 37°C and subsequently analyzed by both phase-contrast (A and C) and epifluorescence (B and D) microscopy. While untreated cells display no fluorescence (B), cells of the HP-shocked sample clearly show expression of green fluorescent protein driven by the *sulA* promoter (D).

pronounced (hyperfilamentation) when the cognate Lon protease, responsible for SulA degradation (38), is inactivated (3) (Fig. 2). Surprisingly, however, even in the absence of SulA, mild cell elongation still takes place after HP stress (3, 31), indicating that a second, redundant elongation pathway is triggered by HP stress. The latter pathway is still obscure but strongly resembles the "transient filamentation" phenomenon observed by Gottesman et al. (19) after SOS induction by UV treatment, which was also shown to be SulA independent.

When the DNA damage is repaired after the HP shock, the SOS response rapidly fades out and allows cell division to proceed. In Lon-deficient cells, however, the persistence of SulA and blocked cell division ultimately leads to the death, or at least nonculturability, of a majority of the population. Additional inactivation of the *sulA* gene can rescue these cells (Fig. 3). Interestingly, this salvage occurred at pressures of >50 MPa and <250 MPa, in-

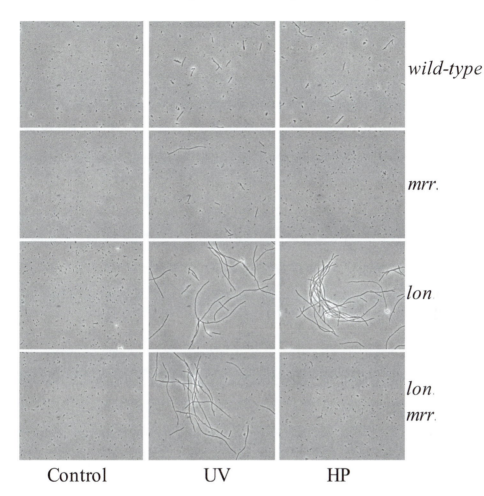

Figure 2. Cells of wild-type *E. coli* MG1655 and its isogenic knockout mutants deprived of Mrr *(mrr)*, Lon *(lon)*, or both Lon and Mrr *(lon mrr)* were either left untreated (control) or shocked by UV (0.1 kJ/m^2) or HP (100 MPa, 15 min, 20°C). After treatment, cultures were incubated for 3 h at 37°C and subsequently analyzed by phase-contrast microscopy. Compared to wild-type cells, *lon* cells display an exaggerated filamentation when exposed to SOS-inducing treatments such as UV or HP shock. Please note that the Mrr protein specifically affects the SOS response induced by HP while leaving the UV response unaltered. Adapted from reference 4.

dicating that at higher pressures the SOS response disappears or its impact becomes inferior to other effects of HP treatment.

Another SOS effect stems from a second activity of the activated RecA complex. In addition to LexA, activated RecA also aids autoproteolysis of the CI repressor encoded by lambdoid phages (33). In lysogens this CI protein keeps the lytic cycle of the prophage repressed. Induction of the SOS response therefore coincides with the production and outburst of resident lambdoid phages. Consequently, it was shown that after HP shock (100 MPa, 15 min) of lysogenic *E. coli* an enormous burst (up to a 1,000-fold increase compared

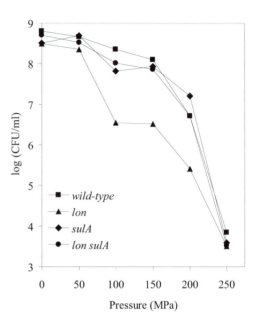

Figure 3. Cells of wild-type *E. coli* MG1655 and its isogenic knockout mutants deprived of Lon, SulA, or both Lon and SulA were treated with different pressures (15 min, 20°C) and survival was expressed as log CFU per milliliter. Due to exaggerated filamentation after HP treatment, *lon* cells are hypersensitive to HP, unless they are rescued by inactivating the *sulA* gene. Adapted from reference 3.

to unpressurized cultures) in the production of infective lambdoid phage particles was measured (2, 5) (Fig. 4). In hindsight, it is noteworthy that HP-mediated prophage induction in lysogenic *E. coli* was already once reported as an obscure phenotype more than 40 years ago by Rutberg (41), well before any notion of the SOS mechanism.

Some of these lambdoid phages, such as Shiga toxin (Stx)-converting bacteriophages, carry virulence genes (45), and their HP-mediated induction during food preservation might contribute to the lateral spread of this pathogenic trait to other *E. coli* strains. Surprisingly, although Stx phages could be induced by HP shock in corresponding lysogens of the laboratory *E. coli* K-12 MG1655 strain (Fig. 4B and C), no such induction was observed in naturally occurring pathogens carrying Stx prophages, such as *E. coli* EDL933 (O157:H7) or H19 (O26:H11) (2). Moreover, these strains did not exhibit HP SOS induction while displaying no intrinsic defects in the SOS cascade (or Stx phage production) triggered by typical DNA damagers such as UV or mitomycin C (2). The observation that HP SOS induction is not a universal trait shared by all *E. coli* strains prompted further research into the molecular mechanism of SOS induction by HP.

How Is the SOS Response Induced after HP Shock?

Although the earlier-mentioned RecA dependence of the HP-induced SOS response clearly indicates involvement of ssDNA and an activated RecA nucleoprotein filament (5), it was not clear how HP treatment would result in the formation of ssDNA. As DNA duplexes are very pressure stable (8, 22), a direct dissociation of both DNA strands could be excluded at HP levels around 100 MPa. Alternatively, DNA replication and transcription are sensitive to pressures up to 100 MPa (56) (see "Effect of HP on Essential Microbial Processes" above), suggesting the possibility that ssDNA somehow could be formed as a result of stalling of the replication or transcription complexes. However, this assumption

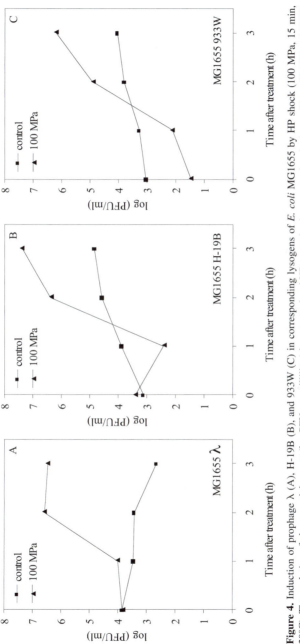

Figure 4. Induction of prophage λ (A), H-19B (B), and 933W (C) in corresponding lysogens of *E. coli* MG1655 by HP shock (100 MPa, 15 min, 20°C). The evolution of phage particle count (log PFU per milliliter) in untreated and HP-treated cultures over time is shown. Phage H-19B and 933W are lamboid phages carrying the genes for Stx1 and Stx2, respectively. Please note that the drop in PFU of phage 933W per milliliter directly after HP treatment is due to the pronounced HP sensitivity of the phage particle. Adapted from references 2 and 5.

also proved unlikely, as not all *E. coli* strains support the HP-induced SOS response despite the fact that this replication and transcription machinery is well conserved.

Finally, the hypothesis was examined that HP stress might incur genuine DNA damage. In that case, such DNA lesions had to be processed by dedicated proteins that sense the ssDNA that originates from DNA damage and pass it on to the RecA protein. Two such dedicated protein complexes convey ssDNA to RecA: the RecFOR complex senses daughter strand gaps, while the RecBCD complex senses double strand breaks (DSBs) (33). A daughter strand gap results from a single lesion on one strand, while a DSB results from opposite lesions on both strands of the double helix. Interestingly, HP activation of the SOS response depended entirely on RecB, indicating DSBs as the trigger for the SOS response (4). Because of thermodynamic constraints, however, pressures of 100 MPa are absolutely incapable of breaking covalent bonds (8, 22), precluding the direct generation of DSBs by HP. This consideration, together with the fact that HP activation of the SOS response can be defective in some *E. coli* strains, led to the speculation of the presence of an HP-activated cellular pathway, still upstream of the RecB-dependent activation of RecA, that couples HP stress to DSBs.

Such a specific upstream pathway became more likely when in a subsequent genetic screen mutants of *E. coli* MG1655 could be isolated that were deficient only in the SOS response triggered by HP, and not in the one triggered by UV or mitomycin C. These mutants led to the identification of the Mrr protein, an endogenous restriction endonuclease (21, 52), as the likely final effector in translating the perception of HP into DSBs and concomitant RecB-dependent SOS induction (Fig. 2). In agreement with these findings, the *mrr* gene is naturally absent in *E. coli* EDL933 (48), explaining the earlier-mentioned lack of HP-induced SOS response in these strains (2). However, how the Mrr protein, in turn, is activated by HP remains unknown.

Is There Also DNA Damage during HP Growth?

As already mentioned earlier in this chapter (see "Pleiotropic Effects of Growth at HP" above), the obvious filamentation phenotype during growth of *E. coli* at HP remains enigmatic because the molecular mechanism delaying cell division is still elusive. However, using immunofluorescence microscopy, Ishii et al. (28) recently observed that FtsZ rings were absent in *E. coli* filaments growing at HP, although they rapidly reappeared and gave rise to cell division as soon as HP was relieved, thereby indicating that the mechanism behind the delay in cell division can reversibly control FtsZ polymerization.

As discussed earlier, reversible prevention of FtsZ ring formation is a typical trait of the SOS response (26), and it could therefore be reasoned that the inhibition of DNA replication during HP growth corresponds with a loss in DNA integrity, which via the SOS response could contribute to filamentation. In this regard, it is tempting to interpret the increased levels of RecA observed by Welch et al. (53) during HP growth (see "Heat and Cold Shock Induction during HP Growth" above) as a sign of DNA damage. However, it is known that RecA is also modestly up-regulated by cold shock (49), and because several other CSPs but no SOS proteins were detected in this proteomic approach, it was assumed that RecA induction fitted in the HP-induced cold shock response. Moreover, FtsZ inhibition during HP growth seems to be independent of this SOS response, since *recA* and *sulA* mutants all keep displaying filamentous growth at permissive pressures (28).

This observation led Ishii et al. (28) to speculate on a direct HP effect on the FtsZ protein, and these authors were able to demonstrate that FtsZ monomers were indeed unable to polymerize in vitro under HP conditions, probably because the increase in volume associated with polymerization is thermodynamically unfavored by HP. This would imply that the inhibition of DNA replication and the prevention of cell division are both direct effects of HP, without the need of a causal link between these mechanisms.

Nevertheless, it remains difficult to verify the extrapolation of this in vitro thermodynamic effect on FtsZ monomers to in vivo conditions during HP growth. Moreover, at this time not all in vivo observations can be straightforwardly reconciled with the proposed direct effect of HP on FtsZ polymerization. It seems, for example, that cell division stops at about the same pressure as DNA synthesis (56). Although this might well be coincidental, it could be indicative of a causal link between inhibition of DNA replication and cell division. Second, it is certainly noteworthy that the initial experiments regarding filamentous HP growth of *E. coli* performed by Zobell and Cobet (58) were carried out with an *E. coli* B strain because of its tendency to readily form filaments during HP growth. It was later shown, however, that the *E. coli* B wild-type strain harbors a deficient *lon* gene (13), which seems to be inactivated by the insertion of an IS*186* element in what appears to be a natural hot spot for this element in the *lon* locus (42). As mentioned earlier, Lon is the cognate protease of SulA, and in the absence of Lon the effects of SulA (i.e., division inhibition and filamentation) become more pronounced. Finally, perhaps the strongest argument was provided during the characterization of a piezosensitive mutant of the piezophile *Photobacterium profundum* SS9. Bidle and Bartlett (11) found that the SS9 RecD protein was indispensable for HP growth of SS9. RecD is part of the RecBCD exonuclease, which is involved in recombinational repair of DNA damage, and *recD* mutants are hyperrecombinogenic and compromised in plasmid maintenance. However, both wild-type *E. coli* and a *recD* mutant react similarly with regard to HP growth, and both form filaments. Surprisingly, however, when an *E. coli recD* mutant was equipped with SS9 RecD, normal rod-like cell morphology was restored during HP growth, indicating that the SS9 *recD* allele is able to confer a piezoadapted phenotype on *E. coli*. The fact that an altered RecD function is somehow able to rescue *E. coli* morphology, and thus FtsZ polymerization, adds complexity to the filamentation phenomenon and is in favor of a yet-to-be-characterized cellular link between inhibition of DNA replication and cell division inhibition.

GENOME-WIDE HP RESPONSE

Some recent studies have embarked on genome-wide expression profiling of *E. coli* subjected to HP by using gene arrays, in order to obtain an integrated view on the impact of this stress. While their findings confirmed induction of members of the heat and cold shock responses, it is currently infeasible to understand and explain every single change in gene expression. Nevertheless, this approach has led to the rapid identification of interesting *E. coli* mutants that are particularly affected by the repercussions of HP stress. For example, it was shown that *E. coli* cells exposed to mild, growth-permitting pressures (<50 MPa) depend heavily on the global regulator H-NS for survival (27). The H-NS protein is an abundant nucleoid structuring factor with a predominant and mostly negative influence on gene expression (14), and its role in HP resistance awaits further clarification.

Another interesting expression study demonstrated that HP shock (100 MPa, 15 min) affected thiol-disulfide redox systems and the Fe-S cluster status, indicating a profound impact of HP shock on the cell's redox homeostasis (36). Moreover, subsequent analysis of specific mutants revealed a surprising contribution of the presence of Fe-S clusters to pressure sensitivity. It was therefore hypothesized that HP is able to affect the Fe-S clusters of some proteins, resulting in the release of Fe, which is disadvantageous to the cell.

CONCLUSIONS AND FUTURE PERSPECTIVES

It is clear that in *E. coli* exposure to HP has pleiotropic effects on physiology and is able to trigger previously characterized and well-defined stress regulons such as the heat shock, cold shock, and SOS response in an anomalous way. Although these regulons may reveal a glimpse of what HP shock is affecting in the cell, an accurate grasp will be obtained only when future research is aimed at elucidating (i) how these regulons are exactly induced, i.e., how the HP signal is molecularly perceived and transduced by the cell, and (ii) how they are embedded in the total physiological response to HP. This approach is essential in understanding and predicting adaptations that enable HP growth or survival, and the microbial behavior under HP or after HP exposure.

Acknowledgment. A.A. is a Postdoctoral Fellow of the Research Fund—Flanders (Belgium) (FWO-Vlaanderen).

REFERENCES

1. **Aertsen, A., P. De Spiegeleer, K. Vanoirbeek, M. Lavilla, and C. W. Michiels.** 2005. Induction of oxidative stress by high hydrostatic pressure in *Escherichia coli. Appl. Environ. Microbiol.* **71:**2226–2231.
2. **Aertsen, A., D. Faster, and C. W. Michiels.** 2005. Induction of Shiga toxin-converting prophage in *Escherichia coli* by high hydrostatic pressure. *Appl. Environ. Microbiol.* **71:**1155–1162.
3. **Aertsen, A., and C. W. Michiels.** 2005. SulA-dependent hypersensitivity to high pressure and hyperfilamentation after high-pressure treatment of *Escherichia coli lon* mutants. *Res. Microbiol.* **156:**233–237.
4. **Aertsen, A., and C. W. Michiels.** 2005. Mrr instigates the SOS response after high pressure stress in *Escherichia coli. Mol. Microbiol.* **58:**1381–1391.
5. **Aertsen, A., R. Van Houdt, K. Vanoirbeek, and C. W. Michiels.** 2004. An SOS response induced by high pressure in *Escherichia coli. J. Bacteriol.* **186:**6133–6141.
6. **Aertsen, A., K. Vanoirbeek, P. De Spiegeleer, J. Sermon, K. Hauben, A. Farewell, T. Nystrom, and C. W. Michiels.** 2004. Heat shock protein-mediated resistance to high hydrostatic pressure in *Escherichia coli. Appl. Environ. Microbiol.* **70:**2660–2666.
7. **Arsene, F., T. Tomoyasu, and B. Bukau.** 2000. The heat shock response of *Escherichia coli. Int. J. Food Microbiol.* **55:**3–9.
8. **Balny, C., P. Masson, and K. Heremans.** 2002. High pressure effects on biological macromolecules: from structural changes to alteration of cellular processes. *Biochim. Biophys. Acta* **1595:**3–10.
9. **Bartlett, D. H.** 2002. Pressure effects on in vivo microbial processes. *Biochim. Biophys. Acta* **1595:**367–381.
10. **Bartlett, D. H., C. Kato, and K. Horikoshi.** 1995. High pressure influences on gene and protein expression. *Res. Microbiol.* **146:**697–706.
11. **Bidle, K. A., and D. H. Bartlett.** 1999. RecD function is required for high-pressure growth of a deep-sea bacterium. *J. Bacteriol.* **181:**2330–2337.
12. **Cordell, S. C., E. J. Robinson, and J. Lowe.** 2003. Crystal structure of the SOS cell division inhibitor SulA and in complex with FtsZ. *Proc. Natl. Acad. Sci. USA* **100:**7889–7894.
13. **Donch, J., and J. Greenberg.** 1968. Ultraviolet sensitivity gene of *Escherichia coli* B. *J. Bacteriol.* **95:**1555–1559.
14. **Dorman, C. J.** 2004. H-NS: a universal regulator for a dynamic genome. *Nat. Rev. Microbiol.* **2:**391–400.

15. **Erijman, L., and R. M. Clegg.** 1995. Heterogeneity of *E. coli* RNA polymerase revealed by high pressure. *J. Mol. Biol.* **253:**259–265.

16. **Erijman, L., and R. M. Clegg.** 1998. Reversible stalling of transcription elongation complexes by high pressure. *Biophys. J.* **75:**453–462.

17. **Forst, S. A., and D. L. Roberts.** 1994. Signal transduction by the EnvZ-OmpR phosphotransfer system in bacteria. *Res. Microbiol.* **145:**363–373.

18. **Friedberg, E. C., G. C. Walker, and W. Siede.** 1995. *DNA Repair and Mutagenesis.* ASM Press, Washington, DC.

19. **Gottesman, S., E. Halpern, and P. Trisler.** 1981. Role of *sulA* and *sulB* in filamentation by *lon* mutants of *Escherichia coli* K-12. *J. Bacteriol.* **148:**265–273.

20. **Hauben, K. J., D. H. Bartlett, C. C. Soontjens, K. Cornelis, E. Y. Wuytack, and C. W. Michiels.** 1997. *Escherichia coli* mutants resistant to inactivation by high hydrostatic pressure. *Appl. Environ. Microbiol.* **63:**945–950.

21. **Heitman, J., and P. Model.** 1987. Site-specific methylases induce the SOS DNA repair response in *Escherichia coli. J. Bacteriol.* **169:**3243–3250.

22. **Heremans, K.** 1995. High pressure effects on biomolecules, p. 81–97. *In* D. A. Ledward, D. E. Johnston, R. G. Earnshaw, and A. P. M. Hastings (ed.), *High Pressure Processing of Foods.* Nottingham University Press, Leicestershire, United Kingdom.

23. **Hildebrand, C. E., and E. C. Pollard.** 1972. Hydrostatic pressure effects on protein synthesis. *Biophys. J.* **12:**1235–1250.

24. **Horikoshi, K.** 1998. Barophiles: deep-sea microorganisms adapted to an extreme environment. *Curr. Opin. Microbiol.* **1:**291–295.

25. **Huisman, O., R. D'Ari, and J. George.** 1980. Further characterization of *sfiA* and *sfiB* mutations in *Escherichia coli. J. Bacteriol.* **144:**185–191.

26. **Huisman, O., R. D'Ari, and S. Gottesman.** 1984. Cell-division control in *Escherichia coli:* specific induction of the SOS function SfiA protein is sufficient to block septation. *Proc. Natl. Acad. Sci. USA* **81:**4490–4494.

27. **Ishii, A., T. Oshima, T. Sato, K. Nakasone, H. Mori, and C. Kato.** 2005. Analysis of hydrostatic pressure effects on transcription in *Escherichia coli* by DNA microarray procedure. *Extremophiles* **9:**65–73.

28. **Ishii, A., T. Sato, M. Wachi, K. Nagai, and C. Kato.** 2004. Effects of high hydrostatic pressure on bacterial cytoskeleton FtsZ polymers in vivo and in vitro. *Microbiology* **150:**1965–1972.

29. **Karatzas, K. A., J. A. Wouters, C. G. Gahan, C. Hill, T. Abee, and M. H. Bennik.** 2003. The CtsR regulator of *Listeria monocytogenes* contains a variant glycine repeat region that affects piezotolerance, stress resistance, motility and virulence. *Mol. Microbiol.* **49:**1227–1238.

30. **Kato, C., T. Sato, M. Smorawinska, and K. Horikoshi.** 1994. High pressure conditions stimulate expression of chloramphenicol acetyltransferase regulated by the *lac* promoter in *Escherichia coli. FEMS Microbiol. Lett.* **122:**91–96.

31. **Kawarai, T., M. Wachi, H. Ogino, S. Furukawa, K. Suzuki, H. Ogihara, and M. Yamasaki.** 2004. SulA-independent filamentation of *Escherichia coli* during growth after release from high hydrostatic pressure treatment. *Appl. Microbiol. Biotechnol.* **64:**255–262.

32. **Knorr, D.** 1999. Novel approaches in food-processing technology: new technologies for preserving foods and modifying function. *Curr. Opin. Biotechnol.* **10:**485–491.

33. **Kuzminov, A.** 1999. Recombinational repair of DNA damage in *Escherichia coli* and bacteriophage lambda. *Microbiol. Mol. Biol. Rev.* **63:**751–813.

34. **Leistner, L.** 2000. Basic aspects of food preservation by hurdle technology. *Int. J. Food Microbiol.* **55:**181–186.

35. **Macdonald, A. G.** 1984. The effects of pressure on the molecular structure and physiological functions of cell membranes. *Philos. Trans. R. Soc. Lond. B* **304:**47–68.

36. **Malone, A. S., Y. K. Chung, and A. E. Yousef.** 2006. Genes of *Escherichia coli* O157:H7 that are involved in high-pressure resistance. *Appl. Environ. Microbiol.* **72:**2661–2671.

37. **Meganathan, R., and R. E. Marquis.** 1973. Loss of bacterial motility under pressure. *Nature* **246:**525–527.

38. **Mizusawa, S., and S. Gottesman.** 1983. Protein degradation in *Escherichia coli:* the *lon* gene controls the stability of SulA protein. *Proc. Natl. Acad. Sci. USA* **80:**358–362.

39. **Mukherjee, A., C. Cao, and J. Lutkenhaus.** 1998. Inhibition of FtsZ polymerization by SulA, an inhibitor of septation in *Escherichia coli. Proc. Natl. Acad. Sci. USA* **95:**2885–2890.

40. **Patterson, M. F.** 2005. Microbiology of pressure-treated foods. *J. Appl. Microbiol.* **98:**1400–1409.
41. **Rutberg, L.** 1964. On the effects of high hydrostatic pressure on bacteria and bacteriophage. 3. Induction with high hydrostatic pressure of *Escherichia coli* K lysogenic for bacteriophage lambda. *Acta Pathol. Microbiol. Scand.* **61:**98–105.
42. **saiSree, L., M. Reddy, and J. Gowrishankar.** 2001. IS*186* insertion at a hot spot in the *lon* promoter as a basis for Lon protease deficiency of *Escherichia coli* B: identification of a consensus target sequence for IS*186* transposition. *J. Bacteriol.* **183:**6943–6946.
43. **San Martin, M. F., G. V. Barbosa-Canovas, and B. G. Swanson.** 2002. Food processing by high hydrostatic pressure. *Crit. Rev. Food Sci. Nutr.* **42:**627–645.
44. **Sato, T., Y. Nakamura, K. K. Nakashima, C. Kato, and K. Horikoshi.** 1996. High pressure represses expression of the *malB* operon in *Escherichia coli. FEMS Microbiol. Lett.* **135:**111–116.
45. **Schmidt, H.** 2001. Shiga-toxin-converting bacteriophages. *Res. Microbiol.* **152:**687–695.
46. **Schwarz, J. R., and J. V. Landau.** 1972. Hydrostatic pressure effects on *Escherichia coli:* site of inhibition of protein synthesis. *J. Bacteriol.* **109:**945–948.
47. **Schwarz, J. R., and J. V. Landau.** 1972. Inhibition of cell-free protein synthesis by hydrostatic pressure. *J. Bacteriol.* **112:**1222–1227.
48. **Sibley, M. H., and E. A. Raleigh.** 2004. Cassette-like variation of restriction enzyme genes in *Escherichia coli* C and relatives. *Nucleic Acids Res.* **32:**522–534.
49. **Thieringer, H. A., P. G. Jone, and M. Inouye.** 1998. Cold shock and adaptation. *Bioessays* **20:**49–57.
50. **VanBogelen, R. A., and F. C. Neidhardt.** 1990. Ribosomes as sensors of heat and cold shock in Escherichia coli. *Proc. Natl. Acad. Sci. USA* **87:**5589–5593.
51. **Vinella, D., and R. D'Ari.** 1995. Overview of controls in the *Escherichia coli* cell cycle. *Bioessays* **17:**527–536.
52. **Waite-Rees, P. A., C. J. Keating, L. S. Moran, B. E. Slatko, L. J. Hornstra, and J. S. Benner.** 1991. Characterization and expression of the *Escherichia coli* Mrr restriction system. *J. Bacteriol.* **173:**5207–5219.
53. **Welch, T. J., A. Farewell, F. C. Neidhardt, and D. H. Bartlett.** 1993. Stress response of *Escherichia coli* to elevated hydrostatic pressure. *J. Bacteriol.* **175:**7170–7177.
54. **Wilson, C. J., H. Zhan, L. Swint-Kruse, and K. S. Matthews.** 2007. The lactose repressor system: paradigms for regulation, allosteric behavior and protein folding. *Cell. Mol. Life Sci.* **64:**3–16.
55. **Yayanos, A. A.** 1995. Microbiology to 10 500 meters in the deep sea. *Annu. Rev. Microbiol.* **49:**777–805.
56. **Yayanos, A. A., and E. C. Pollard.** 1969. A study of the effects of hydrostatic pressure on macromolecular synthesis in Escherichia coli. *Biophys. J.* **9:**1464–1482.
57. **Zobell, C. E., and A. B. Cobet.** 1962. Growth, reproduction, and death rates of *Escherichia coli* at increased hydrostatic pressures. *J. Bacteriol.* **84:**1228–1236.
58. **Zobell, C. E., and A. B. Cobet.** 1964. Filament formation by *Escherichia coli* at increased hydrostatic pressures. *J. Bacteriol.* **87:**710–719.
59. **Zobell, C. E., and F. H. Johnson.** 1949. The influence of hydrostatic pressure on the growth and viability of terrestrial and marine bacteria. *J. Bacteriol.* **57:**179–189.

High-Pressure Microbiology
Edited by C. Michiels, D. H. Bartlett, and A. Aertsen
© 2008 ASM Press, Washington, DC

Chapter 6

Listeria monocytogenes High Hydrostatic Pressure Resistance and Survival Strategies

Marjon Wells-Bennik, Kimon A. Karatzas, Roy Moezelaar, and Tjakko Abee

New trends in farming practices, globalization of markets, and development of new products and production methods are representing new challenges for food safety. The traditional means to control microbial spoilage and safety hazards in foods, such as freezing, blanching, sterilization, curing, and use of preservatives, are being replaced by new, innovative techniques, including mild heating, modified atmosphere and vacuum packaging, and the employment of natural antimicrobial systems. Additionally, the food industry has renewed interest in the use of high hydrostatic pressure (HHP) as a food processing method because of reported quality improvements in specific food products subjected to HHP treatments after packaging (28).

Food preservation techniques are becoming milder in response to the consumers' demands for higher-quality, more convenient foods that are less heavily processed, less heavily preserved (e.g., less acid, salt, and sugar), and less reliant on additive preservatives than hitherto used (e.g., sulfite and nitrite) and that have improved nutritional and organoleptic quality. To this end, food manufacturers apply traditional preservation factors at lower intensities, but also evaluate alternative technologies, including HHP treatment. Pressures of 200 to 600 MPa at ambient temperature reduce the number of microbes present and inactivate enzymes responsible for spoilage but have little effect on nutritional and organoleptic quality. The effects of such treatments on microorganisms are comparable with those of pasteurization: vegetative cells are inactivated but bacterial endospores survive. HHP-treated products must therefore be stored at refrigeration temperatures. Since the introduction of

Marjon Wells-Bennik • Top Institute for Food and Nutrition, P.O. Box 557, 6700 AN, Wageningen, and NIZO Food Research, Kernhemseweg 2, 6718 ZB Ede, The Netherlands. *Kimon A. Karatzas* • Top Institute for Food and Nutrition, P.O. Box 557, 6700 AN, Wageningen, The Netherlands, and School of Clinical Veterinary Science, University of Bristol, Langford House, Langford, Bristol, BS40 5DU, United Kingdom. *Roy Moezelaar* • Top Institute for Food and Nutrition, P.O. Box 557, 6700 AN, and WUR, Wageningen University and Research Centre, Food Technology Centre, Bornsesteeg 59, 6708 PD, Wageningen, The Netherlands. *Tjakko Abee* • Top Institute for Food and Nutrition, P.O. Box 557, 6700 AN, and Laboratory of Food Microbiology, Wageningen University, Bomenweg 2, 6700 EV, Wageningen, The Netherlands.

high-pressure-treated foods in 1990 in Japan, the range of products has gradually increased. This has in particular been the case in the United States, where complete meal kits with several pressure-treated components are currently on the market.

Commonly, mildly preserved and mildly processed food products, including pressure-treated foods, require refrigerated storage and refrigerated distribution to ensure their quality and safety, from both microbial and quality retention standpoints. The major microbiological concerns associated with minimally processed foods are psychrotrophic and mesophilic microorganisms. Psychrotrophic microorganisms can grow at refrigeration temperatures, while mesophilic pathogens can survive under refrigeration and may subsequently grow at higher temperatures (1).

Based on an extensive survey in the United States, Mead et al. (52) estimated that of all illnesses attributable to food-borne transmission, 30% were caused by bacteria, 3% were caused by parasites, and 67% were caused by viruses. Bacteria account for 72% of deaths associated with food-borne transmission, parasites account for 21%, and viruses account for 7%. *Salmonella* and *Listeria* account for 60% of these food-related deaths (52). A detailed overview of food-borne pathogens is provided by Lund et al. (48). Notably, in many cases of food-related illnesses and deaths, the causative agent could not be identified, and these illnesses might be associated with yet-unknown pathogens.

Despite the growing number of pressure-pasteurized products and numerous publications confirming the potential of high pressure to inactivate microorganisms, relatively little is known about the underlying mechanisms that may enable microorganisms to overcome the adverse effects of pressure. In this chapter we highlight aspects of HHP inactivation and survival strategies of the human pathogen *Listeria monocytogenes*.

LISTERIA MONOCYTOGENES

L. monocytogenes is a ubiquitous gram-positive food-borne pathogen that can cause life-threatening illness in immunocompromised and elderly people, pregnant women, and neonates (16, 65, 74). The bacterium has been found in mammals, birds, and some species of fish and shellfish. It can be isolated from soil, silage, and other environmental sources and in food processing plants (9). The resistance of *L. monocytogenes* to the damaging effects of freezing, drying, and heat is remarkable for a non-spore-forming bacterium. Unlike most human pathogens, *L. monocytogenes* can grow at refrigeration temperatures. Its ubiquitous distribution in the environment, its ability to grow at low temperatures, and its pathogenic potential make this bacterium a particular concern with regard to the safety of refrigerated and ready-to-eat foods, which are consumed without reheating and/or cooking. Several outbreaks of listeriosis have been traced to contaminated cold-stored ready-to-eat foods, including dairy, vegetable, and meat products (44).

So far, the genome sequences of four *L. monocytogenes* strains and of one closely related nonpathogenic strain, namely, a *Listeria innocua* strain, have been completed. The genome sequences of *L. monocytogenes* strain EGDe (serovar 1/2a) and *L. innocua* strain CLIP11262 (serovar 6a) were published in 2001 (26), while more recently the sequences of *L. monocytogenes* strains F2365 (serotype 4b, cheese isolate), F6854 (serotype 1/2a, frankfurter isolate), and H7858 (serotype 4b, meat isolate) became available (57). Furthermore, the partial genome sequence of *L. monocytogenes* CLIP80459 (serovar 4b) has been determined (20). The majority of all major food-borne outbreaks of listeriosis appear to be caused by serovar 4b strains. A high degree of colinearity in genome organization (synteny) was found among the *L. monocytogenes* strains, but also with *L. innocua, Bacillus subtilis,* and *Staphylococcus aureus* (10, 57). The burst in genome sequences of microorganisms, including that of *L. monocytogenes*, has opened the way both for functional genomics approaches, including transcriptome, proteome, and metabolome analyses, and for structural genomics (47, 76, 78), and studies are ongoing to develop proteome reference maps of *L. monocytogenes* EGDe for comparison with protein profiles of *Listeria* isolates from food (62).

LISTERIA PATHOGENESIS

L. monocytogenes pathogenesis has been studied extensively, and several high-quality reviews have addressed this topic (16, 23, 74). Surface proteins and secreted proteins play important roles in the interaction of microorganisms with their environment, in particular, in host infection by pathogenic bacteria. In *L. monocytogenes,* many virulence factors are associated with the cell surface or constitute secreted proteins, such as various internalins (Inl [invasion proteins]), phospholipases PlcA (escape from the phagocytic vacuole) and PlcB (cell-to-cell spread), listeriolysin O, and ActA (intracellular actin-based motility). Internalins belong to a family of proteins characterized by leucine-rich repeats in their N-terminal domain, and most of them also contain a so-called C-terminal LPxTG motif for covalent linkage to peptidoglycan (12). Of the 219 predicted surface and secreted proteins encoded in the *L. monocytogenes* EGDe genome, 53 are absent from *L. innocua*, and 20 of the missing surface proteins are LPxTG proteins (10). The *L. monocytogenes* EGDe chromosome has a 10-kb virulence locus that is absent from *L. innocua.* This locus contains the genes *prfA* (encoding the global regulator PrfA), *plcA, plcB, actA, hly* (listeriolysin O precursor), and *mpl* (metalloproteinase precursor). Moreover, many of the genes encoding internalins are absent from the *L. innocua* genome, including a small cluster, *inlG-inlH-inlE* (26). Hybridization analysis showed that all of the well-known virulence factors were present in all 93 *L. monocytogenes* strains tested, but absent in other *Listeria* species (20).

The major regulator of virulence genes in *L. monocytogenes* is PrfA. It activates all genes of the virulence gene cluster, as well as the internalin genes *inlA, inlB,* and *inlC,* and *hpt* (encoding a transporter that mediates rapid intracellular proliferation) (16). PrfA binds to a palindromic recognition sequence (PrfA box) located in the promoter region of regulated genes. More insight into the PrfA regulon was obtained by comparative transcriptome analysis of wild-type *L. monocytogenes* EGDe and a *prfA* deletion mutant (54), in combination with searches for putative PrfA boxes (consensus TTAACANNTGTTAA) in upstream regions of genes. A core set of 12 virulence genes that were preceded by a PrfA box were found to be positively regulated by PrfA.

A second set of more than 50 genes were differentially expressed in the wild type and the *prfA* deletion mutant, but these genes lacked a PrfA box. Most of these genes were predicted to be expressed from a promoter that is σ^B dependent. σ^B in *L. monocytogenes* is an alternative sigma factor that contributes to increased survival under stress conditions such as acid, oxidative stress, and carbon starvation (21, 80). Its role in pressure resistance is discussed below. The activation of a number of (putative) virulence genes by σ^B establishes a link between stress response and virulence. Therefore, stress-resistant variants may have undergone changes in their virulence characteristics compared with unstressed wild-type cells. A risk assessment of exposure to *L. monocytogenes* isolates should therefore include such stress-resistant variants.

HHP INACTIVATION OF BACTERIA

At the surface of the earth the ambient pressure is 0.1 MPa, while at the deepest point of the oceans the pressure is about 100 MPa, and at the center of the earth it is about 360 GPa (60). The term HHP in food applications represents pressures in the range of 50 to 1,000 MPa. Inactivation of bacterial spores for sterilization purposes requires pressures up to 1,000 MPa or higher (67). Typical pressures used to inactivate vegetative cells of bacterial species, and thus pasteurize the product, range from 300 to 700 MPa, with gram-positive bacteria generally being more piezotolerant than gram-negative bacteria (60).

Various factors affect the pressure that is required to inactivate bacteria present in specific foods, such as medium components, temperature, time of inactivation, and pH (11, 53, 60, 61, 67). Increased resistance to HHP has been demonstrated for cultures in the stationary phase of growth and for starved cells (13, 37, 50). For *L. monocytogenes,* it has been demonstrated that stationary-phase cells are more piezotolerant than those of the exponential phase (2- to 3-log cycle differences) (Fig. 1) (43). The growth temperature of a culture can also influence its inactivation by high pressure (51, 79). Another factor that influences HHP sensitivity is the membrane fluidity of cells; the general rule applies that less fluid membranes with similar makeup are more sensitive to HHP (49). Cultures of *L. monocytogenes* that were grown at low temperatures showed a higher survival rate upon pressure treatment than those grown at high temperatures (79).

Great variability in piezotolerance has been reported not only among different bacterial species but also among strains of the same species (4, 29, 53). The different piezotolerances of various strains of the same species, including *L. monocytogenes,* are exemplified in work done by Alpas et al. (5). Several *L. monocytogenes* strains were tested at 345 MPa for 5 min at 25°C, and their piezotolerance varied even by 2.6 log cycles, with *L. monocytogenes* strain CA appearing to be the most piezotolerant and *L. monocytogenes* strain SLR1 appearing to be the least tolerant, showing 0.9- and 3.5-log reductions, respectively. Another study by Tay et al. (71) also highlights the differences in piezotolerances between different *L. monocytogenes* strains. These authors found that the decreases in log CFU per

Figure 1. Reductions in viable numbers of wild-type *L. monocytogenes* cells after exposure to different pressures for 20 min at 20°C. Cells were grown in brain heart infusion broth at 30°C with shaking (160 rpm). Cells were harvested (i) in mid-exponential phase and resuspended in *N*-(2-acetamido)-2-aminoethanesulfonic acid (ACES) buffer before treatment (●), (ii) in stationary phase and resuspended in ACES buffer before treatment (○), and (iii) in mid-exponential phase and resuspended in semiskimmed milk before treatment (▲). Data are also presented in reference 43.

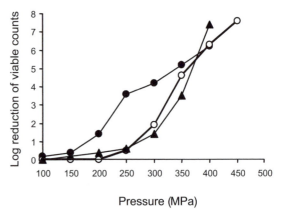

milliliter between nine strains ranged from 1.4 to 4.3 at 400 MPa and from 3.9 to >8 at 500 MPa.

A commonly observed phenomenon with regard to inactivation of *Listeria* upon pressurization is tailing of the survival curves; i.e., exposure during prolonged periods at a certain pressure does not result in a linear inactivation curve (log of number of surviving cells versus time) (11, 71). These observations point toward the presence of a subpopulation of cells that have increased resistance to pressure. Indeed, cells within the same population of *L. monocytogenes* can have different piezotolerances. This became evident upon the isolation of piezotolerant strains of *L. monocytogenes* from a pure culture of *L. monocytogenes* Scott A following pressurization; piezotolerant survivors showed up to 5.2-log-lower inactivation than the wild-type strain under certain conditions (41, 43). The underlying mechanism has been identified for a number of isolates (42) and is described below.

Clearly, the variability in HHP resistance of different species, strains, and even cells within a population makes the proper design of HHP treatments that would allow for adequate reductions of bacteria a challenging task.

INFLUENCE OF FOOD COMPONENTS ON HHP INACTIVATION EFFICIENCY

The composition of the medium or foodstuff in which the microorganisms are dispensed during the pressure treatment can substantially influence the antibacterial effects of pressure.

It is well known that low water activity (a_w) can significantly protect cells from the effects of HHP. It has been suggested that reduced a_w could cause cell shrinkage and thickening of the cellular membrane, reduce the size of the cell, and lead to overall increased piezotolerance (45, 60). Several compounds that lead to reduced a_ws, like salt and sugar, have also been shown to protect cells from the lethal effect of pressure (68, 70). Their effect could be partly explained by the effects of low a_w, but the main underlying mechanism has yet to be elucidated (68). It has been suggested that the trimethyl ammonium compounds glycine betaine and, to a lesser extent, L-carnitine are primarily responsible for the observed increase in piezotolerance at elevated osmolarities. These compounds might affect piezotolerance through stabilization of proteins and increase of the membrane fluidity (49, 68). The suitability of HHP inactivation of bacteria in intermediate-moisture foods with low a_ws, like cheeses, syrups, and cured or fermented meat, is therefore limited. HHP inactivation is furthermore not suitable for inactivation of bacteria in dry products like spices, sugar, starch, or gums. The distribution of pressure within solid particles is not as homogeneous as it is in fluids, and solids could therefore confer physical protection on microorganisms (15).

Another important component that has been shown to influence piezotolerance of microorganisms during high-pressure treatment is fat. Fat can enhance or reduce the lethal effects of pressure, depending on the conditions of HHP treatments (mainly pressure and temperature), percentage of fat, type of fat, and animal species of milk fat (25). Gervilla et al. (25) reported that increased fat content (0, 6, and 50% fat) did not show a protective effect on *Escherichia coli*. Fat had a piezoprotective effect on *Staphylococcus aureus* and *Lactobacillus helveticus,* but no progressive protection was seen between 6 and 50% fat content. In the case of *L. innocua,* the increase of fat content resulted in a progressive protection against inactivation due to HHP. In contrast, pressurization of *Pseudomonas fluorescens* at high fat content (50%) showed stronger pressure inactivation than at intermediate fat content (6%). The mechanism underlying the piezoprotective effect of fat has not been investigated so far. The enhancement of the lethal effects of fat might be related to the increasing concentration of certain liposoluble substances with antimicrobial effect, interchanging triglycerides of milk with lipoproteins of cellular membrane of microorganisms (altering its permeability), and forming crystals of fat (mainly at low temperatures during HHP treatments) (25).

HHP IN COMBINATION WITH OTHER PROCESSES AND FOOD PRESERVATIVES

HHP technology can be used in combination with additional hurdles to reduce the numbers of microorganisms present in food more effectively. Hurdles with synergistic action are particularly effective and could reduce the intensity of the HHP treatment applied, thereby reducing the cost. A variety of antimicrobial treatments or substances have been shown to act synergistically with high pressure. These include organic acids, bacteriocins, chitosan, lysozyme, lactoperoxidase, and essential oils, but also low or high temperatures during growth and during pressurization (2, 24, 39, 75, 79).

The antimicrobial effects of HHP are enhanced in foods with low pH. Pressurization in the presence of the organic acids citric acid and lactic acid results in a reduction of microorganisms higher than that by HHP alone; additional 1.2- to 3.9-log reductions at pH 4.5 for both acids at 345 MPa have been reported (4). The increased pressure inactivation in the presence of acids is believed to relate to a restricted pH range that bacteria can tolerate under pressure, possibly because of the inhibition of ATPase-dependent transfer of protons and cations, or their direct denaturation or the dislocation of bound ATPase in the membrane (58, 81). In addition, HHP has an effect on the pH value itself. HHP causes separation of electrical charges, allowing water molecules to be positioned with a higher range of order, resulting in the decrease of the total volume of the system. This phenomenon is called electrostriction and causes ionic dissociation, resulting in a pH reduction in the presence of weak acids and a pH increase in the presence of weak bases, with numerous implications for various biological functions (15).

It has been shown that HHP acts synergistically with essential oils such as α-terpinene, (R)-($+$)-limonene, carvacrol, or thymol (2, 40). Essential oils are usually lipophilic and affect the cellular membrane by increasing its permeability to cations like H^+ and K^+ (66). HHP also acts on the cellular membrane, inflicting further damage by causing the lipids of the cellular membrane to crystallize; the melting temperature of lipids (triglycerides) increases by more than $10°C$ per 100 MPa. Thus, membrane lipids present in a liquid state at room temperature will crystallize under high pressure, resulting in changes in the structure and permeability of the cell membrane (15). It has also been shown that antimicrobials might be applied following the HHP treatment and not simultaneously, resulting in similar reductions in microbial counts (40).

The temperature at which pressurization is applied can also be adjusted to enhance the effects of HHP. It has been reported that the pressure resistance of microorganisms is highest at pressurization temperatures of 15 to $30°C$ and decreases significantly at higher or lower temperatures (6). The enhancement of the effects of HHP at low temperatures might be related to reduction in the membrane fluidity under pressure. It is known that less fluid membranes are more sensitive to HHP (49). However, the effects of mild heat in combination with HHP treatments might be related to increased misfolding and denaturation of cellular proteins (32, 56).

Another interesting combined processing method involving HHP is manothermosonication. This method involves the simultaneous application of HHP, heat, and ultrasonic waves (63). The effects of the method depend greatly on the species, as it has been found that combinations of these treatments have synergistic effects for some bacterial species, like *Streptococcus faecium,* while only additive effects for others, like *L. monocytogenes* (59).

LISTERIA STRESS RESPONSE AND ADAPTATION CAPACITY AFFECTING HHP SURVIVAL

Since the majority of the earth is covered by sea, high pressure might be a familiar environment for life. In evolutionary terms, it could be more appropriate to talk about adaptation of life to ambient environment rather than adaptation of life to a high-pressure environment (72). Bacteria have the ability to continuously adapt to changes in their environment. This process of adaptation is termed the microbial stress response and is crucial for the growth and survival of the bacterium under unfavorable conditions. The cellular mechanisms which comprise the stress response system can generally be switched on rapidly. This allows the cell to respond to the changes in its environment, either leading to a rapid resumption of its growth or increasing its chances of survival before the cell is irreversibly damaged. The stress response generally confers cross-protection against a wide variety of other stresses to the cell (69). The fact that microbes can adapt to changing environments, which in turn can lead to an increased resistance against a wide variety of stresses, has serious implications for food safety and food processing. *L. monocytogenes* cells that have adapted to acid conditions are, for example, more resistant to extreme acid conditions and to bile salts (33). These cross-protective effects may cause an increased rate of survival in the human gastrointestinal tract upon ingestion of *L. monocytogenes* cells and thus increase the probability of the occurrence of infection of this pathogen (17, 23). Cross-protective effects might also confer increased resistance to high pressure.

Unfortunately, there is only limited information regarding the mechanisms of bacterial adaptation and resistance to high pressure. One of the identified responses that allow for increased survival of *Listeria* upon HHP treatment results from induction of the general stress response mediated by σ^B. Another mechanism that underlies high-pressure resistance involves a mutation in the central regulator of the class III heat shock response, resulting in constitutive expression of the class III heat shock proteins. Strikingly, these mutations were found to occur at relatively high frequencies in wild-type populations of *L. monocytogenes* Scott A. The occurrence of mutations within a population may further contribute to variations in piezotolerance within a culture. The mechanisms mentioned are discussed in more detail below.

σ^B Plays a Role in Piezotolerance

One of the central regulators of the stress response in gram-positive bacteria is the alternative sigma factor σ^B (7, 73). Upon binding of σ^B to core RNA polymerase, genes downstream of a promoter that can be recognized by the σ^B-RNA polymerase complex are transcribed. The role of σ^B and its regulation have been most extensively studied in the gram-positive model organism *Bacillus subtilis,* but also in the food pathogens *L. monocytogenes, Staphylococcus aureus,* and *Bacillus cereus*. In all these organisms σ^B plays a central role in redirecting gene expression under stress conditions, and its activation can confer protection to a wide range of stresses (for a recent review, see reference 73). A surprisingly small number of the proteins that are encoded by σ^B-dependent genes have a direct role in stress response, like catalases and intracellular proteases that can turn over misfolded proteins. A larger group of proteins seem to have a role in metabolic reprogramming of the cell. These proteins have diverse functions, including the generation of vitamins or cofactors, carbon metabolism, and the in- and efflux of amino acids, osmolytes, and ions. These processes may lead to a passive stress resistance (73). *L. monocytogenes* is able to grow at high salt concentrations (up to 10% NaCl), and the accumulation of the naturally occurring osmolytes betaine and L-carnitine allows the organism to adjust to environments of high osmotic strength. Betaine and L-carnitine transporters play an important role in providing osmoprotection to the cell (14, 34, 80). The transporters BetL and GbuABC are dedicated to the uptake of betaine, whereas OpuC transports carnitine (14, 80). All three transporter-encoding genes are osmotically inducible to some extent. *betL* is transcribed from a σ^B-independent promoter, while *gbuA* is transcribed from dual promoters, one of which is σ^B dependent. *opuC* is transcribed exclusively from a σ^B-dependent promoter (14). The presence of accumulated betaine and L-carnitine in the cell has been demonstrated to contribute to increased barotolerance of *L. monocytogenes* at elevated osmolarity (68).

The importance of the σ^B stress response in general survival of HHP was demonstrated by Wemekamp-Kamphuis et al. (80): an *L. monocytogenes* EGDe σ^B deletion mutant was more sensitive to exposure to HHP than the wild type, while induction of σ^B (following preexposure to pH 4.5 for 1 h) resulted in considerable protection against high-pressure treatment compared with the untreated wild type. The function of a considerable number of σ^B-dependent proteins is still unknown. The determination of the function of these proteins is a challenge for future research and may lead to the discovery of novel mechanisms that can contribute to stress resistance, and supply tools for efficient control of these gram-positive food-borne pathogens.

Heat Shock Proteins Play a Role in Piezotolerance

Many stress-induced proteins are molecular chaperones or proteases that play a role in protein folding and degradation (27). The heat shock response in the gram-positive model organism *Bacillus subtilis* involves at least four classes of heat-inducible genes. These classes are distinguished by their regulatory mechanisms.

Class I genes encode classical chaperones such as DnaK, GroES, and GroEL, which are controlled by the HrcA repressor. Their expression involves a σ^A-dependent promoter and the highly conserved CIRCE (controlling inverted repeat of chaperone expression) operator sequence (TTAGCACTC-N9-GAGTGCTAA). The CIRCE sequence is the binding site for the HrcA repressor (30, 55, 83, 84). The GroE chaperonin machine has been shown to be required in vivo to allow HrcA to adopt its active conformation and repress class I heat shock genes efficiently. After heat shock, GroE is thought to be titrated through association with misfolded proteins in the cell, and HrcA becomes inactive and dissociates from its operator sequence, leading to induction of the HrcA regulon (55).

Class II genes are the largest class and encode general stress proteins. Expression of these genes requires the σ^B sigma factor, whose synthesis and activity are increased under stress conditions such as heat shock but also general stress conditions (see above) (8, 31).

Class III heat shock genes are repressed by the regulator CtsR (class III stress gene repressor). In *L. monocytogenes,* CtsR is encoded by the first gene of the *clpC* operon. It contains an amino-terminal helix-turn-helix motif and controls the expression of *clpP, clpE,* and the *clpC* operon by binding to a tandem heptanucleotide direct repeat (A/GGTCAAANANA/GGT CAAA) (18). CtsR-binding sites usually overlap the transcriptional start site or the -35 and -10 sequences of the promoter, suggesting that the repressor acts by interfering with the binding of the RNA polymerase $E\sigma^A$ holoenzyme. An extensive database search indicated that this system is highly conserved among low-G+C gram-positive bacteria, with CtsR-binding sites found mostly upstream from *clp* genes (18). Induction of class III genes is thought to involve targeted degradation of the CtsR repressor by the Clp ATP-dependent protease (19, 46).

Finally, class IV genes are heat shock genes that are not controlled by HrcA, σ^B, or CtsR.

Increased tolerance to HHP was observed for an *L. monocytogenes* spontaneous mutant, strain AK01, compared with the wild type. Interestingly, cells of AK01 lacked flagella, were elongated (Fig. 2), and showed increased resistance to heat, acid, and H_2O_2 compared with the wild type (43). It is noteworthy that the mutant showed reduced virulence (42). Further analysis of this mutant showed constitutive expression of the class III heat shock proteins, which resulted from a codon deletion in *ctsR*, rendering CtsR inactive as a repressor (42). As a result, cells of mutant strain AK01 are preloaded with heat shock proteins, resulting in an increased ability to cope with misfolded and denatured proteins that may be toxic to the cell. Other authors have also reported increased heat resistance of piezotolerant isolates, and it is not inconceivable that these organisms show increased

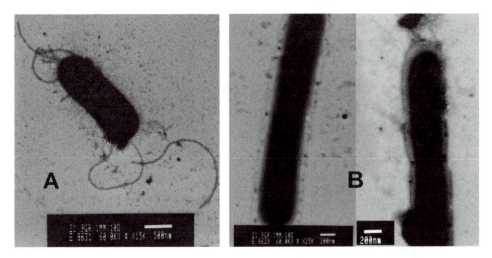

Figure 2. Visualization of exponentially grown cells of wild-type *L. monocytogenes* Scott A (A) and the piezotolerant mutant AK01 (B) with electron microscopy. Bars, 500 nm (A) and 200 nm (B). Reprinted from reference 43.

expression of heat shock proteins (36, 67). In fact, it was demonstrated that heat shock promoters were activated in *Escherichia coli* mutants with increased resistance to pressure compared with the wild-type strains (3). The same study also showed that heat shock rendered *E. coli* significantly more resistant to pressure treatment.

Limited studies are available on the protein expression profiles of bacteria that have been exposed to sublethal pressure treatments. In *E. coli,* induced expression of 55 proteins was observed upon exposure to a sublethal pressure treatment, and many of these proteins were previously identified as heat shock or cold shock proteins (77; see also chapter 5). Sublethal pressurization of the gram-positive bacterium *Lactobacillus sanfranciscensis* resulted in the induction of 16 proteins, 1 of which was specific to the response to HHP. The other proteins were influenced by most of the investigated stresses in a way similar to how they were influenced by HHP. The highest similarity in the HHP proteome was found to be with cold- and NaCl-stressed cells, with 11 overlapping proteins (35; see also chapter 7). Lastly, increased expression of cold shock proteins upon pressurization was demonstrated in *L. monocytogenes* by Western blotting using cold shock protein antibodies (79).

Spontaneous Occurrence of Mutations

Mechanisms that increase genetic variation can allow for a selective advantage under stressful conditions. Bacteria have several stress responses that provide ways to specifically produce mutations and respond to selective pressure. These include the SOS response, the general stress response, the heat shock response, and the stringent response (22, 38). The underlying mechanisms can be a DNA polymerase that synthesizes error-containing DNA, recombination-dependent generation of mutations, or recombination-independent generation of mutation (e.g., strand slippage). The latter two phenomena are facilitated by the presence of hypermutable loci within genetic regions that are important for the expression of stress genes. These loci have been shown to have an overrepresentation in regions related to stress genes (64).

The significance of naturally occurring mutants within a "pure" culture of *L. monocytogenes* for piezotolerance was recently demonstrated by work on spontaneous piezotolerant mutants of *L. monocytogenes* (41, 42). A triple-codon repeat in a glycine-encoding region of the *ctsR* gene of *L. monocytogenes* (Fig. 3) is prone to a codon deletion or insertion, thereby inactivating the repressor function of CtsR. In addition, larger insertions and deletions in this region were observed, leading to truncated or inactive CtsR proteins (41, 42). This results in the occurrence of piezotolerant and stress-tolerant mutants at relatively high frequencies within a wild-type population (41).

amino acid number	59	60	61	62	63	64	65	66	67	68
amino acid	S	R	R	G	G	G	G	Y	I	R
DNA sequence	AGC	AAA	CGT	GGT	GGT	GGT	GGA	TAT	ATT	CGG

Figure 3. Protein and DNA sequences of the glycine-rich region of *ctsR* of *L. monocytogenes*. The piezotolerant mutant strain AK01 lacks a GGT codon sequence in the glycine-rich region (42). Mutations in this region were found at relatively high frequencies in other piezotolerant *L. monocytogenes* isolates (41).

In addition, insight is required in other stress-induced mutagenesis mechanisms. The SOS response is associated with stress-induced mutagenesis triggered by DNA damage and leads to an error-prone DNA replication, resulting in a mutator phenotype (82). An inducible or constitutive mutator phenotype may confer a selective advantage for the population. Heterogeneous, changing, or stressful environments can select for these mutator strains, and it is conceivable that stress-induced rapid evolution can play a role in the development of very stress-resistant strains or the emergence of novel food-borne pathogens.

CONCLUSIONS AND PERSPECTIVES

Variability in piezotolerance occurs between different species, strains, and even cells within a population. Also, the history of the cells at the time of pressurization can influence the inactivation by pressure; e.g., stationary-phase cells or cells that have been exposed to other stresses are generally more resistant to pressure. Furthermore, the composition of the food, the presence of antimicrobial compounds with synergistic effect, and various other factors can affect the inactivation achieved. In general, the intensity and duration of a preserving treatment that a food must undergo to achieve a desired safety level are determined on the basis of inactivation curves of selected target pathogens. For high-pressure inactivation, *L. monocytogenes* is recommended as a target organism by the U.S. Food and Drug Administration. In theory, inactivation curves follow first-order kinetics, and the treatment time is usually calculated by extrapolating the inactivation curve to the desired level of inactivation. When resistant subpopulations are present, an inactivation pattern is obtained that deviates from first-order kinetics: a commonly observed pattern is that of a gradually decreasing rate of inactivation in time (tailing effect). In current practice, safety margins included in the process conditions are generally sufficient to take moderate tailing effects into account.

Extensive application of functional genomics tools may rapidly increase our knowledge of bacterial stress responses and survival mechanisms, including the characterization of stress-induced mutator phenotypes and the occurrence of stable subpopulations of pathogens that are more resistant to inactivation treatments than the wild-type population. It is essential that strains of special epidemiological and food safety relevance are adequately represented in future studies. Identification of key elements in recovery and survival may help the design of new intervention strategies that efficiently target food pathogens, including the use of highly efficient high-pressure treatment, either stand-alone or in combination with other treatments. Moreover, cellular mechanisms of adaptation to environmental and food processing-related stresses, including low pH and (weak) acids, may affect stress resistance and virulence of food pathogens. Therefore, the pathogenic potential of these variants has to be assessed and taken into account in human risk assessment studies concerning exposure to *L. monocytogenes*.

REFERENCES

1. **Abee, T., and J. A. Wouters.** 1999. Microbial stress response in minimal processing. *Int. J. Food Microbiol.* **50:**65–91.
2. **Adegoke, G. O., H. Iwahashi, and Y. Komatsu.** 1997. Inhibition of *Saccharomyces cerevisiae* by combination of hydrostatic pressure and monoterpenes. *J. Food Sci.* **62:**404–405.

3. **Aertsen, A., K. Vanoirbeek, P. De Spiegeleer, J. Sermon, K. Hauben, A. Farewell, T. Nystrom, and C. W. Michiels.** 2004. Heat shock protein-mediated resistance to high hydrostatic pressure in *Escherichia coli. Appl. Environ. Microbiol.* **70:**2660–2666.

4. **Alpas, H., N. Kalchayanand, F. Bozoglu, and B. Ray.** 2000. Interactions of high hydrostatic pressure, pressurisation temperature and pH on death and injury of pressure-resistant and pressure-sensitive strains of food-borne pathogens. *Int. J. Food Microbiol.* **60:**33–42.

5. **Alpas, H., N. Kalchayanand, F. Bozoglu, A. Sikes, C. P. Dunne, and B. Ray.** 1999. Variation in resistance to hydrostatic pressure among strains of food-borne pathogens. *Appl. Environ. Microbiol.* **65:**4248–4251.

6. **Arroyo, G., P. D. Sanz, and G. Préstamo.** 1999. Response to high pressure, low temperature treatment in vegetables: determination of survival rates of microbial populations using flow cytometry and detection of peroxidase activity using confocal microscopy. *J. Appl. Microbiol.* **86:**544–556.

7. **Becker, L. A., S. N. Evans, R. W. Hutkins, and A. K. Benson.** 2000. Role of sigma B in adaptation of *Listeria monocytogenes* to growth at low temperature. *J. Bacteriol.* **182:**7083–7087.

8. **Benson, A. K., and W. G. Haldenwang.** 1993. The σ^B-dependent promoter of the *Bacillus subtilis* sigB operon is induced by heat shock. *J. Bacteriol.* **175:**1929–1935.

9. **Brackett, R. E.** 1988. Presence and persistence of *Listeria monocytogenes* in food and water. *Food Technol.* **4:**162–164.

10. **Buchrieser, C., C. Rusniok, F. Kunst, P. Cossart, and P. Glaser for the Listeria Consortium.** 2003. Comparison of the genome sequences of *Listeria monocytogenes* and *Listeria innocua:* clues for evolution and pathogenicity. *FEMS Immunol. Med. Microbiol.* **35:**207–213.

11. **Buzrul, S., and H. Alpas.** 2004. Modeling the synergistic effect of high pressure and heat on inactivation kinetics of *Listeria innocua:* a preliminary study. *FEMS Microbiol. Lett.* **238:**29–36.

12. **Cabanes, D., P. Dehoux, O. Dussurget, L. Frangeul, and P. Cossart.** 2002. Surface proteins and the pathogenic potential of *Listeria monocytogenes. Trends Microbiol.* **10:**238–245.

13. **Casadei, M. A., and B. M. Mackey.** 1997. The effect of growth temperature on pressure resistance of *Escherichia coli*, p. 281–282. *In* K. Heremans (ed.), *High Pressure Research in the Biosciences and Biotechnology.* Leuven University Press, Leuven, Belgium.

14. **Cetin, M. S., C. Zhang, R. W. Hutkins, and A. K. Benson.** 2004. Regulation of transcription of compatible solute transporters by the the general stress response sigma factor, sigma B, in *Listeria monocytogenes. J. Bacteriol.* **186:**794–802.

15. **Cheftel, C.** 1995. Review: high-pressure, microbial inactivation and preservation. *Food Sci. Technol. Int.* **1:**75–90.

16. **Cossart, P.** 2002. Molecular and cellular basis of the infection by *Listeria monocytogenes*, an overview. *Int. J. Med. Microbiol.* **291:**401–409.

17. **Cotter, P. D., and C. Hill.** 2003. Surviving the acid test: responses of gram-positive bacteria to low pH. *Microbiol. Mol. Biol. Rev.* **67:**429–453.

18. **Derré, I., G. Rapoport, and T. Msadek.** 1999. CtsR, a novel regulator of stress and heat shock response, controls *clp* and molecular chaperone gene expression in Gram-positive bacteria. *Mol. Microbiol.* **31:**117–131.

19. **Derré, I., G. Rapoport, and T. Msadek.** 2000. The CtsR regulator of stress response is active as a dimer and specifically degraded in vivo at 37°C. *Mol. Microbiol.* **38:**335–347.

20. **Doumith, M., C. Cazalet, N. Simoes, L. Frangeul, C. Jacquet, F. Kunst, P. Martin, P. Cossart, P. Glaser, and C. Buchrieser.** 2004. New aspects regarding evolution and virulence of *Listeria monocytogenes* revealed by comparative genomics and DNA arrays. *Infect. Immun.* **72:**1072–1083.

21. **Ferreira, A., C. P. O'Byrne, and K. J. Boor.** 2001. Role of σ^B in heat, ethanol, acid, and oxidative stress resistance and during carbon starvation in *Listeria monocytogenes. Appl. Environ. Microbiol.* **67:**4454–4457.

22. **Foster, P. L.** 2005. Stress responses and genetic variation in bacteria. *Mutat. Res.* **569:**3–11.

23. **Gahan, C. G., and C. Hill.** 2005. Gastrointestinal phase of *Listeria monocytogenes* infection. *J. Appl. Microbiol.* **98:**1345–1353.

24. **Garcia-Graells, C., C. Valckx, and C. W. Michiels.** 2000. Inactivation of *Escherichia coli* and *Listeria innocua* in milk by combined treatment with high hydrostatic pressure and the lactoperoxidase system. *Appl. Environ. Microbiol.* **66:**4173–4179.

25. **Gervilla, R., V. Ferragut, and B. Guamis.** 2000. High pressure inactivation of microorganisms inoculated into ovine milk of different fat contents. *J. Dairy Sci.* **83:**674–682.

26. **Glaser, P., L. Frangeul, C. Buchrieser, C. Rusniok, A. Amend, F. Baquero, P. Berche, H. Bloecker, P. Brandt, T. Chakraborty, A. Charbit, F. Chetouani, E. Couve, A. de Daruvar, P. Dehoux, E. Domann,**

G. Dominguez-Bernal, E. Duchaud, L. Durant, O. Dussurget, K. D. Entian, H. Fsihi, F. Garcia-del Portillo, P. Garrido, L. Gautier, W. Goebel, N. Gomez-Lopez, T. Hain, J. Hauf, D. Jackson, L. M. Jones, U. Kaerst, J. Kreft, M. Kuhn, F. Kunst, G. Kurapkat, E. Madueno, A. Maitournam, J. M. Vicente, E. Ng, H. Nedjari, G. Nordsiek, S. Novella, B. de Pablos, J. C. Perez-Diaz, R. Purcell, B. Remmel, M. Rose, T. Schlueter, N. Simoes, A. Tierrez, J. A. Vazquez-Boland, H. Voss, J. Wehland, and P. Cossart. 2001. Comparative genomics of *Listeria* species. *Science* **294:**849–852.

27. **Gottesman, S., S. Wickner, and M. R. Maurizi.** 1997. Protein quality control: triage by chaperones and proteases. *Genes Dev.* **11:**815–823.

28. **Gould, G. H.** 2000. Emerging technologies in food preservation and processing in the last 40 years, p. 5–23. *In* G. V. Barbosa-Cánovas and G. W. Gould (ed.), *Innovations in Food Processing.* Technomic Publishing Co., Inc. D/B ATP Ltd., Hitchin Herts, United Kingdom.

29. **Hauben, K. J. A., D. H. Bartlett, C. C. F. Soontjens, K. Cornelis, E. Y. Wuytack, and C. W. Michiels.** 1997. *Escherichia coli* mutants resistant to inactivation by high hydrostatic pressure. *Appl. Environ. Microbiol.* **63:**945–950.

30. **Hecker, M., W. Schumann, and U. Völker.** 1996. Heat-shock and general stress response in *Bacillus subtilis. Mol. Microbiol.* **19:**417–428.

31. **Hecker, M., and U. Völker.** 1998. Nonspecific, general and multiple stress resistance of growth-restricted *Bacillus subtilis* cells by the expression of the sigma B regulon. *Mol. Microbiol.* **29:**1129–1136.

32. **Heremans, K.** 1982. High pressure effects on proteins and other biomolecules. *Annu. Rev. Biophys. Bioeng.* **11:**1–21.

33. **Hill, C., P. D. Cotter, R. D. Sleator, and C. G. M. Gahan.** 2002. Bacterial stress response in *Listeria monocytogenes:* jumping the hurdles imposed by minimal processing. *Int. Dairy J.* **12:**273–283.

34. **Holtmann, G., and E. Bremer.** 2004. Thermoprotection of *Bacillus subtilis* by exogenously provided glycine betaine and structurally related compatible solutes: involvement of Opu transporters. *J. Bacteriol.* **186:**1683–1693.

35. **Hormann, S., C. Scheyhing, J. Behr, M. Pavlovic, M. Ehrmann, and R. F. Vogel.** 2006. Comparative proteome approach to characterize the high-pressure stress response of Lactobacillus sanfranciscensis DSM 20451(T). *Proteomics* **6:**1878–1885.

36. **Iwahashi, H., S. Fujii, K. Obuchi, S. C. Kaul, A. Sato, and Y. Komatsu.** 1993. Hydrostatic pressure is like high temperature and oxidative stress in the damage it causes to yeast. *FEMS Microbiol. Lett.* **108:**53–58.

37. **Iwahashi, H., S. C. Kaul, K. Obuchi, and Y. Komatsu.** 1991. Induction of barotolerance by heat shock treatment in yeast. *FEMS Microbiol. Lett.* **80:**325–328.

38. **Joyce, E. A., K. Chan, N. R. Salama, and S. Falkow.** 2002. Redefining bacterial populations: a postgenomic reformation. *Nature* **3:**462–473.

39. **Kalchayanand, N., A. Sikes, C. P. Dunne, and B. Ray.** 1998. Interaction of hydrostatic pressure, time and temperature of pressurization and pediocin AcH on inactivation of foodborne bacteria. *J. Food Prot.* **61:**425–431.

40. **Karatzas, K. A., E. P. W. Kets, E. J. Smid, and M. H. Bennik.** 2001. The combined action of carvacrol and high hydrostatic pressure on *Listeria monocytogenes* Scott A. *J. Appl. Microbiol.* **90:**463–469.

41. **Karatzas, K. A., V. P. Valdramidis, and M. H. Wells-Bennik.** 2005. Contingency locus in *ctsR* of *Listeria monocytogenes* Scott A: a strategy for occurrence of abundant piezotolerant isolates within clonal populations. *Appl. Environ. Microbiol.* **71:**8390–8396.

42. **Karatzas, K. A., J. A. Wouters, C. G. Gahan, C. Hill, T. Abee, and M. H. Bennik.** 2003. The CtsR regulator of *Listeria monocytogenes* contains a variant glycine repeat region that affects piezotolerance, stress resistance, motility and virulence. *Mol. Microbiol.* **49:**1227–1238.

43. **Karatzas, K. A. G., and M. H. J. Bennik.** 2002. Characterization of a *Listeria monocytogenes* Scott A isolate with high tolerance towards high hydrostatic pressure. *Appl. Environ. Microbiol.* **68:**3183–3189.

44. **Kathariou, S.** 2002. *Listeria monocytogenes* virulence and pathogenicity, a food safety perspective. *J. Food Prot.* **65:**1811–1829.

45. **Knorr, D.** 1994. Hydrostatic pressure treatment of foods: microbiology, p. 159–175. *In* G. W. Gould (ed.), *New Methods of Food Preservation.* Blakie, London, United Kingdom.

46. **Krüger, E., D. Zühlke, E. Witt, H. Ludwig, and M. Hecker.** 2001. Clp-mediated proteolysis in Grampositive bacteria is autoregulated by the stability of a repressor. *EMBO J.* **20:**852–863.

47. **Kuipers, O. P.** 1999. Genomics for food biotechnology: prospects of the use of high-throughput technologies for the improvement of food microorganisms. *Curr. Opin. Biotechnol.* **10:**511–516.

48. **Lund, B. M., T. C. Baird-Parker, and G. W. Gould.** 2000. *The Microbiological Safety and Quality of Food,* vol. 2. Aspen Publishers, Gaithersburg, MD.

49. **MacDonald, A. G.** 1992. Effects of high hydrostatic pressure on natural and artificial membranes, p. 67–74. *In* C. Balny, R. Hayashi, K. Heremans, and P. Masson (ed.), *High Pressure and Biotechnology*. Paris INSERM/John Libbey, Paris, France.

50. **Mackey, B. M., K. Forestiére, and N. Isaacs.** 1995. Factors affecting the resistance of *Listeria monocytogenes* to high hydrostatic pressure. *Food Biotechnol.* **9:**1–11.

51. **McClements, J. M., M. F. Patterson, and M. Linton.** 2001. The effect of growth stage and growth temperature on high hydrostatic pressure inactivation of some psychrotrophic bacteria in milk. *J. Food Prot.* **64:** 514–522.

52. **Mead, P. S., L. Slutsker, V. Dietz, L. F. McCaig, J. S. Bresee, C. Shapiro, P. M. Griffin, and R. V. Tauxe.** 1999. Food-related illness and death in the United States. *Emerg. Infect. Dis.* **5:**607–625.

53. **Metrick, C., D. G. Hoover, and D. F. Farkas.** 1989. Effects of high hydrostatic pressure on heat-resistant and heat-sensitive strains of *Salmonella. J. Food Sci.* **54:**1547–1549.

54. **Milohanic, E., P. Glaser, J. Y. Coppee, L. Frangeul, Y. Vega, J. A. Vazquez-Boland, F. Kunst, P. Cossart, and C. Buchrieser.** 2003. Transcriptome analysis of Listeria monocytogenes identifies three groups of genes differently regulated by PrfA. *Mol. Microbiol.* **47:**1613–1625.

55. **Mogk, A., G. Homuth, C. Scholz, L. Kim, F. X. Schmid, and W. Schumann.** 1997. The GroE chaperonin machine is a major modulator of the CIRCE heat shock regulon of *Bacillus subtilis. EMBO J.* **16:**4579–4590.

56. **Mozhaev, V. V., K. Heremans, J. Frank, P. Masson, and C. Balny.** 1996. High pressure effects on protein structure and function. *Proteins Struct. Funct. Genet.* **24:**81–91.

57. **Nelson, K. E., D. E. Fouts, E. F. Mongodin, J. Ravel, R. T. DeBoy, J. F. Kolonay, D. A. Rasko, S. V. Angiuoli, S. R. Gill, I. T. Paulsen, J. Peterson, O. White, W. C. Nelson, W. Nierman, M. J. Beanan, L. M. Brinkac, S. C. Daugherty, R. J. Dodson, A. S. Durkin, R. Madupu, D. H. Haft, J. Selengut, S. Van Aken, H. Khouri, N. Fedorova, H. Forberger, B. Tran, S. Kathariou, L. D. Wonderling, G. A. Uhlich, D. O. Bayles, J. B. Luchansky, and C. M. Fraser.** 2004. Whole genome comparisons of serotype 4b and 1/2a strains of the food-borne pathogen *Listeria monocytogenes* reveal new insights into the core genome components of this species. *Nucleic Acids Res.* **32:**2386–2395.

58. **Pagán, R., S. Jordan, A. Benito, and B. Mackey.** 2001. Enhanced acid sensitivity of pressure-damaged *Escherichia coli* O157 cells. *Appl. Environ. Microbiol.* **67:**1983–1985.

59. **Pagán R., P. Mañas, J. Raso, and S. Condón.** 1999. Bacterial resistance to ultrasonic waves under pressure at nonlethal (manosonication) and lethal (manothermosonication) temperatures. *Appl. Environ. Microbiol.* **65:**297–300.

60. **Palou, E., A. López-Malo, G. V. Barbosa-Cánovas, and B. G. Swanson.** 1999. High-pressure treatment in food preservation, p. 533–576. *In* M. S. Rahman (ed.), *Handbook of Food Preservation*. Dekker, New York, NY.

61. **Patterson, M. F., M. Quinn, R. Simpson, and A. Gilmour.** 1995. The sensitivity of vegetative pathogens to high hydrostatic pressure in phosphate buffered saline and foods. *J. Food Prot.* **58:**524–529.

62. **Ramnath, M., K. B. Rechinger, L. Jansch, J. W. Hastings, S. Knochel, and A. Gravesen.** 2003. Development of a *Listeria monocytogenes* EGDe partial proteome reference map and comparison with the protein profiles of food isolates. *Appl. Environ. Microbiol.* **69:**3368–3376.

63. **Raso, J., R. Pagán, S. Condon, and F. J. Sala.** 1998. Influence of temperature and pressure on the lethality of ultrasound. *Appl. Environ. Microbiol.* **64:**465–471.

64. **Rocha, E. P. C., I. Matic, and F. Taddei.** 2002. Over-representation of repeats in stress response genes: a strategy to increase versatility under stressful conditions? *Nucleic Acids Res.* **30:**1886–1894.

65. **Schlech, W. F., III.** 2001. Food-borne listeriosis. *Clin. Infect. Dis.* **32:**1518–1519.

66. **Sikkema, J., J. A. M. de Bont, and B. Poolman.** 1994. Interactions of cyclic hydrocarbons with biological membranes. *J. Biol. Chem.* **269:**8022–8028.

67. **Smelt, J. P. P. M.** 1998. Recent advances in the microbiology of high pressure processing. *Trends Food Sci. Technol.* **9:**152–158.

68. **Smiddy, M., R. D. Sleator, M. F. Patterson, H. Hill, and A. L. Kelly.** 2004. Role for compatible solutes in glycine betaine and L-carnitine in listerial barotolerance. *Appl. Environ. Microbiol.* **70:**7555–7557.

69. **Sofos, J. N.** 2002. Stress-adapted, cross-protected: a concern? *Food Technol.* **56:**22.

70. **Takahashi, K., H. Ishii, and H. Ishikawa.** 1993. Sterilization of bacteria and yeast by hydrostatic pressurization at low temperature: effect of temperature, pH and the concentration of proteins, carbohydrates and lipids, p. 244–249. *In* R. Hayashi (ed.), *High Pressure Bioscience and Food Science*. San-Ei Publishing Co., Kyoto, Japan.

71. **Tay, A., T. H. Shellhammer, A. E. Yousef, and G. W. Chism.** 2003. Pressure death and tailing behavior of *Listeria monocytogenes* strains having different barotolerances. *J. Food Prot.* **66:**2057–2061.

72. **Turley, C.** 2000. Bacteria in the cold deep-sea benthic boundary layer and sediment-water interface of the NE Atlantic. *FEMS Microbiol. Ecol.* **33:**89–99.

73. **Van Schaik, W., and T. Abee.** 2005. The role of sigma B in the stress response of Gram-positive bacteria—targets for food preservation and safety. *Curr. Opin. Biotechnol.* **16:**218–224.

74. **Vásquez-Boland, J. A., M. Kuhn, P. Berche, T. Chakraborty, G. Domínguez-Bernal, W. Goebel, B. Gonzalez-Zorn, J. Wehland, and J. Kreft.** 2001. *Listeria* pathogenesis and molecular virulence determinants. *Clin. Microbiol. Rev.* **14:**584–640.

75. **Vurma, M., Y. K. Chung, T. H. Shellhammer, E. J. Turek, and A. E. Yousef.** 2006. Use of phenolic compounds for sensitizing *Listeria monocytogenes* to high-pressure processing. *Int. J. Food Microbiol.* **106:**263–269.

76. **Washburn, M. P., and J. R. Yates III.** 2000. Analysis of the microbial proteome. *Curr. Opin. Microbiol.* **3:**292–297.

77. **Welch, T. J., A. Farewell, F. C. Neidhardt, and D. H. Bartlett.** 1993. Stress response of *Escherichia coli* to elevated hydrostatic pressure. *J. Bacteriol.* **175:**7170–7177.

78. **Wells, J. M., and M. H. Bennik.** 2003. Genomics of food-borne bacterial pathogens. *Nutr. Res. Rev.* **16:**21–35.

79. **Wemekamp-Kamphuis, H. H., A. K. Karatzas, J. A. Wouters, and T. Abee.** 2002. Enhanced levels of cold shock proteins in *Listeria monocytogenes* LO28 upon exposure to low temperature and high hydrostatic pressure. *Appl. Environ. Microbiol.* **68:**456–463.

80. **Wemekamp-Kamphuis, H. H., J. A. Wouters, P. P. L. A. de Leeuw, T. Hain, T. Chakraborty, and T. Abee.** 2004. Identification of sigma factor σ^B-controlled genes and their impact on acid stress, high hydrostatic pressure, and freeze survival in *Listeria monocytogenes* EGD-e. *Appl. Environ. Microbiol.* **70:**3457–3466.

81. **Wouters, P. C., E. Glaasker, and J. P. P. M. Smelt.** 1998. Effects of high pressure on inactivation kinetics and events related to proton efflux in *Lactobacillus plantarum. Appl. Environ. Microbiol.* **64:**509–514.

82. **Yang, H., E. Wolff, M. Kim, A. Diepand, and J. H. Miller.** 2005. Identification of mutator genes and mutational pathways in *Escherichia coli* using a multicopy cloning approach. *Mol. Microbiol.* **53:**283–295.

83. **Yuan, G., and S. L. Wong.** 1995. Isolation and characterization of *Bacillus subtilis groE* regulatory mutants: evidence for *orf39* in the *dnaK* operon as a repressor gene in regulating the expression of both *groE* and *dnaK. J. Bacteriol.* **177:**6462–6468.

84. **Zuber, U., and W. Schumann.** 1994. CIRCE, a novel heat shock element involved in regulation of heat shock operon *dnaK* of *Bacillus subtilis. J. Bacteriol.* **176:**1359–1363.

High-Pressure Microbiology
Edited by C. Michiels, D. H. Bartlett, and A. Aertsen
© 2008 ASM Press, Washington, DC

Chapter 7

Effects of Pressure on Lactic Acid Bacteria

Rudi F. Vogel and Matthias A. Ehrmann

WHY STUDY PRESSURE STRESS IN LAB?

Lactic acid bacteria (LAB) harbor a large variety of species, with their broadest diversity within the lactobacilli (69). With few exceptions, namely, within the genus *Streptococcus,* the vast majority of these bacteria are harmless to humans, animals, and plants. Originating from a long tradition in artisanal food fermentation, a constantly growing number of strains are deliberately used as starter cultures in food biotechnology to produce stable, sensorially desirable foods of high commercial value and hygienic quality (51). Although their primary contribution centers on rapid acid production, they also contribute to the flavor, texture, and nutritional value of foods. Also, LAB are considered to be important components of the normal intestinal microbiota and are also found naturally in human and animal habitats, including the gastrointestinal tract and the oral and vaginal cavities (72). Selected strains of *Lactobacillus* are widely used as probiotics, primarily in dairy products and dietary supplements.

The selection of starter strains fulfilling all metabolic, technical, and handling requirements is the result of a multidisciplinary approach, i.e., to analyze, monitor, and direct the microbial ecology in food fermentations by molecular biology tools, genomics, biochemical and physiological analyses, pilot trials, and modeling of behavior and metabolism. Beyond their applications in food fermentation, these bacteria are gaining growing attention as probiotics, therapeutic agents in chronic diseases, cellular factories, and live oral vaccines. On the other hand, the same bacteria can be potent food spoilage organisms in many types of food, particularly if fermentable carbohydrates are present or released by endogenous enzymes.

The analysis of the tolerance to environmental parameters focused on conditions prevailing during food processing and storage will promote both the application of LAB in food fermentations and their killing or growth inhibition as undesirable spoilage microbiota. Thorough knowledge of stress responses of these bacteria not only allows determination of minimal processing conditions for their inactivation but also enables the design

Rudi F. Vogel and Matthias A. Ehrmann • Technische Mikrobiologie, Technische Universität München, 85350 Freising, Germany.

of stable preparations of desirable strains which therefore have to undergo stresses, including heat and/or cold shock as well as desiccation. Yet, the use of many metabolically interesting strains is hampered by the lack of knowledge of their stress response and tolerance. Among these are strains from cereal fermentations, namely, sourdoughs, which therefore were a focus of the work described here.

Consumer demand for freshness and minimally processed foods along with a prolonged shelf life has driven the development of novel technologies in food preservation, with high hydrostatic pressure (HHP) being the most promising. Therefore, the response to HHP of LAB in their ambivalent role as factories in biotechnology and in biomedical applications versus food-spoiling organisms is most interesting.

STRESS RESPONSE OF LAB

The ability to quickly respond to stressful environments is essential for the survival of bacteria and involves sensing systems, regulatory networks, and the variation of gene expression or cellular structures. In this context the systematic use of adaptive stress responses and cross-protection is a matter of growing interest. Although adaptive stress responses follow general rules and reveal common aspects in bacteria, their specific outcome appears to be species or even strain specific and relies on different molecular bases (78). To understand HHP stress, a knowledge of mechanisms of responses to other stresses is essential. Although mesophilic bacteria may have faced high pressure during their evolution in ancient times (67), high-pressure stress can be considered "unknown" to most of them. Based on cross-protection studies it can be anticipated that mechanisms of cellular pressure stress response use or overlap with those of "known" stresses, including pH, temperature, and osmotic changes, or are simply general. At first glance, this may be a general difference from the pressure response of piezophiles adapted to deep-sea environments. However, the response of deep-sea bacteria to high pressure has been determined at pressure levels they can adapt to, and their responses may reveal patterns of stress response which are similar to those of mesophilic bacteria when they are exposed to pressure levels which present stress to them. Furthermore, primary HHP stress responses which result from the thermodynamic effects of pressure on a (bio)molecule, must be distinguished from secondary responses, which are triggered by pressure-induced environmental and intracellular changes (e.g., pH drop or ribosomal dysfunction). To cope with such effects, cells may use regular signal cascades, which need not directly be affected by pressure.

Stress responses, including responses to acid, heat, cold, and osmotic stress, have been studied in a variety of LAB species, including *Lactococcus lactis* (18), *Streptococcus* (52), and *Lactobacillus sanfranciscensis* (15). LAB react to metabolic stresses with the induction of at least 100 general stress proteins (78). These various stress responses are characterized by induction of sets of general traits, including DNA repair mechanisms (UvrA), enzymes of the chaperone/protease family (DnaK, GroEL, and Clp), and typical metabolic changes (arginine metabolism upon acid stress and trehalose metabolism upon osmotic stress). However, the primary sensors for the signal cascades mediating these "classical" stress responses within the LAB mostly remain putative. In *Bacillus subtilis,* the closest LAB relative studied with respect to stress response, and in many other bacteria, typical categories of stress signaling are found, including (i) HrcA-repressed genes like *dnaK* and the *groESL* operon, (ii) genes regulated by the alternative sigma B factor, (iii) CtsR-repressed genes, (iv) genes regulating caseinolytic proteases *(clp)*, and (v) genes regulated by other factors. However, HrcH-repressed and CtsR-regulated genes cannot be separated in many gram-positive organisms (13), and in the genome of *L. lactis* (6) and other sequenced LAB no sigma B ortholog could be found.

When LAB are subjected to HHP, membrane fluidity (21) and tension (64) and translational efficiency (77) are likely to be the primary targets for high-pressure stress sensing. While the membrane fluidity and tension may directly affect function of membrane proteins or initiate signal cascades (5), the ribosome as stress sensor is tightly coupled with the quality control of proteins by expression of heat shock proteins and the process of *trans*-translation (82). This mechanism prohibits accumulation of stalled ribosomes by releasing them from the mRNA upon binding of transfer mRNA (tmRNA). At the same time, the truncated nascent protein is tagged with a short signature peptide, making it a target for selective proteolytic degradation by Clp proteases. The function of tmRNA is not fully understood, and its role in HHP stress remains to be proven not only for the LAB but also for other bacteria. As HHP via its thermodynamic force is able to influence the physical status of bacterial membranes and ribosomal function in vitro, the use of HHP to study the stress response of these organisms also fosters a basic understanding of stress responses and other cellular events involving macromolecular interaction. This chapter focuses on the HHP-mediated stress responses of selected strains of LAB, which were used unless stated otherwise; the selected strains included *Lactococcus lactis* subsp. *cremoris* MG1363, *Lactobacillus plantarum* TMW1460, and *Lactobacillus sanfranciscensis* DSM20451[T].

HIGH-PRESSURE-MEDIATED CELL DEATH, SUBLETHAL INJURY, AND RESISTANT FRACTIONS

As observed with a variety of other microorganisms (47), LAB residing as suspensions in buffered media at physiological pH are inactivated at pressures between 200 and 600 MPa at holding times of 5 to 60 min, with the decrease of CFU following asymmetric sigmoid declines. Between 200 and 300 MPa the sigmoid shape is pronounced and the curve usually ends in a plateau, indicating the survival of a pressure-resistant fraction. A typical example of a pressure inactivation curve is given for *L. plantarum* in Fig. 1. With some LAB, e.g., strains of *Lactobacillus brevis,* an increase in CFU was detected at low pressure with short holding times, which could be explained by a dissociation of doubles and short chains (unpublished observation). With other strains, e.g., *Enterococcus faecalis,* pressure-mediated cell aggregation may occur which is often overlooked, as it simulates a strong pressure sensitivity with sharply decreasing CFU numbers (unpublished observation). The determination of CFU is inevitably done ex situ, and therefore, only irreversible inactivation is determined. Still, the sigmoid curve shape is indicative for bacteria residing in various states of injury within the population, some of which may be able to recover and some of which may not.

The number of reversibly injured cells can be estimated by plating on rich and on stress media (e.g., containing 4% NaCl) and calculating the ratio. In Fig. 1 the number of these sublethally injured cells, which can only recover under optimal conditions, is plotted as a time course along the holding time. This way, the condition producing the maximal number of sublethally injured cells can be determined by calculating the ratio of uninjured to injured cells along the holding time course and used in studies of the stress response of the

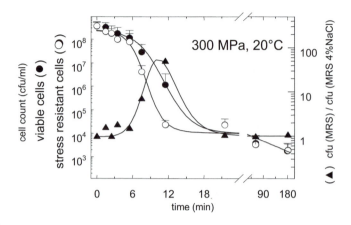

Figure 1. Time course of HHP inactivation of *Lactobacillus plantarum* demonstrating areas of sublethal injury.

respective strain. Figure 2a demonstrates for two strains of *L. sanfranciscensis* that studies on cellular responses induced by high pressure are best performed within a pressure range of 30 to 100 MPa. In this range, cells are able to grow and metabolize maltose. However, filamentation of the cells may be observed under these conditions, as shown in Fig. 2b. As shown in Fig. 2a and 3 for *L. sanfranciscensis* and *L. lactis,* respectively, the loss of metabolic activity and membrane damage occurred prior to or concomitant with cell death. The cellular basis of this sublethal injury was demonstrated to reside in the impaired or rerouted metabolism, membrane integrity and functionality, and effects on ribosomal biosynthesis, which are discussed later in this chapter.

The pressure-resistant fraction can be regrown and HHP treated again repeatedly, giving rise to similar inactivation curves. Ulmer et al. (75) were unable to isolate piezoresistant mutants from *L. plantarum* after four such cycles, indicating that tailing of pressure inactivation reflects a phenotypic diversity within the population rather than the presence of piezoresistant mutants. Specific conclusions resulting from the behavior of a given strain under HHP cannot necessarily be extrapolated to other strains of the same species. However, the differences observed among lactobacilli so far are not as large as observed with strains of *Escherichia coli* (see chapters 4 and 5).

PROTECTIVE ENVIRONMENTS AND CROSS-RESISTANCE

HHP inactivation of LAB is affected by the environment and the matrix they are residing in, which in foods are dictated by process and product parameters such as temperature, food constituents, and pH. Furthermore, the physiological status of the target cells as well as their recent history can influence pressure tolerance. Therefore, the protective and also the synergistic effects towards pressure inactivation are hardly reflected by determination of CFU alone but are further illuminated by techniques providing insight into basic cellular functions and/or structures during or after HHP treatment. Methods involving fluorescent dyes, namely, propidium iodide and tetrazolium salts, have been successfully adapted to study membrane integrity and metabolic activity in *L. lactis* (54), *L. plantarum* (75), and

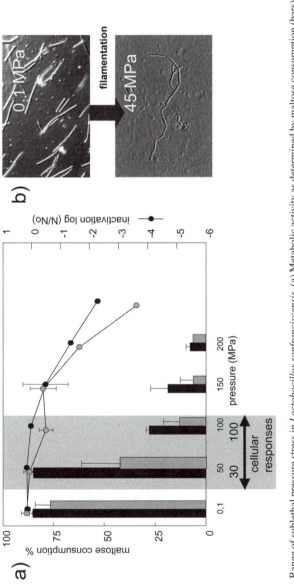

Figure 2. Range of sublethal pressure stress in *Lactobacillus sanfranciscensis*. (a) Metabolic activity as determined by maltose consumption (bars) and survival (lines) upon 20 min of HHP treatment of two different strains (black and grey). (b) Filamentation upon growth at 45 MPa (10% of maximal growth velocity).

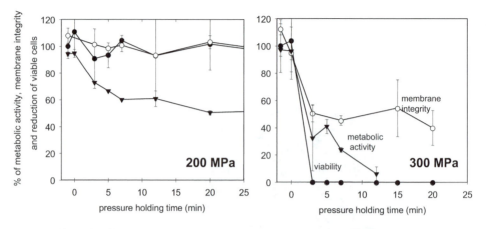

Figure 3. Effect of pressure treatment at 20°C on metabolic activity (▼), membrane integrity (○), and viability (●) of *Lactococcus lactis* MG1363.

Lactobacillus rhamnosus (4). Cells of *L. rhamnosus* could also be sorted upon staining in a flow cytometer, and subpopulations of sublethally injured bacteria were quantified. In contrast to the tests with tetrazolium salts, propidium iodide probing could not be adopted to in situ measurements during pressurization, because its fluorescence requires intercalation into DNA, which is not possible above 100 MPa (unpublished data).

As with many other bacteria, it is generally observed with LAB that the presence of a (food) matrix has a protective effect on their inactivation by HHP. Many food compounds and parameters may have an influence on inactivation but are poorly studied and therefore beyond the scope of this chapter. Our focus is on the effects of selected food components and environmental factors, which are experimentally accessible in food model systems and provide principal ideas on the mechanisms involved in synergisms and antagonisms.

Antagonistic (Protective) Effects

The protective effects of NaCl and sucrose on HHP-mediated inactivation were studied in a reconstituted milk system with *L. lactis* by Molina-Gutierrez et al. (54). *L. lactis* responded differently to the same water activity (a_w) if the reduction was achieved with an ionic or nonionic solvent. While the piezoprotective effect of the solutes was proportional to their concentrations, it was not proportional to the a_w. Normally, high salt concentrations are detrimental to most LAB (57); however, NaCl and KCl could protect *L. lactis* against HHP, and cells did survive otherwise lethal salt concentrations. The bacterium survives low a_w only because this is a transient stress, as it is able to equilibrate extra- and intracellular concentrations of sucrose and lactose or pick up a variety of compatible solutes. One and 1.5 mol of sucrose liter^{-1} provided complete protection against cell death at either 400 or 600 MPa. At 300 MPa a protective effect of 2.0, 3.0, and 4.0 mol of NaCl or KCl liter^{-1} was also observed. With ionic solutes no complete piezoprotection could be achieved. The cytoplasmic accumulation of exogenous glycine betaine upon an osmotic upshift restores the cellular volume and increases the hydration of the cytoplasm of *L. lactis* (53) and also *L. plantarum* (31). Consequently, when this compatible solute was added to 2.5 mM, the piezoprotective effect of 2 mol of NaCl liter^{-1} was increased in HHP treatments up to 400 MPa. As *L. lactis* cannot synthesize glycine betaine, it is likely that it accumulates other solutes from the medium to compensate for decreased a_w.

An example of these protective functions is given in Fig. 4, which demonstrates that the mechanisms of protection are different with ionic and nonionic solutes. The different types of protective effects of the solutes appear to be mediated by their different capabilities to stabilize metabolism and membrane integrity. In the presence of sucrose the metabolic activity and membrane integrity were retained upon pressurization to 90 and 50%, respectively. Disaccharides stabilize proteins in their native state (48) due to a preferential exclusion of water, thus raising the energy barrier for unfolding (74). Also, the transition temperature of the membrane is lowered by replacing water between the lipid head groups (14).

NaCl protects the membrane to the same level, but the metabolic activity decreases rapidly. The protective effect of ionic solutes relies on the intracellular accumulation of compatible solutes as a response to the osmotic stress. Thus, ionic solutes provide indirect protection, and piezoprotection with ionic solutes requires higher concentrations of the osmolytes compared to disaccharides.

Synergistic Effects

While many foods contain protective compounds, some foods have natural preservatives, which may act synergistically with HHP inactivation of bacteria. An interesting example thereof is beer, which is low in nutrients and pH and contains carbon dioxide, ethanol, and hop acids. The HHP-mediated inactivation of hop-tolerant strains of *L. plantarum* (29) and *L. brevis* (unpublished data) was investigated in model beer systems. Ethanol (5 and 10%) enhanced HHP effects, whereas hop extract (50 and 100 ppm) was less effective. Conversely, hop extracts killed pressurized cells during subsequent storage in beer but ethanol did not. HHP-treated beer-spoiling lactobacilli failed to survive under conditions of cold storage at acidic conditions. This could be related to membrane damage and the inactivation of multiple drug resistance (MDR) transporters, such as HorA, which

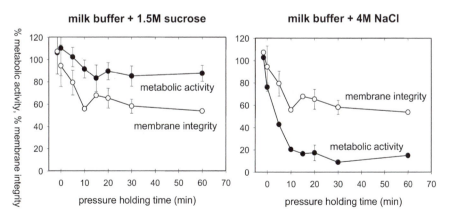

Figure 4. Different effects of solutes on the metabolic activity and membrane integrity of *Lactococcus lactis* MG1363 after treatment with 300 MPa at 20°C. Metabolic activity and membrane integrity were determined as described by Ulmer et al. (75).

mediates part of the hop resistance (75). In *L. lactis* investigated in milk model systems, sublethal pressure treatments led to inactivation of the MDR transporter LmrP, and cells also lost their inability to maintain their internal pH (55). The latter appears to be a very fast, initially reversible process occurring during the pressure ramp above threshold levels of 150 to 200 MPa.

In foods harboring adverse conditions, it may be sufficient to prevent spoilage by sublethally injuring cells by HHP treatments in the range of 200 to 300 MPa, making them unable to survive upon storage. This is primarily true in the presence of low pH or compounds dissipating membrane gradients, as the presence of an impaired membrane structure and membrane transport proteins leaves the cell unable to maintain transmembrane gradients and internal pH.

Cross-Resistance

Typical pictures from general and specific stress responses result from individual stresses, including low pH, ethanol, heat, osmotic, or oxidative stress. Both general and specific stress responses are known to enhance survival in harsh environments (28) and are prone to interact with HHP stress responses and cross tolerance. The study of such cross tolerances provides some insight into the sensors, pathways, and effectors of HHP stress. Cross-resistances have frequently been demonstrated in LAB for "conventional" stresses with different combinations of successional stresses (7–9, 35). Scheyhing et al. (63) demonstrated for *L. sanfranciscensis* that stresses reducing growth velocity by 90% through the action of cold (12.5°C), acid (pH = 3.7), salt (1.9%), and starvation in the stationary phase induced piezotolerance in *L. sanfranciscensis* when it was exposed to a subsequent, lethal stress of 300 MPa for 30 min. Heat shock (43°C) could not increase pressure tolerance but left cells with increased sensitivity to HHP. However, cells pretreated with 80 MPa were more tolerant to lethal heat (50°C) and less tolerant to low pH. A summary of the cross-protection as observed with *L. sanfranciscensis* is given in Fig. 5. Similar findings were reported for other species of *L. plantarum* in which an increased high-pressure sensitivity was observed after preincubation at elevated temperatures (73), and for *L. rhamnosus* GG, for which pressure-induced thermotolerance was reported (4). Therefore, in LAB and in contrast to the findings with other bacteria, heat shock proteins cannot generally be considered to cause piezotolerance. As not all stresses induce piezotolerance, a general stress response induced by all kinds of stresses, including heat, is also not suitable to explain the increase in piezotolerance by various prestresses. De novo protein biosynthesis must be involved in acquired piezotolerance (cross tolerance) of LAB, because it cannot be observed in the presence of protein biosynthesis inhibitors (unpublished data). In addition, other mechanisms are likely to participate in cross tolerance to HHP. In *Saccharomyces* the accumulation of trehalose in response to high osmotic pressure is more important for piezotolerance than accumulation of heat shock proteins, which act as molecular chaperones (27). We have also observed this effect for *L. brevis* upon osmotic shock and desiccation (unpublished data). There is also strong evidence for a contribution of an altered membrane composition, permeability, and interference with membrane-bound transport systems to the mechanisms of high-pressure-mediated inactivation (59, 75). The composition of lipid membranes is known to be altered in response to variations of pH, external osmolality, and low temperature to retain membrane fluidity and function-

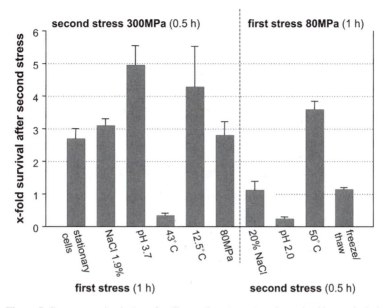

Figure 5. Cross-protection in *Lactobacillus sanfranciscensis* as determined by survival of prestressed cells to a second stress of the same or another type.

ality (25, 33, 55, 79). Some of these aspects are revisited in more detail elsewhere in this chapter.

Despite the high complexity and the fact that underlying physiological details are not fully understood, the occurrence of specific cross-resistances should be taken into account when high-pressure processes are introduced into multistep food production and storage schemes.

MODELING INACTIVATION AND SUBLETHAL INJURY STATES

Many attempts have been made to characterize the inactivation kinetics of bacteria in HHP treatments to assist the development of food processes. Current modeling of HHP in-activation mostly relies on differential equation models, where kinetics of surviving cells or kinetics of the inactivation of important enzymes as a function of pressure, temperature, and time (22, 23, 49) are used to describe the whole inactivation process. These approaches do not regard the physiological activities of the examined population, e.g., as reflected in membrane integrity, metabolic or transport activity, or ability to maintain membrane potential and internal pH. Models based on artificial neuronal networks have also been used and are well suited to describe the complex behavior that is often seen in HHP inactivation. However, this is a black-box approach which does not provide any biological or physiological insight in the mechanisms of microbial inactivation during HHP (30).

For *L. lactis* a fuzzy-logic model was developed by Kilimann et al. (45) which is able to predict as a function of pressure level, pressure holding time, and several substrate additives the viability and sublethal injury reflected in membrane integrity and metabolic activity of *L. lactis* as well as the activity of LmrP, a proton motive force-dependent multidrug transport enzyme. It is based on experimental data covering the pressure range of 200 to 600 MPa, pH values between 4.0 and 6.5, and the presence of additives known to exert a protective effect on pressure-induced inactivation. Furthermore, they applied multivariate statistical analysis to large inactivation data sets obtained with *L. lactis* (28a) to demonstrate that pressure inactivation of *L. lactis* is temperature dependent, and the piezoprotective effects of sucrose and NaCl strongly depend on the temperature. To date these are the most powerful mathematical approaches describing the HHP inactivation of a bacterium, including several sublethal states.

MEMBRANE PHYSIOLOGY AND CELL DIVISION

Membrane Physiology

The cytoplasmic membrane consists of many lipids and proteins representing a complex macromolecular structure (79) which is very sensitive to thermodynamic changes introduced by temperature or HHP. As a result, all vital functions the membrane offers to the living system are easily impaired upon HHP treatment. Lipids in biological membranes are in the fluid (liquid crystalline) phase, allowing fast lateral movement of molecules. Pressure upshift and/or temperature downshift in pure one-component phospholipid bilayers induces a phase transition from the liquid crystalline phase to a gel phase, characterized by an increased rigidity and reduced conformational degrees of freedom for the acyl chains (12). In natural membranes with a complex lipid composition, a more or less wide coexistence region of gel and liquid crystalline phases is observed at the phase transition. LAB take up and incorporate into the membrane a wide range of phospholipids, depending on their availability in the medium and growth temperature, to adjust membrane fluidity and maintain its functionality. Ulmer et al. (76) found that this clearly influences the phase transition of their respective membranes upon HHP exposure. In addition to the fatty acids present in the bacterial membrane, the piezoresistance of the same strain was also reported to be related to the phospholipid head group composition (73).

For *L. plantarum* and *L. lactis* it has been shown that the membrane becomes permeabilized for ions and also larger molecules, e.g., propidium iodide (55, 75, 83). The HHP-induced permeability of the membrane for molecules with the size of propidium iodide has been described to be a point of no return, coinciding with HHP-induced cell death (45). Its permeability to protons and potassium ions occurs above very low threshold levels, from 50 MPa, as a very fast process observed during the ramp of HHP treatments. It results in the inability of the cell to maintain a transmembrane proton gradient and therefore its internal pH, which causes a number of secondary effects. The latter result from the inability of the cell to distinguish the inside from the outside. While ions and other compounds leak out of the cell, environmental compounds come in and exogenous factors can elicit detrimental changes to the metabolism and integrity of cellular macromolecules, namely, proteins. The factors involved in the loss of pH homeostasis of pressure-treated *L. plantarum* and *L. lactis* were investigated by Wouters et al. (83) and Molina-Gutierrez et al. (55). Sublethal pressure treatment of *L. plantarum* resulted in a loss of the cell's ability to maintain the internal pH, in decreased acid efflux, and in inactivation of the F_oF_1 ATPase, whereas the ability to generate ATP remained unaffected. As shown in Fig. 6, a drop of the intracellular pH occurs very fast during the ramp of the HHP treatment when cells are in an acidic

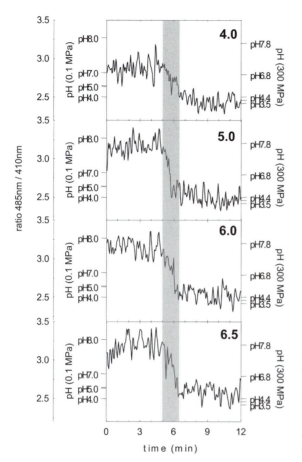

Figure 6. Changes in the intracellular pH of *Lactococcus lactis* subsp. *cremoris*. Samples were treated at 300 MPa and 20°C in milk buffer. The compression rate was 200 MPa min^{-1}; the ramp-up time was 90 s (shaded area). The measurement of the intracellular pH was performed as described by Molina-Gutierrez et al. (55).

environment. This effect may be forced in phosphate-buffered (food) systems, as HHP decreases their pH.

Due to the changes in the membrane function and structure, the functionality of membrane proteins can be impaired. The effects of composition and phase behavior of the membrane on HorA, an ATP-binding cassette MDR transporter, during sublethal HHP treatment (200 MPa) of cultures of *L. plantarum* were investigated by Ulmer et al. (76). With the information obtained by Fourier transform-infrared spectroscopy and Laurdan fluorescence spectroscopy on the thermodynamic phase state of the cytoplasmic membrane, they established a pressure-temperature diagram for cell membranes. Cells grown at 37°C and pressure treated at 15°C lost >99% of their HorA activity and viable cell counts within 36 and 120 min, respectively. The membranes of these cells were in the gel phase region at ambient pressure. In contrast, cells grown at 15°C and pressure treated at 37°C lost >99% of HorA activity and viable cell counts within 4 and 8 min, respectively. The membranes of these cells were in the liquid crystalline phase region at ambient pressure. The kinetic analysis of inactivation of *L. plantarum* provided evidence that inactivation of membrane enzymes and subsequent cell death depend on the thermodynamic properties of the membrane before and their changes during HHP treatment. Phase transitions from a liquid crystalline phase to a gel phase appeared to be most detrimental to membrane enzymes, and thus to support cell death. However, the role of water in this system should not be underestimated. It is highly likely, but not yet demonstrated, that both the presence of water in the membrane and the hydration of the membrane head groups change upon HHP treatment, produce compact lipid stacking, and influence hydration and thus conformation and functionality of the membrane proteins.

Apart from the great potential of HHP for study of the structure and function of membranes and membrane proteins, there is a significant impact of these findings when it comes to the incorporation of HHP treatment into a food process. As pressure treatment of cells with a liquid crystalline membrane at 0.1 MPa resulted in HorA inactivation and cell death more rapidly than those of cells with a gel phase membrane at 0.1 MPa, HHP treatment should hit cells when their membrane is in a liquid crystalline phase, i.e., before cooling of a product. However, treatment at low, and particularly at subzero, temperatures is another story which is not completely explored yet. Also, if a food offers conditions adverse to microbial growth and survival, sublethal injury caused by HHP may be sufficient to ensure inability to survive upon storage in the respective food. This is especially true if a cell with an impaired membrane functionality resides at low pH.

Cell Division

Cell division in bacteria is a multifactorial process which depends on the interaction of macromolecules along a strictly defined time scale. In the early stage of this process the FtsZ protein assembles into a ring (20, 70) which can be visualized with fluorescent antibodies. As the principal action of HHP on a cellular system appears to reside in its ability to affect interaction of macromolecules in the first place, cellular division, including FtsZ ring assembly, is very sensitive to HHP. Sato et al. (62) showed for *Escherichia coli* that the macromolecular assembly of FtsZ into a ring was inhibited by high-pressure conditions. They suggested that the cell division might be "frozen" at an early stage in the cell cycle by pressure application. In contrast, the FtsZ protein of the deep-sea isolate *Shewanella violacea* is functional at 50 MPa (38). *L. lactis* harbors an FtsZ homolog whose assembly was studied under ambient and high pressures by Molina-Höppner et al. (56). Under high-pressure conditions, cell segmentation and segregation of chromosomal DNA were apparent, indicating that the cytoskeletal assembly apparatus involved in cell division retained some of its functionality. Thus, cells recovering from such a process could not be considered as "synchronous" in a strict sense. While rapid reassembly of the FtsZ ring after decompression, followed by fragmentation of filamentous cells within minutes, was demonstrated in *E. coli* (62), this was not observed in *L. lactis*. However, a strong reduction of the number of FtsZ rings in exponentially growing cells of *L. lactis* MG1363 under high-pressure conditions indicated that aggregation of the FtsZ protein was inhibited and that cytoskeleton assembly during cell division contributes to the growth arrest of *L. lactis*. The cell morphology of *L. lactis* grown at elevated pressure resembles that of cells grown at supraoptimal temperatures, i.e., under growth conditions of FtsZ overexpression. Therefore, inhibition of FtsZ assembly rather than a reduced expression of the *ftsZ* gene is likely to explain the altered cell morphology. While HHP-mediated inhibition of cell division in *E. coli* may include reduced FtsZ expression (40), it appears to be dominated in *L. lactis* by impaired FtsZ ring assembly.

Recently, in *Bacillus subtilis* ClpX, the substrate recognition subunit of the ClpXP protease was identified as an inhibitor of FtsZ assembly. Both in vitro and in vivo studies support a model in which ClpX helps to maintain the cytoplasmic pool of unassembled FtsZ that is required for the dynamic nature of the cytokinetic ring functions as a general regulator of FtsZ assembly in bacteria (81). In this context it is worth mentioning that we found increased amounts of ClpX in high-pressure-treated cells of *L. sanfranciscensis* (see paragraph 6).

HIGH-PRESSURE-SENSITIVE GENE EXPRESSION
Proteome Level

Many aspects of the cellular response to HHP, including metabolic changes, cell membrane composition, transport mechanisms, and acquired stress tolerance and cross-resistance to other stresses, indicate that changes in the proteome are induced by HHP. It has also been shown for *L. sanfranciscensis* and *L. rhamnosus* that cross-resistance is not inducible in the presence of protein biosynthesis inhibitors (4, 63). Furthermore, in vitro studies on the influence of HHP on the ribosomes indicate that protein biosynthesis is affected. Therefore, the changes in the proteome induced by HHP can be assigned to altered and impaired protein biosynthesis on the one side and stress response and adaptation on the other. The effects of HHP stress on the proteome level can only be defined upon comparison with other stresses in differential proteomics. As a prerequisite for such studies, the stress response at the physiological level of the organism must be well known with respect to its tolerance as measured by survival or growth rates. Still, studies of the stress response, not only on the proteome level, often seem to give contradictory results at first glance even if they are done with the same bacterium. This is because two types of stress responses must be distinguished with respect to the experimental setup. First, bacteria can be exposed to a shock for minutes, which they would not survive for long, and are subsequently held under optimum conditions in a recovery phase. Proteins synthesized as a reaction on this lethal stress are synthesized during the recovery phase, and a general stress response always overlaps with the specific stress applied. Second, bacteria can be grown at suboptimal growth conditions, typically at 10% of their maximal growth rate, for several hours. Proteins are synthesized during growth under these stress conditions.

For LAB there are only a few studies on changes in the proteome under HHP stress. All of them use the second approach, i.e., growth under sublethal stress and analysis of the proteome of these cells in two-dimensional electrophoresis along the isoelectric points and molecular weights of the soluble proteins. Among the LAB, the most detailed proteome studies on the HHP response were performed with *L. sanfranciscensis*. In this organism Drews et al. (17) studied the stepwise-altered protein expression at various pressures (20 to 200 MPa) to find a picture with few proteins being affected upon pressurization. Most remarkable was the finding of a protein which apparently switched between two forms as a result of pressure stress. This protein, identified as a cold shock protein homolog, would appear with the same molecular weight at a higher isoelectric point (pI = 4.2) in the proteome of cells treated with 150 MPa for 60 min than for cells grown under optimal conditions (pI = 4.0). The molecular background of this apparently conformational change

remains unclear. Hörmann et al. (37) chose a comparative approach with organisms growing for 60 min at 10% of their maximal growth rate upon stresses including HHP (80
MPa), cold (12.5°C), heat (43°C), acid (pH = 3.7), salt (1.9%), and starvation in the stationary phase. Sixteen proteins were found to be affected by HHP, and they were identified
by using N-terminal amino acid sequencing and mass spectrometry (Fig. 7). Nine of them
were up-regulated (P1 to P9), whereas seven (P10 to P16) appeared to be repressed under
these conditions. *L. sanfranciscensis* ribokinase (P1) is one of the strongest increased en-

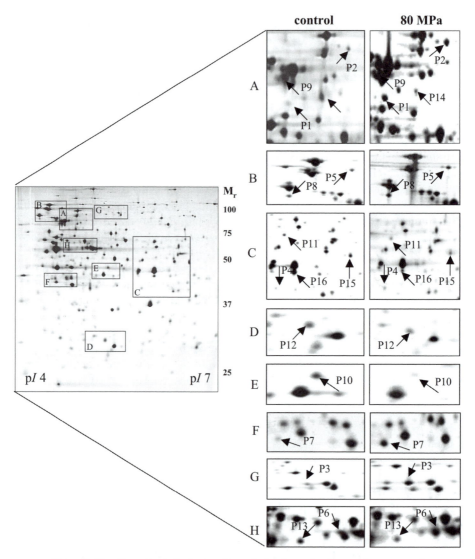

Figure 7. Two-dimensional electrophoretic analysis of cytoplasmic protein extracts of
Lactobacillus sanfranciscensis left untreated and after incubation for 1 h at 80 MPa. Proteins were silver stained. High-pressure-affected proteins are indicated by arrows. Isoelectric points and molecular weights are indicated.

zymes after exposure to the different stress conditions and thus appears to be useful for the cell to cope with stress. Through the formation of ribose-5-phosphate, ribokinase plays a central role in the initiating steps of synthesis of purine and pyrimidine nucleotides and also for the amino acids histidine and tryptophan. A specific role for a trehalose and maltose hydrolase (P5) in the stress response could reside in energy metabolism or maintenance of an osmotic balance. However, their kinetics and therefore preferential direction of the catalyzed reaction have not been studied under these conditions. The other up-regulated proteins are identified as chaperones and proteases, which are found in a variety of other stress responses. The comparison with the proteomes from cells subjected to the other stresses revealed a Clp homolog as the only HHP-specific inducible protein. At first glance it appears to be nearly a contradiction that there is an overlap of the HHP response with cold- and NaCl-stressed cells (11 of 16 proteins) and heat-stressed cells (10 of 16 proteins). On the other hand, this indicates that the HHP stress response uses subsets of other stress responses rather than eliciting an HHP-specific answer. A similar conclusion can be made for *E. coli,* in which HHP induces an SOS response (2) and several heat shock proteins (3). The different type of HHP stress response in *L. sanfranciscensis* could be due to the lack of different sigma factors in lactobacilli compared to the ones found in *B. subtilis* and *E. coli* (78).

Transcriptome Level

The limits of this proteome approach can be seen in the limited number of membrane proteins, which are not all efficiently solubilized and included in the analysis, and in the detection limit, which usually does not allow detection of low-copy-number regulatory proteins. Some of these drawbacks can be solved by comparison of the proteome data with analyses on the transcriptome level. Ishii et al. (40) investigated the HHP response of *E. coli* at the transcriptome level to get insight into regulatory mechanisms mediating the stress response. For gram-positive bacteria no such comprehensive study based on full genomic data has been published. In the absence of a genome sequence, Pavlovic et al. (58) performed a transcriptome analysis with a redundancy-cleared shotgun microarray of *L. sanfranciscensis* allowing the reliable readout of 750 spots. Upon HHP treatment with 45 MPa for 30 min, the intensity of 42 spots was increased, and that of 6 was decreased. The expression of the most strongly responsive genes was quantified with real-time PCR to confirm the array data. An overview of the identified HHP-responsive genes is given in Table 1. A significant overlap with the proteome data is observed, which helps the interpretation of the data. For *L. sanfranciscensis,* the *xpk* gene, encoding phosphoketolase, was found to be unaffected by HHP and was used for normalization of the data.

Generally, transcriptome analyses quantify the amount of specific mRNAs present in the cell at a specific time. Thus, the up-regulation by HHP identified in transcriptome analyses could be caused by overexpression of a gene or by enhancement of the stability of its mRNA under HHP conditions. Kaarniranta et al. (41) concluded from the use of the transcription blocker actinomycin with a chondrocytic cell line that mRNA stability was indeed affected by HHP. They correlated the *hsp60* induction to HHP-mediated stabilization of the mRNA, rather than to increased expression. However, their data apparently cannot be generalized for other systems, and, also, they did not investigate the effect of a transcription blocker (e.g., rifampin) in the absence of HHP. In a similar approach we could

Table 1. Genes responsive to high pressure (45 MPa for 30 min) as determined by array hybridization and/or real-time PCR[a]

Gene	Protein	Similarity to known genes	Induction/repression			Accession no.
			MA 45[b]	RT 45[c]	RT 80[d]	
Stress response						
hsp60	GroEL	ACQ93GO7 (75%)	2.9	2.9 ± 0.8	2.8 ± 0.4	AY922321
clpL	ATP-dependent Clp protease	AAD 34338 (55%)	2.1	4.2 ± 0.5	15.2 ± 2.3	AY912122
guaA	GMP synthetase	ACO85192 (66%)	3.0	ND[e]	ND	AY912128
gyrA	DNA gyrase, A subunit	ACQ89OK3 (65%)	2.0	ND	ND	AY912125
ppK	Polyphosphatekinase	ACQ88YD2 (99%)	2.5	ND	ND	AY912130
ppx	Exopolyphosphatase	ACQ88YD1 (98%)	2.3	ND	ND	AY912130
ORF1	Hypothetical protein (similar DEAD ATP-dependent RNA helicase)	ACQ88Z45 (73%)	3.9	ND	ND	AY912129
pepO	Endopeptidase	AL935254 (60%)	2.4	ND	ND	AY917120
Translation factors and ribosomal proteins						
rplK	Ribosomal protein L11	ACQ88YX0 (82%)	2.7	ND	ND	AY912111
rplF	Ribosomal protein L6	ACQ88XX1 (66%)	2.1	ND	ND	AY912116
rpsB/tsf	Ribosomal protein S2 and translation elongation factor Tsf	ACQ88VJ4 (86%) ACP34831 (71%)	2.2	ND	ND	AY912117 AY912118
tuf	EF-Tu	ACQ8KMR6 (83%)	2.2	1.9 ± 0.2	1.8 ± 1.0	AY912112
fusA	EF-G	ACQ88XY8 (80%)	-2.5	-3.3 ± 0.0	1.2 ± 0.1	AY912119
lepA	GTP-binding translation elongation factor	NP_785542.1 (65%)	ND	0.89	4.0	AY912120
prfB	Peptide chain release factor 2	AE017202 (70%)	-2.5	ND	ND	AY912121
prfC	Peptide chain release factor 3	ACCP000023 (65%)	1.0	1.0 ± 0.2	1.1 ± 0.3	AY917120
infB	Translation initiation factor 2	ACCP000023_334 (65%)	ND	4.5 ± 3.85[f]	4.5 ± 4.9	AY912123
hisS	Histidyl-tRNA synthetase	ACQ88VQ7_1447 (45%)	2.5	ND	ND	AY912113
ORF2	GTPase with unknown function	ACQ88WT7 (96%)	2.9	3.3 ± 0.3	ND	AY912115
tRNA-modifying enzymes						
trmA	tRNA-methyltransferase, TrmA family	ACQ73EJ5 (55%)	2.0	ND	ND	AY912114
gidA	Glucose-inhibited cell division protein	ACQ88RX6 (80%)	2.3	2.5 ± 0.7	2.7 ± 0.4	AY912127
thdf/trmE	tRNA-modifying GTPase	ACQ88RX5 (61%)	ND	1.4 ± 0.2	1.9 ± 0.6	AY912124
trmA	tRNA-methyltransferase, TrmA family	ACQ73EJ5 (55%)	2.0	ND	ND	AY912114

[a]The real-time PCR analysis of the investigated genes was normalized against the expression of phosphoketolase (AJ586560).
[b]MA, x-fold induction data from microarray assay (45 MPa, 30 min).
[c]RT 45, x-fold induction data from real-time-PCR analyses (45 MPa, 30 min).
[d]RT 80, x-fold induction data from real-time-PCR analyses (80 MPa, 30 min).
[e]ND, not determined.
[f]Real-time PCRs for infB were poorly reproducible and varied between 2.0- and 8.3-fold induction.

demonstrate that in *L. sanfranciscensis,* increased mRNA stability was not the cause of the presence of more *hsp60* mRNA. In this case, the presence of the transcription blocker was responsible for *hsp60* mRNA stabilization in the absence and presence of HHP to the same extent.

More than 20% of the HHP-responsive genes were found to encode either translation factors (EF-G and EF-Tu) or ribosomal proteins (S2, L6, and L11) or to be genes changing translational accuracy or molecular chaperones (GroEL and ClpL). The strongest HHP response was observed with an open reading frame with a conserved DEAD-ATP-dependent helicase motif, GroEL and a GTPase most similar to Era and EngA, all of which are involved in ribosomal processes (16, 71, 77).

This provides strong in vivo evidence that the translational machinery is a major target for HHP and that the cell tries to counteract the decrease in translational capacity by (i) regulating translational factors, (ii) regulating genes controlling translational accuracy, and (iii) inducing stress proteins. Genes for the translation elongation factors (EF) *tuf* and *tsf,* both responsible for the binding of aminoacyl-tRNA to the ribosomal A site, were induced under high pressure. On the other hand, the absence of induction or even the repression of genes concerning translocation or release of the peptide chain (*fusA, prfB,* and *prfC*), respectively, points to a rate-limiting reaction before translocation. The translation initiation factor encoded by *infB,* responsible for binding of the initiator tRNA and GTP to the 30S subunit, was also induced under HHP conditions (Table 1). InfB was also induced in *E. coli* upon cold shock (10, 26) and described as a chaperone (11), indicating its role in general stress responses.

The view of the ribosome as a major HHP target that appears from this transcriptome analysis with *L. sanfranciscensis* is supported by in vitro data provided by Schwarz and Landau (65, 66) and Smith et al. (68) obtained with ribosomes of *E. coli.* According to their results, the inhibition of translation by HHP takes place at aminoacyl-tRNA binding or translocation. As the inability of aminoacyl-tRNAs to bind to the ribosome under high pressure is quantitatively identical to the inhibition of protein synthesis in the whole cell, the inhibition of protein synthesis was attributed to the 30S ribosomal subunit, which undergoes conformational changes concomitant with an increase in volume by binding aminoacyl-tRNA.

The apparent paradox observed in the proteome analysis is also seen at the transcriptome level. Genes encoding proteins involved in heat shock (*hsp60, clpL, ppk,* and *ppx*) or cold shock (*infB, gyrA,* and *ORF1*) were up-regulated under HHP. However, a common principle of cold shock and high pressure residing in the decrease of the translational capacity was repeatedly suggested (10, 24, 26). Apparently, it also fits the HHP stress response in *L. sanfranciscensis,* which shows preferred induction of translation factors and ribosomal proteins to compensate for this decrease. At the same time it reacts to impaired translational accuracy, the induction of several molecular chaperones (GroEL, ClpL, and EF-Tu) and aminopeptidases (PepC and PepO) is observed.

The picture of the ribosome as a sensor and mediator of the HHP stress response in *L. sanfranciscensis* receives another convincing addition through the overexpression of *ssrA,* encoding tmRNA (82), a molecule which recycles stalled ribosomes and tags truncated proteins. This gene was not part of the shotgun array and therefore did not participate in the transcriptome analysis. It was detected in the analysis of a mutant and is therefore discussed in paragraph 7.

Expression Studies and Reporter Systems

It is a fascinating idea to identify and study HHP-sensitive promoters, which may offer new perspectives for biotechnology for the LAB. One may start with the idea that a promoter or other regulatory element represents a three-dimensional structure prone to be affected even in a living cell by the thermodynamic changes occurring upon HHP treatment. Also, it is tempting to speculate that RNA polymerase or effector binding to a specific motif in an operator region is HHP sensitive and expression may be stimulated, e.g., by repressor dissociation.

To study the expression of HHP-responsive genes and therefore the HHP responsiveness of their promoters, reporter systems can be used which generally put the corresponding promoter in front of a reporter gene, the expression of which is easily monitored. The influence of high pressure on expression of single proteins was first detected in *E. coli* that expressed green fluorescent protein (GFP) under the control of defined promoters. GFP was placed downstream of the T7 RNA polymerase-binding site. Expression of T7 RNA polymerase was controlled by an isopropyl-β-D-thiogalactopyranoside (IPTG)-inducible *lac*UV5 promoter. In non-IPTG-induced cells, fluorescence increased three- to fourfold during a 2-h expression at 70 MPa compared to that at atmospheric pressure (19). Additional induction with IPTG showed a strong increase (seven to eight times) for both 50 and 70 MPa compared to that in IPTG-induced cells under atmospheric conditions. The HHP-induced increase in fluorescence is consistent with previous reports that gene expression initiated from the *lac* and *tac* promoters was greatly enhanced by growth at 30 and 50 MPa (44, 61).

The *lac* promoter and its derivatives *tac* and *lac*UV5 are induced by binding of the chemical inducer IPTG to the repressor protein LacI, releasing it from its binding site on the DNA. As inactivation of LacI expression itself can be excluded, Kato et al. favored the hypothesis that the effects relate entirely to a change in plasmid supercoiling that incidentally interferes with the binding of LacI (44).

Hörmann et al. (37) have developed a reporter system for *L. sanfranciscensis* using the *melA* gene, encoding an alpha-galactosidase, to study stress reactions. Its function was validated with the heat- and NaCl-dependent expression of the *dnaK* operon. With this system, pressure-responsive promoters located upstream of *rbsK/rki* and *clpL* were studied and compared with the results obtained in the transcriptome and proteome analyses. A pressure-induced increase in expression of the *melA* coupled to the *rbk* promoter (15-fold) and *clpL* promoter (2.5-fold) after 1 h at 80 MPa was demonstrated. The comparison of the promoter sequences of so-far-identified pressure-sensitive genes of *L. sanfranciscensis* does not reveal any common sequence or secondary-structure motifs.

GENOME FLEXIBILITY, MUTATION, AND ADAPTATION

HHP introduces a broad spectrum of changes in a cellular system which can result in direct or indirect cell death or leave cells with impaired functions. These survivors, described as sublethally injured cells, mostly have defects of a transient nature; i.e., they are reversible when the pressure is released and repaired by the use of regular cellular functions. When these survivors grow up to a new population it will be unchanged. However, HHP can also introduce permanent changes in the genome, which are passed on to the progeny. These changes can be random, but they are likely to change the ability of the cell to cope with HHP and other stresses in terms of survival or ability to grow. As with many other cellular functions, the analysis of mutants provides insight into the mechanisms of HHP tolerance and stress response. As HHP is also used as a food preservation technology, the occurrence of tolerant mutants may be of further interest with respect to determination of proper process conditions.

A few reports are available on the characterization of HHP-resistant mutants of different bacteria. As the HHP response patterns and signaling pathways are apparently different, it is expected that mutations leading to HHP tolerance need not be in the same or related loci. Only in a few cases was the molecular background of HHP-tolerant mutants characterized. Some mutants showed altered expression of heat shock proteins DnaK, GroEL, GroES, GrpE, and ClpB in *E. coli* MG1615, or they overexpressed global regulatory elements, e.g., the master stress regulator sigma^S, encoded by *rpoS* in *E. coli* O157:H7 (3, 60). Karatzas and Bennik (42) isolated a spontaneous HHP-tolerant mutant of *Listeria monocytogenes* and demonstrated that codon deletion in the gene of the global class III stress regulator CtsR could be linked to higher-pressure resistance of *L. monocytogenes*. Thus, overproduction of chaperones and proteases as well as up-regulation of their regulators overcome the accumulation of truncated and misfolded proteins as the general problem cells face during growth under high pressure. As mature (successfully translated) proteins and enzymes can sustain significant damage under moderate pressures (below 150 MPa), the origin of proteinaceous "waste" resides in their incomplete synthesis via impaired ribosomal translation. For the LAB, Marquis and Bender (50) suggested that a regulatory defect in the arginine dihydrolase system of *Streptococcus faecalis* (now called *Enterococcus faecalis*) causes both acid tolerance and piezotolerance. They gradually increased the growth pressure for agar stab subcultures in steps of 5 MPa starting from 75 MPa (the maximum growth pressure for unadapted cultures) to 100 MPa. The dihydrolase system produced ammonia during glycolysis, even in the presence of high concentrations of glucose. The ammonia acted to neutralize metabolic acids and to confer a type of acid tolerance. This acid tolerance may be involved in piezotolerance, since pressure is known to render *E. faecalis* (and other LAB) hypersensitive to acid conditions. In all, it appeared that pressure could be used as a selection agent for isolation of rare piezotolerant variants in normal bacterial populations.

This could be easily explained on the basis of the observations of Molina-Gutierrez et al. (55), who found that HHP causes an adaptation of the intracellular pH to the extracellular pH as a result of instant membrane damage. Thus, a cell which loses an important system for pH homeostasis will be less tolerant to pH drop, which can be seen as a major indirect HHP effect.

We have obtained HHP-tolerant mutants of *L. sanfranciscensis* upon growth for 25 cycles (including approximately 100 generations) at 50 MPa (unpublished results). Compared to the wild type, these strains showed a twofold increase in growth when incubated under 50 MPa for 15 h. Under Mg^{2+}- and Mn^{2+}-limited conditions, a 2-log increase was also observed in resistance against lethal pressure conditions (150 MPa). Interestingly, an altered sensitivity was recorded against antibiotics influencing ribosomal action (Table 2). Temperature-dependent growth at ambient pressure was significantly altered. While the growth rate at the optimal temperature of 30°C was increased, the strains were no longer able to grow at 35°C.

Table 2. Susceptibilities of wild-type *Lactobacillus sanfranciscensis* and the pressure-adapted mutant to various antibiotics acting on the ribosomes or on translational processes

Antibiotic	Zone diam (mm)[a]		Function affected
	Wild type	Mutant	
Increase in resistance			
Kasugamycin	6.7 ± 0.8	14 ± 0.7	Initiation (fMet-tRNA binding)
Tetracycline	20.2 ± 0.3	26.2 ± 0.3	aa-tRNA–EF-TU–GTP binding
Puromycin	20.2 ± 0.3	25.5 ± 1.0	Elongation
Erythromycin	41.0 ± 0.7	46.8 ± 0.9	Elongation
Decrease in resistance			
Kanamycin	13.0 ± 0.0	11.5 ± 1.0[b]	Translation accuracy
Paromomycin	8.4 ± 0.8	11.9 ± 0.2[b]	Translation accuracy
Tobramycin	15.7 ± 0.4	16.9 ± 1.0[b]	Translation accuracy translocation
Spectinomycin	15.0 ± 0.7	11.4 ± 0.5	(EF-G–GTP interaction)
Spiramycin	24.8 ± 0.4	20.5 ± 1.0	Peptide bond formation
Streptomycin	27.8 ± 0.4[c]	25.2 ± 0.4	Translation accuracy
No effect			
Amikacin	20.4 ± 0.5	20.4 ± 0.5	Translation accuracy
Gentamicin	14.0 ± 0.0	15.3 ± 0.6	Translation accuracy
Neomycin	15.0 ± 0.0	15.6 ± 0.5	Translation accuracy

[a]The disk diameter is 6 mm. A value of 6 mm indicates no zone of inhibition.
[b]Spontaneous resistant colonies occur in the zone of inhibition of the mutant in contrast to that of the wild type.
[c]Spontaneous resistant colonies occur in the zone of inhibition of the wild type in contrast to that of the mutant.

To address the molecular basis of the mutation, the genotypic characterization was focused on alterations of ribosomal components. No mutations were detected in 16S rRNAs or in ribosomal proteins S4, S5, and S12, most often involved in streptomycin resistance. Northern analysis revealed constitutive overexpression of *ssrA* (tmRNA) in the mutant. A 2.2-fold induction after pressure shock indicated the tmRNA as the genetic determinant of a piezotolerance response in the wild type. In the mutant the basal expression of tmRNA was 3.5-fold higher than in the HHP-induced wild type. Still, its expression was further increased severalfold upon pressure treatment. A mutation in the regulation of the tmRNA gene leading to increased amounts of tmRNA might help to prevent accumulation of truncated, potentially harmful proteins and making proteolysis more efficient. Nevertheless, it should be specified that a direct link between overexpression of *ssrA* and HHP resistance remains to be demonstrated. Thus, the finding of a tmRNA-overproducing, piezotolerant mutant fits well with a picture of ribosomal sensing of a high-pressure stress response in *L. sanfranciscensis,* because the tmRNA-directed tag targets the unfinished proteins for proteolysis via the Clp protease system (34). By using a proteome and a transcriptome approach to characterize the HHP response of *L. sanfranciscensis,* a remarkable increase was found in ClpL (discussed above) (37, 58). Thus, we propose *trans*-translation and peptide tagging, processes that promote recycling of stalled ribosomes and prevent accumulation of abortively synthesized polypeptides, to be involved in combating HHP damage and conferring moderate piezotolerance. In such a model, the ribosome would be the primary target and sensor for the thermodynamic changes induced by pressure, while the expression of specific enzymes, chaperones, and stress proteins appears to be a secondary step. This is in accordance with the absence of common pressure-sensitive motifs in pressure-inducible promoters.

The occurrence of HHP-tolerant mutants has been described as potentially relevant in high-pressure food processing (36). The finding of a close connection with changes in antibiotic resistance adds another aspect to this discussion. However, the relatively low pressures, long incubation times, and necessity of growth to establish such a mutation suggest that such mutations are unlikely to arise in HHP food processes.

While pressure-tolerant mutants were obtained upon repeated pressure shock and recovery or prolonged growth under sublethal conditions, there are indications that genome flexibility may occur under HHP stress already during one cycle of pressurization only. Microorganisms can enhance their chances of survival under stress by creating genetic diversity at the population level (1).

Insertion sequence (IS) elements and transposons are known to be induced under stress conditions in a variety of bacteria; however, no such report is available on the induction of such elements by HHP. We could demonstrate for several IS elements in *L. sanfranciscensis* that HHP induces transcription of their transposases and that this indeed increases mobility of IS elements under sublethal HHP conditions. The HHP-induced generation of genetic diversity by the loss of one copy of an IS element is shown in Fig. 8. The rate of ISLsf6 transposition may be even higher, because only those IS jumps can be detected that do not negatively affect the fitness of the cell.

MACROMOLECULAR INTERACTION IN FOOD AND BIOSCIENCE

The driving force to study the influence of HHP on behavior of food-borne bacteria originates from the need to develop HHP as a novel tool for mild food processing aimed at bacterial inactivation. In this system the LAB have an ambivalent function as spoilage organisms versus their deliberate use in food biotechnology. The identification of pressure-sensitive cellular targets contributes both to the understanding of HHP-induced inactivation mechanisms and to the development of novel strategies for improving survival of bacteria when they need to survive stressful preparation technologies and exhibit optimal performance in food biotechnological processes (80). In Fig. 9, HHP-sensitive cellular targets identified and studied in LAB are shown. These targets, which may serve to initiate cell death or resistance development, must be differentiated in primary targets and secondary ones. In this view a primary target would be a macromolecule that is itself sensitive to thermodynamic changes induced by HHP discussed elsewhere in this book (see chapter 1). A secondary target would be a macromolecule that is changed by a cellular reaction or response to a change in a primary target. This response may use any of the regulatory pathways a particular cell offers or be a result of environmental conditions. Thus, an HHP stress response observed at the secondary level will always remain different from one cellular system to another, while the primary targets always remain very similar, if not the same. Keeping this in mind, many seemingly contradictory results obtained with various bacteria under HHP treatment can fit together nicely. As LAB lack most of the general stress signaling pathways residing in SOS responses or use of alternative sigma factors of other bacteria, they are nice models to identify and study the primary targets of HHP.

Obviously, the interaction of macromolecules can be a primary target depending on the pressure height. The most sensitive vital processes appear to be translation and membrane

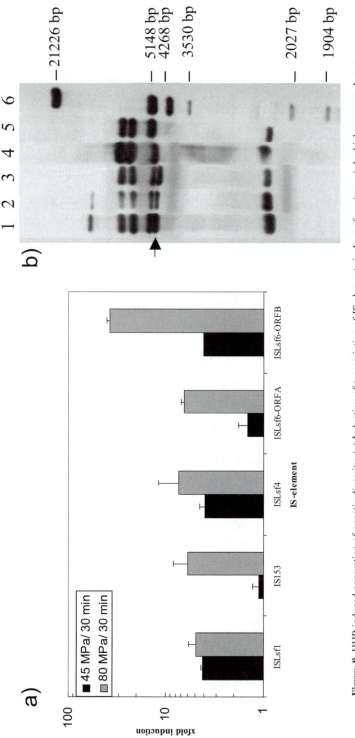

Figure 8. HHP-induced generation of genetic diversity. (a) Induction of transcription of IS elements in *L. sanfranciscensis* by high pressure. *L. sanfranciscensis* was subjected to 45 MPa (black bars) or 80 MPa (grey bars) for 30 min, and transcription ratios (high pressure to atmospheric pressure) of several transposable elements were determined. Data were normalized against *xpk* transcription; shown are the means of three independent experiments. (b) High-pressure-induced changes of hybridization patterns of ISLsf6 of *L. sanfranciscensis*. EcoRV-digested DNA was hybridized with a probe for *orfB* of ISLsf6. Lane 1 shows genomic DNA of the ancestor. Lanes 2 through 5 show genomic DNA after 1 growth cycle at 0.1 MPa (lane 2), 25 growth cycles at 0.1 MPa (lane 3), 1 growth cycle at 50 MPa (lane 4), and 25 growth cycles at 50 MPa (lane 5). Lane 6 shows DNA markers. Hybridization with a probe for *orfA* of ISLsf6 resulted in the same pattern. The arrow indicates the loss of a fragment in high-pressure-treated cells.

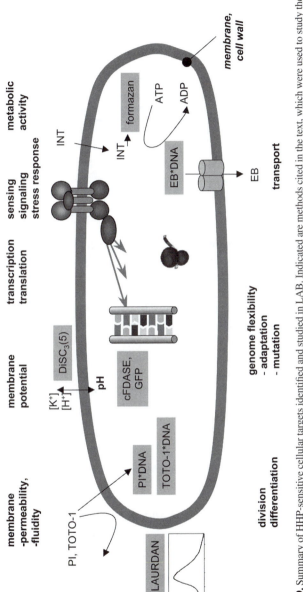

Figure 9. Summary of HHP-sensitive cellular targets identified and studied in LAB. Indicated are methods cited in the text, which were used to study these cellular functions. PI, propidium iodide; PI*DNA, fluorescent complex of PI with DNA; TOTO, 1,1′-(4,4,8,8-tetramethyl-4,8-diazaundecamethylene) bis[4-(3-methyl-2,3-dihydrobenzo-1,3-thiazolyl-2-methylidene)quinolinium] tetraiodide; TOTO*DNA, fluorescent complex of TOTO with DNA; DISC₃(5), dipropylthiadicarbocyanin iodide; cFDASE, carboxyfluorescein diacetate lyase; INT, 2-(-iodophenyl)-3-(p-nitrophenyl)-5-phenyltetrazolium chloride; EB, ethidium bromide; EB*DNA, fluorescent complex of EB with DNA. Laurdan was used for measurements of membrane fluidity.

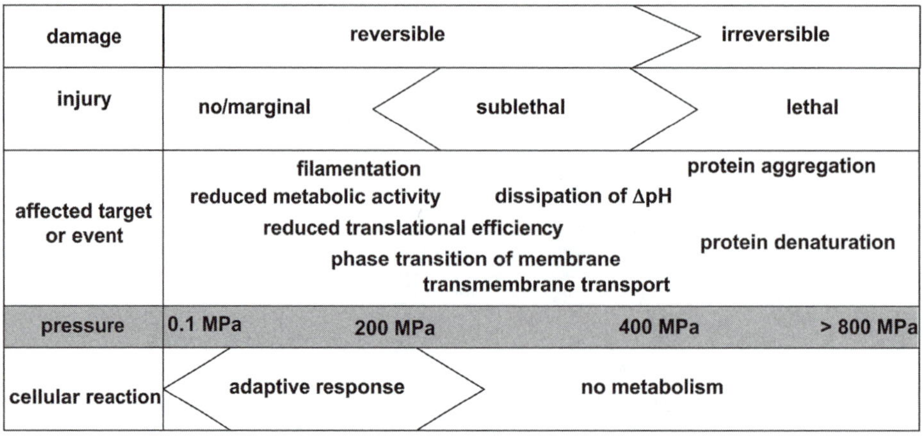

damage	reversible		irreversible
injury	no/marginal	sublethal	lethal
affected target or event	filamentation reduced metabolic activity dissipation of ΔpH reduced translational efficiency phase transition of membrane transmembrane transport		protein aggregation protein denaturation
pressure	0.1 MPa 200 MPa	400 MPa	> 800 MPa
cellular reaction	adaptive response	no metabolism	

Figure 10. Cellular processes along increasing pressure.

functions (e.g., barrier functions, maintenance of proton potential, and functionality of transporters or ATPases). On the other hand, any imagination of pressure-induced intracellular protein refolding or altered promoter structures lacks proof and should therefore be reserved to much higher pressures used in studies of protein structure and dynamics and may not even be reached in HHP food processing. The overview given in Fig. 10 may add some clarity to this view depicting various cellular processes along increasing pressure.

The current understanding of HHP-induced microbial inactivation of vegetative cells enables establishment of HHP food processes from a microbial food safety point of view. The challenge to use HHP in microbial studies has, however, changed its focus, looking at HHP as a tool to study macromolecular interaction in cellular systems or models thereof. These studies have not much in common with previous views, because it is not the HHP response which is in question. The questions arise from the macromolecular system itself, and HHP serves as a tool only, to study what it is most suitable for: macromolecular interaction.

REFERENCES

1. **Aertsen, A., and C. W. Michiels.** 2005. Diversify or die: generation of diversity in response to stress. *Crit. Rev. Microbiol.* **31**:69–78.
2. **Aertsen, A., R. Van Houdt, K. Vanoirbeek, and C. W. Michiels.** 2004. An SOS response induced by high pressure in *Escherichia coli. J. Bacteriol.* **186**:6133–6141.
3. **Aertsen, A., K. Vanoirbeek, P. De Spiegeleer, J. Sermon, K. Hauben, A. Farewell, T. Nyström, and C. W. Michiels.** 2004. Heat shock protein-mediated resistance to high hydrostatic pressure in *Escherichia coli. Appl. Environ. Microbiol.* **70**:2660–2666.
4. **Anata, E., and D. Knorr.** 2003. Pressure induced thermotolerance of *Lactobacillus rhamnosus* GG. *Food Res. Int.* **36**:991–997.
5. **Beney, L., and P. Gervais.** 2001. Influence of the fluidity of the membrane on the response of microorganisms to environmental stresses. *Appl. Microbiol. Biotechnol.* **57**:34–42.
6. **Bolotin, A., P. Wincker, S. Mauger, O. Jaillon, K. Malarme, J. Weissenbach, D. Ehrlich, and A. Sorokin.** 2001. The complete genome sequence of the lactic acid bacterium *Lactococcus lactis* ssp. *lactis* IL1403. *Genome Res.* **11**:731–753.
7. **Boutibonnes, P., B. Gillot, and Y. Auffray.** 1991. Heat-shock induces thermotolerance and inhibition of lysis in a lysogenic strain of *Lactococcus lactis. Int. J. Food Microbiol.* **14**:1–9.

8. **Broadabent, J. R., and C. Lin.** 1999. Effect of heat shock or cold shock treatment on the resistance of *Lactococcus lactis* to freezing and lyophilization. *Cryobiology* **39:**88–109.

9. **Broadbent, J. R., C. J. Oberg, and H. Wang.** 1997. Attributes of the heat shock response in three species of dairy *Lactobacillus*. *Syst. Appl. Microbiol.* **20:**12–19.

10. **Broeze, R. J., C. J. Solomon, and D. H. Broeze.** 1978. Effects of low temperature on in vivo and in vitro protein synthesis in *Escherichia coli* and *Pseudomonas fluorescens. J. Bacteriol.* **134:**861–874.

11. **Caldas, T., S. Laalami, and G. Richarme.** 2000. Chaperone properties of bacterial elongation factor EF-G and initiation factor IF2. *J. Biol. Chem.* **275:**855–860.

12. **Casadei, M. A., P. Manas, G. Niven, E. Needs, and B. M. Mackay.** 2002. Role of membrane fluidity in pressure resistance of *Escherichia coli* NCTC 8164. *Appl. Environ. Microbiol.* **68:**5965–5972.

13. **Chastanet, A., and T. Msadek.** 2003. *clpP* of *Streptococcus salivarius* is a novel member of the dually regulated class of stress response genes in gram-positive bacteria. *J. Bacteriol.* **185:**683–687.

14. **Crowe, J. H., A. E. Oliver, F. A. Hoekstra, and L. M. Corwe.** 1997. Stabilization of dry membranes by mixtures of hydroxyethyl starch and glucose: the role of vitrification. *Cryobiology* **35:**20.

15. **De Angelis, M., L. Bini, V. Pallini, P. S. Cocconcelli, and M. Gobbetti.** 2001. The acid-stress response in *Lactobacillus sanfranciscensis* CB1. *Microbiology* **147:**1863–1873.

16. **De la Cruz, J., D. Kressler, and P. Linder.** 1999. Unwinding RNA in *Saccharomyces cerevisiae:*DEAD-box proteins and related families. *Trends Biochem. Sci.* **24:**192–198.

17. **Drews, O., W. Weiss, G. Reil, H. Parlar, R. Wait, and A. Görg.** 2002. High pressure effects step-wise altered protein expression in *Lactobacillus sanfranciscensis. Proteomics* **2:**765–774.

18. **Duwat, P., S. D. Ehrlich, and A. Gruss.** 1999. Effects of metabolic flux on stress response pathways in *Lactococcus lactis. Mol. Microbiol.* **31:**845–858.

19. **Ehrmann, M. A., C. H. Scheyhing, and R. F. Vogel.** 2001. *In vitro* stability and expression of green fluorescent protein under high pressure conditions. *Lett. Appl. Microbiol.* **32:**230–234.

20. **Erickson, H. P.** 1997. FtsZ, a tubulin homologue in prokaryote cell division. *Trends Cell Biol.* **7:**362–367.

21. **Eriksson, S., R. Hurme, and M. Rhen.** 2002. Low-temperature sensors in bacteria. *Philos. Trans. R. Soc. Lond. B* **357:**887–893.

22. **Erkmen, O.** 2001. Mathematical modelling of *Escherichia coli* inactivation under high pressure carbon dioxide. *J. Biosci. Bioeng.* **92:**39–43.

23. **Fachin, D., A. van Loey, A. Indrawati, L. Ludikhuyze, and M. Hendrickx.** 2002. Thermal and high-pressure inactivation of tomato polygalacturonase: a kinetic study. *J. Food Sci.* **67:**1610–1615.

24. **Farewell, A., and F. C. Neidhardt.** 1998. Effect of temperature on *in vivo* protein synthetic capacity in *Escherichia coli. J. Bacteriol.* **180:**4704–4710.

25. **Foster, J. W.** 2000. Microbial response to acid stress, p. 99–115. *In* G. Storz and R. Hengge-Aronis (ed.), *Bacterial Stress Responses*. ASM Press, Washington, DC.

26. **Friedman, H., P. Lu, and A. Rich.** 1969. An *in vivo* block in the initiation of protein synthesis. *Cold Spring Harbor Symp. Quant. Biol.* **34:**255–260.

27. **Fujii, S., H. Iwahashi, K. Obuchi, and Y. Komatsu.** 1996. Characterization of a barotolerant mutant of the yeast *Saccharomyces cerevisiae:*importance of trehalose content and membrane fluidity. *FEMS Microbiol. Lett.* **141:**97–101.

28. **Gaidenko, T. A., and C. W. Price.** 1998. General stress transcription factor rB and sporulation transcription factor rH each contribute to survival of *Bacillus subtilis* under extreme growth conditions. *J. Bacteriol.* **180:**3730–3733.

28a.**Gänzle, M. G., K. V. Kilimann, C. Hartmann, R. F. Vogel, and A. Delgado.** 2007. Data mining and fuzzy modelling of high pressure inactivation pathways of *Lactococcus lactis. Innovative Food Sci. Emerg. Technol.* **8:**461–468.

29. **Gänzle, M. G., H. M. Ulmer, and R. F. Vogel.** 2001. High pressure inactivation of *Lactobacillus plantarum* in a model beer system. *J. Food Sci.* **66:**1175–1181.

30. **Geeraerd, A. H., C. H. Herremanns, L. R. Ludikhuyze, M. E. Hendrickx, and J. F. van Impe.** 1998. Modeling the kinetics of isobaric-isothermal inactivation of *Bacillus subtilis* α-amylase with artificial neural networks. *J. Food Eng.* **36:**263–279.

31. **Glaasker, E., W. N. Konings, and B. Poolman.** 1996. Glycine betaine fluxes in *Lactobacillus plantarum* during osmostasis and hyper- and hypo-osmotic shock. *J. Biol. Chem.* **271:**10060–10065.

32. **Gualerzi, C. O., A. M. Giuliodori, and C. L. Pon.** 2003. Transcriptional and post-transcriptional control of cold-shock genes. *J. Mol. Biol.* **331:**527–539.

33. **Guillot, A., D. Obis, and M.-Y. Mistou.** 2000. Fatty acid membrane composition and activation of glycine-betaine transport in *Lactococcus lactis* subjected to osmotic stress. *Int. J. Food Microbiol.* **55:**47–51.

34. **Haebel, P. W., S. Gutmann, and N. Ban.** 2004. Dial tm for rescue: tmRNA engages ribosomes stalled on defective mRNAs. *Curr. Opin. Struct. Biol.* **14:**58–65.

35. **Hartke, A., S. Bouche, J. C. Giard, A. Benachour, P. Boutibonnes, and Y. Auffray.** 1996. The lactic acid stress response of *Lactococcus lactis. Curr. Microbiol.* **33:**194–199.

36. **Hauben, K. J., D. H. Bartlett, C. C. Soontjens, K. Cornelis, E. Y. Wuytack, and C. W. Michiels.** 1997. *Escherichia coli* mutants resistant to inactivation by high hydrostatic pressure. *Appl. Environ. Microbiol.* **63:**945–950.

37. **Hörmann, S., C. Scheyhing, J. Behr, M. Pavlovic, M. Ehrmann, and R. F. Vogel.** 2006. Comparative approach to characterize the high pressure stress response of *Lactobacillus sanfranciscensis* DSM 20451T. *Proteomics* **6:**1878–1885.

38. **Ishii, A., K. Nakasone, T. Sato, M. Wachi, M. Sugai, K. Nagai, and C. Kato.** 2002. Isolation and characterization of the dcw cluster from the piezophilic deep-sea bacterium *Shewanella violacea. J. Biochem.* **132:**183–188.

39. **Ishii, A., T. Oshima, T. Sato, K. Nakasone, H. Mori, and C. Kato.** 2005. Analysis of hydrostatic pressure effects on transcription in *Escherichia coli* by DNA microarray procedure. *Extremophiles* **9:**65–73.

40. **Ishii, A., T. Sato, M. Wachi, K. Nagai, and C. Kato.** 2004. Effects of high hydrostatic pressure on bacterial cytoskeleton FtsZ polymers in vivo and in vitro. *Microbiology* **150:**1965–1972.

41. **Kaarniranta, K., M. A. Elo, R. K. Sironen, H. M. Karjalainen, H. J. Helminen, and M. J. Lammi.** 2003. Stress responses of mammalian cells to high hydrostatic pressure. *Biorheology* **40:**87–92.

42. **Karatzas, K. A., and M. H. Bennik.** 2002. Characterization of a *Listeria monocytogenes* Scott A isolate with high tolerance towards high hydrostatic pressure. *Appl. Environ. Microbiol.* **68:**3183–3189.

43. **Kato, C., and K. Horikoshi.** 1996. Gene expression under high pressure, p. 59–66. *In* R. Hayashi and C. Balmy (ed.), *High Pressure Bioscience and Biotechnology.* Elsevier Science, Amsterdam, The Netherlands.

44. **Kato, C., T. Sato, M. Smorawinska, and K. Horikoshi.** 1994. High pressure conditions stimulate expression of chloramphenicol acetyltransferase regulated by the lac promoter in *Escherichia coli. FEMS Microbiol. Lett.* **122:**91–96.

45. **Kilimann, K. V., C. Hartmann, A. Delgado, R. F. Vogel, and M. G. Gänzle.** 2005. A fuzzy logic-based model for the multistage high-pressure inactivation of *Lactococcus lactis* ssp. *cremoris* MG1363. *Int. J. Food Microbiol.* **98:**89–105.

46. Reference deleted.

47. **Lee, D.-U., V. Heinz, and D. Knorr.** 2001. Biphasic inactivation kinetics for *Escherichia coli* in liquid whole egg by high hydrostatic pressure treatments. *Biotechnol. Progr.* **17:**1020–1025.

48. **Leslie, S. B., E. Israeli, B. Lighthart, J. H. Crowe, and L. M. Crowe.** 1995. Trehalose and sucrose protect both membranes and proteins in intact bacteria during drying. *Appl. Environ. Microbiol.* **61:**3592–3597.

49. **Ludikhuyze, L. R., I. Van den Broech, C. A. Weemaes, C. H. Herremans, J. F. Van Impe, M. E. Hendrickx, and P. P. Tobback.** 1997. Kinetics for isobaric-isothermal inactivation of Bacillus subtilis α-amylase. *Biotechnol. Progr.* **13:**532–538.

50. **Marquis, R. E., and G. R. Bender.** 1980. Isolation of a variant of *Streptococcus faecalis* with enhanced barotolerance. *Can. J. Microbiol.* **26:**371–376.

51. **Mckay, L. L., and K. A. Baldwin.** 1990. Applications for biotechnology—present and future improvements in lactic acid bacteria. *FEMS Microbiol. Rev.* **87:**3–14.

52. **Mechold, U., and H. Malke.** 1997. Characterization of the stringent and relaxed responses of *Streptococcus equisimilis. J. Bacteriol.* **179:**2658–2667.

53. **Molenaar, D., A. Hagting, H. Alkema, A. Driessen, and W. N. Konings.** 1993. Characteristics and osmoregulatory roles of uptake systems for proline and glycine betaine in *Lactococcus lactis. J. Bacteriol.* **175:**5438–5444.

54. **Molina-Gutierrez, A., B. Rademacher, M. G. Gänzle, and R. F. Vogel.** 2002. Effect of sucrose and sodium chloride on the survival and metabolic activity of *Lactococcus lactis* under high-pressure conditions. *Trends High Pressure Biosci. Biotechnol.* **19:**295–302.

55. **Molina-Gutierrez, A., V. Stippl, A. Delgado, M. G. Gänzle, and R. F. Vogel.** 2002. In situ determination of the intracellular pH of *Lactococcus lactis* and *Lactobacillus plantarum* during pressure treatment. *Appl. Environ. Microbiol.* **68:**4399–4406.

56. **Molina-Höppner, A., T. Sato, C. Kato, M. G. Gänzle, and R. F. Vogel.** 2003. Effects of pressure on cell morphology and cell division of lactic acid bacteria. *Extremophiles* **7:**511–516.

57. **Padan, E., and T. A. Krulwich.** 2000. Sodium stress, p. 117–130. *In* G. Storz and R. Hengge-Aronis (ed.), *Bacterial Stress Responses.* ASM Press, Washington, DC.

58. **Pavlovic, M., S. Hörmann, R. F. Vogel, and M. A. Ehrmann.** 2005. Transcriptional response reveals translation machinery as target for high pressure in *Lactobacillus sanfranciscensis. Arch. Microbiol.* **184:** 11–17.

59. **Perrier-Cornet, J. M., M. Hayert, and P. Gervais.** 1999. Yeast cell mortality related to a high-pressure shift: occurrence of cell membrane permeabilization. *J. Appl. Microbiol.* **87:**1–7.

60. **Robey, M., A. Benito, R. H. Hutson, C. Pascual, S. F. Park, and B. M. Mackey.** 2001. Variation of resistance to high hydrostatic pressure in natural isolates of *Escherichia coli* O157:H7. *Appl. Environ. Microbiol.* **67:**4901–4907.

61. **Sato, T., C. Kato, and K. Horikoshi.** 1995. The effect of high pressure on gene expression by the lac and tac promoters in *Escherichia coli. J. Mar. Biotechnol.* **3:**89–92.

62. **Sato, T., T. Miwa, A. Ishii, C. Kato, M. Wachi, M. Nagai, M. Aizawa, and K. Horikoshi.** 2002. The dynamism of *Escherichia coli* under high hydrostatic pressure repression of the FtsZ-ring formation and chromosomal DNA condensation, p. 233–238. *In* R. Hayashi (ed.), *Trends in High Pressure Bioscience and Biotechnology.* Elsevier, Amsterdam, The Netherlands.

63. **Scheyhing, C. H., S. Hörmann, M. A. Ehrmann, and R. F. Vogel.** 2004. Barotolerance is inducible by preincubation under hydrostatic pressure, cold-, osmotic- and acid-stress conditions in *Lactobacillus sanfranciscensis* DSM 20451^T. *J. Appl. Microbiol.* **39:**284–289.

64. **Schwab, C., and M. G. Gänzle.** 2006. Effect of membrane lateral pressure on the expression of fructosyltransferases in *Lactobacillus reuteri. Syst. Appl. Microbiol.* **29:**89–99.

65. **Schwarz, J. R., and J. V. Landau.** 1972. Hydrostatic pressure effects on *Escherichia coli*: site of inhibition of protein synthesis. *J. Bacteriol.* **109:**945–948.

66. **Schwarz, J. R., and J. V. Landau.** 1972. Inhibition of cell-free protein synthesis by hydrostatic pressure. *J. Bacteriol.* **112:**1222–1227.

67. **Sharma, A., J. H. Scott, G. D. Cody, M. L. Fogel, R. M. Hazen, R. J. Hemley, and W. T. Huntess.** 2002. Microbial activity at gigapascal pressures. *Science* **295:**1514–1516.

68. **Smith, W., D. Pope, and J. V. Landau.** 1975. Role of bacterial ribosome subunits in barotolerance. *J. Bacteriol.* **124:**582–584.

69. **Stiles, M. E., and W. H. Holzapfel.** 1997. Lactic acid bacteria of foods and their current taxonomy. *Int. J. Food Microbiol.* **36:**1–29.

70. **Sun, Q., and W. Margolin.** 1998. FtsZ dynamics during the division cycle of live *Escherichia coli* cells. *J. Bacteriol.* **180:**2050–2056.

71. **Tanner, N. K., and P. Linder.** 2001. DEAD/H box RNA helicases: from genetic motors to specific dissociation functions. *Mol. Cell* **8:**251–262.

72. **Tannock, G. W.** 1999. Analysis of the intestinal microflora, a renaissance. *Antonie Leeuwenhoek* **76:**265–278.

73. **ter Steeg, P. F., J. C. Hellemons, and A. E. Kok.** 1999. Synergistic actions of nisin, sublethal ultrahigh pressure, and reduced temperature on bacteria and yeast. *Appl. Environ. Microbiol.* **65:**4148–4154.

74. **Timasheff, S. N., J. C. Lee, E. P. Pittz, and N. Tweedy.** 1976. The interaction of tubulin and other proteins with structure-stabilizing solvents. *J. Colloid Interface Sci.* **55:**658–663.

75. **Ulmer, H. M., M. G. Gänzle, and R. F. Vogel.** 2000. Effects of high pressure on survival and metabolic activity of *Lactobacillus plantarum* TMW1.460. *Appl. Environ. Microbiol.* **66:**3966–3973.

76. **Ulmer, H. M., H. Herberhold, S. Fahsel, M. G. Gänzle, R. Winter, and R. F. Vogel.** 2002. Effects of pressure-induced membrane phase transitions on inactivation of HorA, an ATP-dependent multidrug resistance transporter, in *Lactobacillus plantarum. Appl. Environ. Microbiol.* **68:**1088–1095.

77. **Van Bogelen, R. A., and F. C. Neidhardt.** 1990. Ribosomes as sensors of heat and cold shock in *Escherichia coli. Proc. Natl. Acad. Sci. USA* **87:**5589–5593.

78. **van de Guchte, M., P. Serror, C. Chervaux, T. Smokvina, S. D. Ehrlich, and E. Maguin.** 2002. Stress responses in lactic acid bacteria. *Antonie Leeuwenhoek* **82:**187–216.

79. **van den Boom, T., and J. E. Cronan.** 1989. Genetics and regulation of bacterial lipid metabolism. *Annu. Rev. Microbiol.* **43:**317–343.

80. **Vogel, R. F., M. A. Ehrmann, M. G. Gänzle, C. Kato, M. Korakli, C. H. Scheyhing, A. Molina-Gutierrez, H. M. Ulmer, and R. Winter.** 2002. High pressure response of lactic acid bacteria, p. 249–255. *In Proceedings of 2nd International Conference on High Pressure Bioscience and Biotechnology.*

81. **Weart, R. B., S. Nakano, B. E. Lane, P. Zuber, and P. A. Levin.** 2005. ClpX chaperone modulates assembly of the tubulin-like protein FtsZ. *Mol. Microbiol.* **57:**238–249.
82. **Withey, J., and D. Friedman.** 2002. The biological roles of *trans*-translation. *Curr. Opin. Microbiol.* **5:** 154–159.
83. **Wouters, P., E. Glaasker, and J. P. P. M. Smelt.** 1998. Effects of high pressure on inactivation kinetics and events related to proton efflux in *Lactobacillus plantarum. Appl. Environ. Microbiol.* **64:**509–514.

High-Pressure Microbiology
Edited by C. Michiels, D. H. Bartlett, and A. Aertsen
© 2008 ASM Press, Washington, DC

Chapter 8

Saccharomyces cerevisiae Response to High Hydrostatic Pressure

Patricia M. B. Fernandes

Life is within a Euclidean space (with the usual Euclidean metric), as there are only three essential axes for sustaining life. The first is the source of energy, the second is water, and the third is a range of conditions that the organism can tolerate, meaning how much of a stress an organism can cope with.

First, living organisms must burn fuel for metabolism, which drives growth, reproduction, and maintenance of the structure and integrity of their bodies. The major source for energy and carbon in the yeast *Saccharomyces cerevisiae* is glucose, and glycolysis is the general pathway for conversion of glucose to pyruvate. Both the baking and brewing industries rely upon the ability of *Saccharomyces cerevisiae* to convert glucose to ethanol and carbon dioxide. Although this yeast can use pyruvate in further energy production by respiration, fermentation is its main lifestyle. Those two processes are mainly regulated by environmental factors, the best documented being the availability of glucose and oxygen. Even when *S. cerevisiae* is grown aerobically in high concentrations of glucose, fermentation will occur, and only when the levels of glucose decline is the respiratory pathway induced, resulting in oxidative consumption of ethanol.

The second axis for sustaining life, water, does not constitute a nutrient per se, but it is vital for microbial growth, as most essential metabolic reactions occur in an aqueous solution. Water influences the majority of the structures of the macromolecules on which life is based (proteins, nucleic acids, lipids, and sugars) and provides the medium in which the chemical reactions, hence the metabolism, of living organisms take place. The unique solvent properties of water derive from its polarity as well as its hydrogen bonding. It regulates osmotic pressure and acts as a thermoregulator due to its high specific heat and also as a dispersant of inorganic ions. On the other hand, the osmolarity of a growing yeast cell is maintained at a higher level inside than outside the cell. The resulting osmotic gradient drives water transport across the plasma membrane, which causes cell expansion and creates turgor.

To understand the third axis, we need to define what a normal environmental condition is. Let us consider that there is an optimum condition at which activity, growth rate, and

---

Patricia M. B. Fernandes • Biotechnology Core, Federal University of Espirito Santo, Vitória, ES Brazil, 29040-090.

metabolism are at their greatest. Stress conditions are those deviating from the normal, and they invoke particular cell responses to address the hazardous consequences of cell damage. The main stresses on organisms are high or low temperature, high pressure, desiccation, acidity or alkalinity, high or low osmotic and ionic stress, and low or high oxygen levels (74). Moreover, unicellular organisms, such as yeasts, must cope with a natural environment that does not always provide all the conditions needed for constant growth. An abrupt change in the environmental conditions leads to a rapid adjustment of metabolism and the genomic expression program to adapt to the new situation.

Saccharomyces cerevisiae is a unicellular fungus and therefore a true yeast, and the cell divides by budding. It also exhibits sexual reproduction: mating, having two opposite mating types, **a** and α. The cells can grow as haploids or can mate and vegetatively grow as diploids or sporulate (undergo meiosis) and form haploid gametes. Moreover, the basic cellular mechanisms of replication, recombination, cell division, and metabolism are generally conserved between *S. cerevisiae* and higher eukaryotes. For that reason, besides its industrial importance, *S. cerevisiae* is a useful model system for the study of many important problems faced by eukaryotic cells. Moreover, this yeast was the first eukaryote whose genome has been completely sequenced, giving insight into many different aspects of life (http://genome-www.stanford.edu/Saccharomyces/).

This chapter describes the effects of short exposures of *S. cerevisiae* to lethal and sublethal pressures, with a focus on cellular inactivation/resistance and stress response, respectively.

YEAST CELL SURVIVAL OF HHP

High hydrostatic pressure (HHP) exerts a broad effect within eukaryotic cells, with characteristics similar to common stresses, such as temperature, ethanol, and oxidative stresses. However, HHP represents an interesting form of stress, as it is a way of changing only one variable, the reaction volume. In this context, when comparing with temperature stress that involves both volume and thermal changes, results obtained by HHP experiments are straightforward. Moreover, it is important to consider that biochemical reactions are accompanied by changes in volume. If a reaction results in an increase in volume, it will be inhibited by pressure, whereas if it is associated with a decrease in volume, it will be enhanced (47).

Hydrostatic pressure has a strong effect on a variety of cellular structures and functions (32, 80). Cytokinetic and mitotic activities of dividing cells are delayed or inhibited by pressure, depending on the magnitude and duration of treatment. High pressure interferes with the processes of polymerization and depolymerization of proteins which are essential for the formation and functioning of the mitotic structure and its stability. In addition, reactivity of enzymes and other proteins has been shown to be affected by hydrostatic pressure (81).

Yeast viability during hydrostatic pressure treatment decreases with increasing pressure, and this effect is more pronounced when cells are submitted to pressures above 100 MPa, while at 220 MPa all wild-type cells are killed. A pressure of 50 MPa is sufficient neither to kill nor to alter the yeast cell morphology. Yeast cells in stationary phase, which undergo growth arrest and a variety of morphological and physiological changes, are more resistant to pressure than proliferating cells (25).

My group recently found that wild-type yeast strains with different genetic backgrounds display different hydrostatic pressure sensitivities. The results were obtained for two brewing strains isolated from Brazilian distilleries (BT0501 and BT0502), one commercial baker's strain (Fleishmann Yeast; AB Brasil Indústria e Comércio de Alimentos Ltda.), and four laboratory strains (Y440, *MAT* **a** *leu2;* BY4741, *MAT* **a** *his3Δ1 leu2Δ0 met15Δ0 uraΔ0;* W303, *MAT* **a** *ade2-1 trp1-1 leu2-3,112 can1-100 ura3-1 his3-11,115;* and S228C, *MAT* α *SUC2 gal2 mal mel flo1 flo8-1 hap1*). The similar profiles of the inactivation curves of the different strains suggest a universal mechanism of cell survival after subjection to HHP stress (Fig. 1). Nevertheless, the onset of inactivation seems to be different for different strains, and according to Hartmann and Delgado (31), it is correlated with cell volume. An obvious difference in piezosensitivity, or resistance, can be seen in the range of 50 to 100 MPa.

Application of HHP in the food processing industry has stimulated the effort to understand the impact of pressure in combination with other parameters, such as high or low temperature, on cell viability. The effects of high pressure and high temperature on the survival of yeast cells appear to be additive. *S. cerevisiae* becomes more sensitive to pressure at higher sublethal temperatures, and this effect tends to increase stepwise as pressure increases. Several factors can cause a nonlogarithmic decrease in survival fractions of a microbial population; e.g., the presence of injured cells might change the homogeneity of a cell population, which eventually causes nonlinearity in the death curve (12). For subzero temperatures, a linear pressure-temperature dependence of microorganism isoinactivation is observed. Increasing pressure and/or decreasing temperature will lead to a decrease in membrane fluidity and a phase transition (from liquid crystalline to gel phase) of phospholipids. The pressure

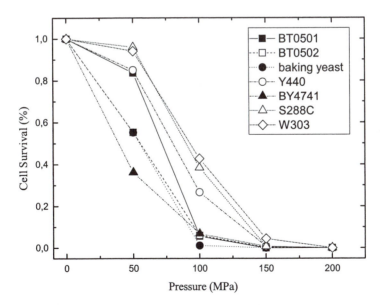

Figure 1. Effect of HHP on different wild-type yeast cells. Cells from logarithmic phase were submitted to various hydrostatic pressures for 30 min each time. Cell survival is expressed as percentage of viable cells. The standard deviations are smaller than the symbols used.

dependence of the phase transition temperature is about 0.2°C/MPa and is almost independent of the type of phospholipids (44). Pressurization of *S. cerevisiae* at subzero temperatures without freezing significantly enhanced the effect of pressure. For example, at a pressure of 150 MPa, the decrease in temperature from ambient to −20°C allowed an increase in the pressure-induced inactivation from less than 1 log up to 7 to 8 log units (55).

In the budding yeast *Saccharomyces cerevisiae* the presence of a bud reflects the cell cycle position; cells in the G_1 interval of the cell cycle are nonbudded, and the bud appears at approximately the same time as the cell enters S phase. When *S. cerevisiae* cells are exposed to a nutrient limitation, mating pheromone, or mild stress, HHP included, they arrest in the G_1 phase of the cell cycle (17, 51). Yeast cells pressurized at 50 MPa for 30 min and then incubated at atmospheric pressure for 120 min show a decrease in the number of budded cells after the stress condition, reaching a minimum value 45 min postpressurization. The cells start recovering after 60 min but are fully active only after 2 h. Comparing yeast cell growth responses to pressure of 50 MPa for 30 min and to the classical stress of heat shock of 40°C for 30 min, both sublethal stresses, shows that pressurized cells have a slower response and also take longer to recover normal growth (51). This suggests that after the yeast cells have been relieved from pressure stress they need a relatively long time and sufficient resources to restore cellular homeostasis.

EFFECT OF HHP ON CELLULAR STRUCTURES

Cell Wall

Pressure interferes with cell architecture and cell division, affecting the structure of the cell wall, membrane fluidity, and several intracellular organelles (46). The yeast cell wall is remarkably thick (100 to 200 nm), with major structural constituents being polysaccharides (80 to 90%), mainly glucans and mannans, and a minor percentage of chitin. HHP-treated *S. cerevisiae* shows an abnormal distribution of the calcofluor white fluorescent stain in the cell wall corresponding to a delocalized deposition of chitin, which normally appears only at bud necks and in bud scars.

Transmission electron microscopic images of yeast cells suggest that hydrostatic pressure induces changes in the cell wall and cytoskeleton, and thus in the cell membranes and organellar dynamics, as shown in Fig. 2 (24). Indeed, pressure up-regulates *HSP12* (23), which codes for a hydrophilic 12-kDa protein that has been shown to be present adjacent to the plasma membrane (61) as well as in the cell wall (49). Deletion of *HSP12* reduced the changes in cell size on exposure to hypo- or hyperosmotic stress and increased the sensitivity to rapid pressure changes. Hsp12p was therefore considered to act as a cell wall plasticizer. This was confirmed by using the carbohydrate polymer agarose as a model system to represent the yeast cell wall glucan. Solutes known to up-regulate HSP12 expression when present in the medium were found to decrease the flexibility of agarose when incorporated into the gel; normal flexibility was restored upon simultaneous incorporation of Hsp12p (40). These data therefore suggest a model whereby pressure, hydrostatic or osmotic, directly affects cell wall flexibility, and the cell responds by production of Hsp12p. Hsp12p would interrupt the hydrogen bonding and ionic interactions between adjacent polysaccharide polymers to ensure a flexible structure. This has recently been tested in yeast cells using atomic force microscopy (41). While wild-type cells expressing HSP12 required a deformation force of 17 mN·m^{-1}, rhsp12 cells in which the HSP12 gene was disrupted required a deformation force of 72 mN·m^{-1}, confirming the plasticizing effect of Hsp12p. Additionally, wild-type cells in the presence of mannitol in the medium required a deformation force of 141 mN·m^{-1}. Recent work has suggested that this flexibility might be brought about by an interaction between Hsp12p and chitin (G. Lindsey, personal communication). Thus, Hsp12p yeast has been shown to be more sensitive to the cell wall-destabilizing agent Congo red (49), and Hsp12p was found to inhibit Congo red binding to chitin in an in vitro experiment (G. Lindsey, personal communication). A possible hypothesis is that Congo red binds to cell wall carbohydrates, forming noncovalent cross-links which reduce cell wall flexibility. The situation must be far more complicated than this simple model, as Hsp12p has been shown to be required for biofilm formation (79) and Congo red induces invasive growth in a normally noninvasive yeast strain providing that HSP12 has not been disrupted (G. Lindsey, personal communication).

Membranes

Hydrostatic pressure also interferes with cellular membrane structure, increasing the order of lipid molecules, especially in the vicinity of proteins. This phenomenon is driven by the smaller volume associated with a more ordered, tighter packing. The consequence is a decrease in cell membrane fluidity followed by an increase in thickness (47). Furthermore, the effect of pressure on the membranes might be due to the fact that lipids are particularly sensitive to pressure effects, being an order of magnitude more compressible than proteins (73). This phenomenon also occurs at low temperature (57). An increase in the extent of fatty acid unsaturation can compensate for these effects and maintain the membrane in a functional liquid crystalline state (homeoviscous adaptation). In fact, many deep-sea organisms modulate their membrane fluidity by increasing the proportions of unsaturated fatty acids in response to pressure (7). Nevertheless, unsaturated membranes have a higher degree of disorder than the saturated bilayers, but their motionally disordered regions are barely accessible to water molecules (47), leading to an increase in the piezoresistance or low-temperature resistance.

Wild-type yeast cells subjected to 200 MPa for 30 min and analyzed by transmission electron microscopy demonstrate a cell outer shape that is almost unaffected, but the cell membrane exhibited an increase in undulations and profuse invaginations (Fig. 2c and d). There was also evidence of nuclear membrane dissolution (Fig. 2c) (24). Pressure treatment for shorter periods (10 min) showed the same effect if higher pressure values (400 to 600 MPa) or subzero temperatures ($-20°$C) were used. Analysis of the external parts of *S. cerevisiae* by scanning electron microscopy indicates that the appearance of the cell surface is almost unaffected at pressures up to 300 MPa for 10 min. But at pressures higher than 500 MPa (again, for 10 min) there is visible damage and disruption in the bud scar area of the cell wall. Nevertheless, transmission electron microscopy revealed that already at 300 to 400 MPa most of the intracellular organelles in the cell, such as the nucleus, mitochondria, endoplasmic reticulum, and vacuole, are completely deformed or disrupted. Furthermore, after pressure treatment at above 500 MPa, the nuclei and other intracellular organelles are no longer recognizable and membranous fragments can be seen. Surprisingly, the pattern of nuclear staining (with 4′,6′-diamidino-2-phenylindole [DAPI]) is not greatly changed even above 400 MPa, in spite of total disruption of nuclear membrane.

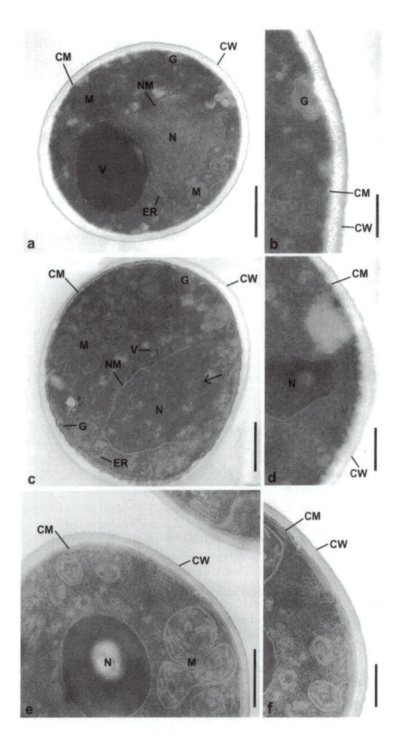

Also, chromosomal DNA samples from pressure-treated cells reveal 13 distinct bands with almost the same mobility as those of untreated cells in pulsed field electrophoresis analysis, suggesting that chromosomal DNA within the cell is as resistant as were extracted DNA fractions subjected to HHP (62, 68).

Actually, the irreversible decrease in cell volume with HHP treatment that leads to cell mortality has been attributed to a mass transfer across the cell membrane, and this mass transfer reflects a change in membrane permeability. The cell membrane certainly becomes more permeable during pressure and allows leakage of intracellular solutes. As pressure increases, cell volume shrinkage is observed, up to 15% at 250 MPa. A phase transition of cell membranes could be expected to occur between 80 and 150 MPa at room temperature. Nevertheless, no extreme change in the yeast cell volume is observed at these pressures, and this phenomenon might be due to the cell wall stiffness (54).

Stresses that cause membrane destabilization, such as osmotic pressure (76), ethanol (11), cold (59), and hydrostatic pressure (24, 54), show a discrete induction of the *ERG25* gene (3, 21, 57, 73). Erg25p is a sterol desaturase involved in the biosynthesis of ergosterol, a molecule that is structurally and functionally similar to cholesterol in animal cells, and is suggested to be membrane bound (5). During stress exposure, the cell must maintain membrane integrity by activating mechanisms that are capable of minimizing these deleterious effects. On the other hand, heat shock down-regulates the *ERG25* gene in yeast cells (28). Yet, in *S. cerevisiae* it has been shown that the addition of ergosterol induces thermal and ethanol tolerance in a sterol-auxotrophic strain (69), and vesicles with membranes containing cholesterol prove to be more resistant to HHP than cholesterol-free ones (47). Ergosterol is a sterol containing an unsaturated side chain, and cholesterol has a saturated side chain. *S. cerevisiae* cells enriched with ergosterol seem to be more resistant to ethanol than cells enriched with cholesterol (47). Thus, biological membranes have been implicated as a primary sensor of environmental stress, and membrane sterols appear to be important in stress tolerance.

Another interesting feature is the piezoinduction of the *OLE1* gene, which encodes a Δ9-desaturase in yeast. There is considerable evidence that an increased proportion of unsaturated fatty acids in membrane lipids is strongly correlated with bacterial resistance under high pressure as well as at low temperature (7). A higher proportion of unsaturated fatty acids would help to maintain favorable fluidity and viscosity of biological membranes under high pressure or at low temperature. Furthermore, microarray analysis performed with *S. cerevisiae* submitted to low temperatures also demonstrated the induction of the *OLE1* gene (59), indicating that the cell can sense the fluidity state of the membrane and also possesses mechanisms to compensate for the deleterious effects of environmental conditions.

Figure 2. Transmission electron micrographs of a thin section through *Saccharomyces cerevisiae* Y440 wild-type cells. (a and b) Typical *S. cerevisiae* cell at atmospheric pressure. (b) Detailed image illustrating the appearance of the cell wall, cell membrane, and Golgi apparatus. (c and d) Cell submitted to 200 MPa for 30 min. The cell outer shape was almost unaffected. The arrow points to a broken nuclear membrane. (d) Detail of the cell membrane. (e and f) Heat shock-pretreated cell (40°C for 60 min) submitted to HHP of 200 MPa. (f) Detailed image showing a lamellar structure flanking the cell membrane. CM, cell membrane; CW, cell wall; NM, nuclear membrane; N, nucleus; V, vacuoles; M, mitochondria; G, Golgi apparatus; ER, endoplasmic reticulum. The bar in panel a represents 0.8 μm; the bars in panels b, d, and f represent 0.3 μm; and the bars in panels c and e represent 0.5 μm. Reprinted from *Letters in Applied Microbiology* (24) with permission of the publisher.

Cytoskeleton and Organelles

The cytoskeleton is a complex network of polymers and associated proteins that plays a role in many aspects of cellular physiology. The actin cytoskeleton has been implicated in maintenance of cell polarity, changes in cell shape, resistance against osmotic forces, and responses to environmental signals (8). Additionally, experiments with *S. cerevisiae* containing mutations in actin metabolism have shown, among other phenotypes, delocalization of chitin deposition that normally appears only at bud necks and bud scars (26), presenting a clear link between these cell components.

Microfilaments have a cell cycle-specific organization in *S. cerevisiae*. Fluorescence microscopy analysis of F-actin-stained yeast cells (Fig. 3) shows that actin cables are oriented along the long axis of the mother cell during the cell cycle. During cytokinesis, actin dots cluster in the neck region between the separating cells. When cells at various stages are pressurized for 10 min, the actin cables in the mother cells disappear and the cell cycle-specific actin organization is lost at 100 MPa. Short and thick fragmented actin cables are seen in both buds and mother cells at 150 MPa; they become vague at 250 MPa, and no fluorescence is visible due to the depolymerization of F-actin above 300 MPa (42).

Transmission electron microscopy suggests that the impact of hydrostatic pressure on yeast cells occurs directly on the membrane system, particularly the nuclear membrane

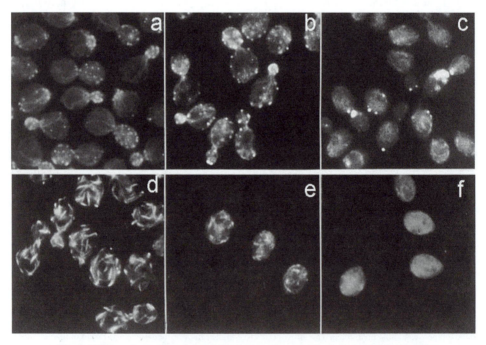

Figure 3. Fluorescent micrographs of *Saccharomyces cerevisiae* cells stained for F-actin with rhodamine-conjugated phalloidin. (a) Untreated cells evidencing the cell cycle-specific organization of microfilaments; (b to f) cells treated with 100 (b), 150 (c), 200 (d), 250 (e), and 300 (f) MPa. Reprinted from *High Pressure Bioscience and Biotechnology* (42) with permission of the publisher.

(24, 50). Cells submitted to piezotreatment of 200 MPa exhibit an alteration in the structure of the Golgi apparatus (Fig. 2c). In *Schizosaccharomyces pombe,* disruption of the microtubular network causes dissociation of the Golgi cisternae (4). Hence, hydrostatic pressure affects the stack of cisternal structures, leading to a system of flattened tubules on the side of the plasma membrane. The effect of the hydrostatic pressure on the actin cytoskeleton may also be noticed by the mitochondrial agglomeration (Fig. 2c). Mitochondria are organized along actin fibers, as shown by colocalization of actin cables and mitochondria in wild-type yeast cells and the disruption of this organization in certain actin mutants (20). Besides the abnormal distribution, approximately 40% of pressured cell mitochondria are elongated or burst. Another specialized organelle, the vacuole, is broken and disappears after the pressure treatment (24).

Another feature of HHP is the promotion of cytoplasm and vacuole acidification in a manner dependent on the magnitude of pressure applied. Pressure-induced internal acidification is connected to carbon dioxide production through ethanol fermentation and is caused by the dissociation of protons from H_2CO_3 or sugar phosphoesters (glycolytic intermediates, such as glucose-6-phosphate or fructose-6-phosphate) facilitated by the decrease in volume driven by pressure (2). Stresses like hydrostatic pressure, heat shock, ethanol exposure, and osmostress in *S. cerevisiae* promote a cytoplasmic acidification (56). The maintenance of neutral intracellular pH is essential for the viability of yeast cells. Microarray analysis of pressurized yeast cells revealed the induction of *HSP30* (23), which codes for a hydrophobic stress-responsive protein that negatively regulates the plasma membrane H^+-ATPase. H^+-ATPase pumps H^+ at the expense of ATP depletion. Hsp30p thus plays a role in energy conservation during stress conditions (63), and it is a physiological response of *S. cerevisiae* to increasing hydrostatic pressure.

RESISTANCE TO PRESSURE AND CROSS-PROTECTION

The cellular stress response is evolutionarily conserved in all living organisms, a major role being attributed to the induced heat shock proteins, or stress proteins, and other molecules that confer stress protection. The molecular responses elicited by the cells dictate whether the organism adapts, survives, or, if injured beyond repair, undergoes death. In the baker's yeast *S. cerevisiae,* suboptimal environmental conditions lead to achieved tolerance to high temperature, osmotic pressure, dehydration, cryotreatment, and HHP (17, 24, 25, 62, 72). Generally, yeast cells are better able to withstand severe stress after they have been exposed to a mild form of stress, and this phenomenon is evidence for the existence of a general stress response.

The heat shock response in *S. cerevisiae* is one of the best-studied pathways of eukaryotic cells, and pretreatment with a mild heat stress leads to protection against more severe heat shock and several other stresses, including HHP (24, 34, 56). There is an increase in viability of approximately a factor of 500 for cells that have been subjected to a heat pretreatment compared to cells that had been directly subjected to HHP (24). However, a mild heat shock is only one of a number of preconditioning treatments known to induce pressure tolerance. Mild treatments with hydrogen peroxide, ethanol, and cold shock also induce piezoresistance (52). Although all of these stresses affect the cell membrane, each of them has its own key target. Protein denaturation is the main cause of death of cells exposed to high temperature (56), but the major problem that yeast cells face at low temperature is the reduction of membrane fluidity (59), while ethanol (16) and hydrogen peroxide (15) treatments cause oxidative stress.

S. cerevisiae cells submitted to a mild, sublethal pressure do not immediately acquire resistance to a subsequent and more severe HHP treatment (25). Piezoresistance after pressure treatment is only acquired if the cells are incubated at ambient pressure for a short period before the severe HHP stress. The short postrecovery period after the pressure pretreatment is also necessary to enhance survival following a subsequent severe stress shock of high or low temperature. The protection effect is seen after 15 min of incubation at atmospheric pressure, and with regard to pressure and ultracold severe stresses, this effect persists for 1 h. In contrast, the observed protection against heat shock is shorter (51).

It is worth noting that most pressure-up-regulated genes encode proteins related to membrane protection, and pressure does not induce the larger heat shock proteins related to chaperone activity. Pressure, like low temperature, induces two small heat shock proteins (encoded by *HSP12* and *HSP26*) related to membrane stabilization (23, 59). These two heat shock proteins are also induced when cells are pressurized at low temperature (36).

There is a straight correlation between trehalose content and cellular tolerance to stress, hydrostatic pressure included. Mutants of *S. cerevisiae* with a deletion of the trehalose-6-phosphate synthase gene *(tps1),* unable to accumulate trehalose, are more sensitive to hydrostatic pressure than the wild-type cells. Nevertheless, *tps1* cells at the stationary phase are more resistant to pressure than proliferating ones. Addition of 10% trehalose to the yeast culture subjected to hydrostatic pressure improved survival about 10-fold for both wild-type and *tps1* cells (25). In yeast, high trehalose content induced by heat stress protects enzymes at high temperatures but inhibits them at optimum temperature. Trehalose protects cellular proteins against denaturation and subsequent aggregation but inhibits the solubilization of protein aggregates and the refolding of the partially denatured proteins during recovery from stress. The presence of neutral trehalase, the enzyme responsible for breaking down trehalose in cells, is therefore essential for recovery after heat shock (66). This has also proven to be true for pressure stress, as a neutral trehalase mutant strain is less piezotolerant than the wild type at stationary phase (35). Despite the important role of trehalose in protecting the cell membrane and proteins during heat stress, it is probably not the main protector during pressure exposure, given that the *S. cerevisiae tps1* mutant can acquire piezotolerance when at stationary phase or after a mild temperature pretreatment (25). Moreover, trehalose metabolism genes were neither induced nor repressed under hydrostatic pressure (23), and yeast cells do not accumulate trehalose after a pressure treatment of 75 MPa for 30 min at 30°C (71).

Yeast cells can sense oxidative stress and are able to build a response at the molecular level involving the induction of primary and secondary antioxidant defenses. The response to oxidative stress is characterized by a strong induction of genes involved in the detoxification of active oxygen species (such as superoxide dismutases, glutathione peroxidases, and thiol-specific antioxidants), as well as genes involved in oxidative and reductive reactions within the cell (thioredoxin, thioredoxin reductases, glutaredoxin, and glutathione reductase) (28). Microarray analysis of *S. cerevisiae* submitted to HHP revealed a >2-fold induction of genes that encode proteins involved with oxidative stress, such as cytoplasmic catalase and CuZn superoxide dismutase (23). The superoxide anion (O_2^-), hydrogen peroxide (H_2O_2), and the hydroxyl radical (OH) are the most important reactive oxygen species (ROS) produced by cells. ROS cause oxidative damage on nucleic acids, lipids, proteins, carbohydrates, and other cellular components (15). Hydrogen peroxide pretreatment (0.4 mM) is capable of inducing piezoresistance in a time-dependent way, with a maximum after 45 min of treatment and a decrease after 60 min. Another feature of H_2O_2 stress is the induction of *GSH1*, encoding γ-glutamylcysteine synthetase, an enzyme required for biosynthesis of glutathione (GSH), the most abundant low-molecular-mass intracellular thiol compound. GSH acts as a radical scavenger, with the redox-active sulfhydryl group reacting with oxidants (37), and *S. cerevisiae* possesses a specific GSH transport system for uptake of GSH from the extracellular environment (53). When yeast cells are subjected to HHP in the presence of GSH, they exhibit piezoresistance. These results reinforce the idea that HHP causes oxidative stress within the cell and confirm the importance of an oxidative defense mechanism, as induced by the H_2O_2 treatment, to diminish the damage caused by pressure.

The role of ethanol as an agent affecting the physicochemical state and biological functions of various membranes is well known (38), and ethanol increases the amount of unsaturated fatty acyl residues in the membrane phospholipids (3, 13, 39, 60, 78). Although the highly unsaturated membranes have a higher degree of disorder than saturated ones, they are less sensitive to HHP due to the fact that their motionally disordered regions are barely accessible to water molecules (47). The fact that an unsaturated bilayer is important to pressure resistance is supported by the induction by pressure of the gene responsible for a desaturase, an enzyme that introduces a double bond in the fatty acid (23). Sublethal ethanol exposure induces the up-regulation of genes involved in the production of heat shock proteins, antioxidant enzymes, and changes in the plasma membrane composition (3). These features most probably are responsible for the piezoresistance induction after ethanol pretreatment. A 3-log increase in HHP tolerance is observed after 1.5 to 2 h of ethanol treatment. The protection persists, although it diminishes, for 4 h. An important feature of ethanol-induced protection is that ethanol has to be present during the pressurization, as ethanol-pretreated cells that have been washed immediately before pressure treatment do not show piezotolerance. Nevertheless, the simple presence of ethanol in the pressurization media without the pretreatment does not induce pressure resistance (52). A possible explanation of this feature is that during ethanol treatment the cells had adapted to the presence of ethanol, changing some of their physical properties, due to the ethanol effect to reduce the hydrating ability of water and to compete with water for hydration sites (38). After ethanol removal a new situation occurs in the cell due to the presence of water molecules occupying the hydration sites previously occupied by ethanol. Even the transcriptional changes induced by the ethanol pretreatment are not enough to lead to piezoresistance, since macromolecule-associated water constitutes a target particularly sensitive to pressure forces (13).

In conclusion, the induction of genes related to preservation of the cell membrane integrity, ROS detoxification, or protein folding will protect yeast cells against the deleterious effects of HHP.

GENE EXPRESSION UNDER HHP

Yeast cells have mechanisms to sense stress and subsequently induce a specific gene expression response. The influence of HHP on gene and protein expression has been widely recognized (6, 29, 70). Actually, even though in most cases gene expression is inhibited by pressure, some specific pressure shock proteins are synthesized (47). For example, some genes (e.g., *ompH* for a porin protein) from high-pressure-tolerant deep-sea bacteria cloned into an atmospheric living bacterium are only expressed if this bacterium is pressurized (7), indicating a mechanism to sense pressure.

As DNA microarray technology has become available, the expression of yeast genes in response to a variety of physical stimuli has been investigated (3, 23, 28, 36, 59, 76). Even though there are some common genes that respond to a variety of stresses, there is not a uniform response for all kinds of stress situations. Microarray analysis of *S. cerevisiae* submitted to HHP treatment of 200 MPa for 30 min at room temperture revealed a stress response expression profile. Analysis of the 6,200 known or predicted genes of *S. cerevisiae* shows that approximately 5% are affected by hydrostatic pressure treatment. Among the genes that undergo a >2-fold change in expression, 131 are induced, while 143 are repressed (23). Table 1 presents the major genes induced by pressure and the corresponding proteins, with their physiological function. The analysis reveals that most of the pressure-regulated mRNAs correspond to uncharacterized open reading frames (ORFs) or ORFs with unknown function. In fact, one of the most up-regulated genes is an uncharacterized ORF, *YER067W*, followed by two genes that code for the small heat shock proteins, *HSP30* and *HSP12*.

Overall, genes that are involved in stress defense and carbohydrate metabolism are highly induced by pressure, while several genes involved in cellular transcription, protein synthesis and targeting, and cell cycle regulation are down-regulated by pressure treatment.

Further functional analyses reveal some interesting features: categories such as transport facilitation, control of cellular organization, and transcription have approximately the same number of genes induced and repressed; conversely, other classes show a distinguished up- or down-regulation. Most of the known pressure-induced genes are among energy metabolism and stress defense categories, resembling a typical stress response pattern. In addition, genes related to protein synthesis and fate (folding, modification, and destination) together with genes involved in cell cycle progression are strongly repressed, leading to the characteristic yeast growth arrest caused by stressful conditions (65).

A schematic representation of the global gene expression profile after HHP in functional categories, according to the MIPS database (http://mips.gsf.de/proj/yeast /CYGD/ db/index.html), is shown in Fig. 4.

Another microarray study has been carried out for 1 h at 25°C with cells in fresh rich yeast extract-peptone-dextrose medium after a pressure shock of 180 MPa at 4°C (36). Pressure has an effect that can be compared with lowering the temperature, and combining the treatments could lead to a synergistic effect in living cells. But the combination of pressure and cold gives a different pattern of gene expression; nevertheless, some common features could be seen in both the arrays, such as induction of small heat shock proteins and genes involved in protein degradation. Actually, in chondrocytic cells, cDNA array changes are different in cells submitted to continuous, intermittent, or static hydrostatic pressure (43), demonstrating the high complexity of in vivo systems. An interesting feature of combining cold and pressure is the up-regulation of the *HSP104* gene 1 h after the treatment (36), which was not seen upon treatment with pressure or cold stress individually (23, 59). Hsp104p is mainly involved in the rescue of proteins from insoluble aggregates after heat stress. The overexpression of the corresponding gene has been implicated in thermotolerance and cell resistance at stationary phase (17, 65). So it seems that *HSP104* is not

Table 1. Up-regulated characterized genes after 30 min of HHP (200 MPa)[a]

Gene	Protein	Molecular function
Stress response (cell rescue, defense, and regulation of or interaction with cellular environment)		
HSP30	Heat shock protein	Stress-responsive protein that negatively regulates the H^+-ATPase Pma1p; induced by heat shock, ethanol treatment, weak organic acid, glucose limitation, and entry into stationary phase
HSP12	Heat shock protein 12	Protects membranes from desiccation; induced by heat shock, oxidative stress, osmostress, entry into stationary phase, glucose depletion, oleate, and alcohol
TFS1	Lipid-binding protein; (putative) suppressor of a *cdc25* mutation	Lipid binding, protease inhibitor activity
CTT1	Catalase T	Catalase activity
DDR2	Multistress response protein	Activated by a variety of xenobiotic agents and environmental or physiological stresses
HSP26	Heat shock protein 26	Chaperone activity that is regulated by a heat-induced transition from an inactive oligomeric (24-mer) complex to an active dimer; induced by heat, upon entry into stationary phase, and during sporulation
YRO2	Putative plasma membrane protein	Transcriptionally regulated by Haa1p; green fluorescent protein-fusion protein localizes to the cell periphery and bud
CUP1-1	Copper-binding metallothionein	Copper ion binding
CUP1-2	Copper-binding metallothionein	Copper ion binding
SOD2	Manganese-containing superoxide dismutase	Manganese superoxide dismutase activity
HOR7	Hyperosmolarity-responsive protein	Overexpression suppresses Ca^{2+} sensitivity of mutants lacking inositol phosphorylceramide mannosyltransferases Csg1p and Csh1p; transcription is induced under hyperosmotic stress and repressed by alpha factor
MCR1	NADH-cytochrome b_5 reductase	Cytochrome b_5 reductase activity
Metabolism and energy		
HXT4	Hexose transporter	Glucose, fructose, and mannose transporter activity
STF2	ATPase-stabilizing factor	Regulation of the mitochondrial F_oF_1-ATP synthase
HXT7	Hexose transporter	Fructose, mannose, and glucose transporter activity
ERG25	C-4 sterol methyl oxidase	C-4 methyl sterol oxidase activity
GPM2	Phosphoglycerate mutase	Phosphoglycerate mutase activity
PHO12	Acid phosphatase	Acid phosphatase activity
ALD4	Aldehyde dehydrogenase	Aldehyde dehydrogenase (NAD) activity
HXK1	Hexokinase I	Hexokinase activity
GAC1	Glc7p regulatory subunit	Heat shock protein binding (IDA), protein phosphatase type 1 regulator activity (TAS), structural molecule activity (TAS)
ALD3	Aldehyde dehydrogenase	Aldehyde dehydrogenase activity
GLK1	Glucokinase	Glucokinase activity
GDB1	Glycogen debranching enzyme	4-Alpha-glucanotransferase activity, amylo-alpha-1,6-glucosidase activity

(Table continues)

Table 1. *Continued*

Gene	Protein	Molecular function
GSY2	Glycogen synthase (UDP-glucose-starch glucosyltransferase)	Glycogen (starch) synthase activity
EMI2	Nonessential protein	Required for transcriptional induction of the early meiosis-specific transcription factor IME1, also required for sporulation
OLE1	Delta-9-fatty acid desaturase	Stearoyl-coenzyme A 9-desaturase activity
NOG1	Nucleolar G-protein (putative)	GTPase activity
YPC1	Alkaline ceramidase with reverse activity	Ceramidase activity
CYB2	L-Lactate cytochrome *c* oxidoreductase *cytochrome b₂*	L-Lactate dehydrogenase (cytochrome) activity
YDC1	Alkaline dihydroceramidase with minor reverse activity	Ceramidase activity
GAD1	Glutamate decarboxylase	Glutamate descarboxylase activity
TDH1	Glyceraldehyde-3-phosphate dehydrogenase	Glyceraldehyde-3-phosphate dehydrogenase (phosphorylating) activity
MAM33	Mitochondrial acidic matrix protein	Oxidative phosphorylation
HXT6	Hexose transporter	Fructose, mannose, and glucose transporter activity
PGM2	Phosphoglucomutase	Phosphoglucomutase activity
FAU1	5,10-Methenyltetrahydrofolate synthetase	5-Formyl-tetrahydrofolate cycloligase activity
GUT2	Glycerol-3-phosphate dehydrogenase	Glycerol-3-phosphate dehydrogenase activity
PIG2	Type 1 protein phosphatase regulatory subunit	Protein phosphatase type 1 regulator activity
DOG1	2-Deoxyglucose-6-phosphate phosphatase	2-Deoxyglucose-6-phosphatase activity
URA10	Orotate phosphoribosyltransferase 2	Orotate phosphoribosyltransferase activity
TEL1	Protein kinase	Protein kinase activity
RGT2	Glucose receptor	Glucose binding, glucose transporter, and receptor activities
Cell cycle		
RNR3	Ribonucleotide reductase	Ribonucleoside-diphosphate reductase activity
SOL4	6-Phosphogluconolactonase	6-Phosphogluconolactonase activity
BUD14	Protein phosphatase	Protein phosphatase type 1 regulator activity
SAE3	Meiosis-specific protein	Meiosis-specific protein involved in DMC1-dependent meiotic recombination
SPC24	Spindle pole component	Structural constituent of cytoskeleton
SRD1	Zinc finger motif protein	Processing of pre-rRNA to mature rRNA
Transcription		
STP1	Zinc finger motif protein	Specific RNA polymerase II transcription factor activity
STP2	Transcription factor	Specific RNA polymerase II transcription factor activity
RLM1	MADS box transcription factor	DNA bending, DNA binding, and transcriptional activator activities
HUL5	Ubiquitin ligase	Ubiquitin-protein ligase activity
Protein synthesis and fate		
VAM6	Vacuolar protein	Rab-guanyl nucleotide exchange factor activity
PAI3	Inhibitor of proteinase Pep4p	Endopeptidase inhibitor activity
SRT1	*Cis*-Prenyltransferase	Dehydrodolichyl-diphosphate-sintase activity
PMT5	Dolichyl phosphate-D-mannose:protein O-D-mannosyltransferase	Dolichyl-phosphate-mannose-protein mannosyltransferase activity
RPL31A	Ribosomal protein L31A	Structural constituent of ribosome

[a]Adapted from *FEBS Letters* (23) with permission of the publisher.

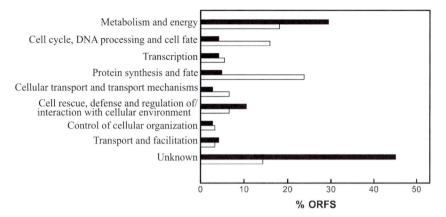

Figure 4. Global gene expression profile in functional categories. Black bars and white bars represent the percentages of induced and repressed genes, respectively. The classification is based on the MIPS database, available on the web. Reprinted from *FEBS Letters* (23) with permission of the publisher.

immediately regulated by pressure or cold, but it is induced as a late response to the cellular damages caused by those stresses.

FROM GENE TO PROTEIN

Life at extreme environmental conditions requires a number of molecular adaptations, some of which are just beginning to be understood. Nevertheless, protein stability is certainly one of the most important factors, and most cells rely on the accumulation of solutes and/or of a special class of proteins to maintain this stability, mainly during a stress condition. The mechanisms are not fully known, but many stabilizing solutes do not bind to proteins but are excluded from the protein's hydration layer, leading the protein to fold up more compactly. Other stabilizers have a more direct interaction and enhance native protein conformation through electrostatic interactions. Cells accumulate "compatible solutes" in response to stress as a way to maintain cell volume or to stabilize macromolecules, or as antioxidants. Many different small molecules are known to serve as organic osmolytes and other compatible solutes, which are small carbohydrates (sugars, polyols, and derivatives), amino acids and derivatives, etc. In nature, stabilizing ability seems to be used only when there are stresses that directly destabilize macromolecules and membranes, such as perturbing solutes, anhydrobiosis, high temperature, freezing, and HHP. However, as far as studies suggest, some solutes ("piezolyte") can enhance survival of deep-sea organisms (77), yet this is not conclusive for yeast cells.

The synthesis of many proteins transiently decreases after a stress. Nevertheless, a special class of proteins, formerly known as heat shock proteins but widely called stress proteins, are specifically translated in response to an alteration in the environmental condition. Many stress-inducible proteins are also synthesized at lower levels under optimal growth conditions, but some are expressed only under suboptimal conditions. Several conserved heat shock proteins fall into a class of protein referred to as molecular chaperones, which function to facilitate protein folding or to maintain conformation. Others have functions that include ubiquitin- and proteasome-dependent degradation and the synthesis of the thermoprotectant trehalose (17, 58).

Protein synthesis is indeed one of the most piezosensitive cellular functions. As HHP dissociates and inactivates ribosomes, this has been considered the major reason for protein synthesis inhibition by pressure. As far as protein synthesis is completely blocked at 67 MPa in organisms that are not adapted to high-pressure environments, high pressures induce the transcription and translation of a specific set of genes encoding proteins related to stress response. Also, besides protein synthesis inhibition, high pressure causes protein denaturation and dissociation, both of which are reversible after pressures up to 100 to 300 MPa (47).

Results from cross-protection experiments, specifically with pressure pretreatment inducing resistance to a subsequent more severe stress, led my group to suggest that genes responsible for stress-inducible proteins that are up-regulated after HHP are unable to synthesize their proteins due to the inhibition of the protein synthesis apparatus during pressurization. When the cells return to atmospheric pressure after a sublethal pressure treatment, the alterations of organelles and of biological processes are rapidly reversed, and only then will the newly transcribed mRNAs be translated (51).

It is important to note that the presence of proteins that protect cells against the deleterious effect of a stress is not solely dependent on transcription and translation. The analysis of mRNA may not be a direct reflection of protein content or of protein activation in the cell. Actually, many studies had shown that there is no clear correlation between mRNA and protein expression levels (1, 30, 33).

Regarding the signaling pathways implicated in pressure response, indeed, they are immediately activated, leading to the induction and repression of a variety of genes, as mentioned above, whose products confer protection on the yeast cell. This feature can be seen not only by the microarray results but also from cross-protection studies. Nevertheless, exactly which pathways are involved is still under investigation.

As previously reported (24, 52), the pressure protection induced by heat pretreatment is substantially higher than the one achieved by a previous mild pressure condition. One explanation of this observation is that high temperature stress triggers a broader cell response, which might contribute to the pressure protection. In *S. cerevisiae* there are three major transcriptional control elements that are activated under different stressful conditions: heat shock elements, which bind the heat shock transcription factor (Hsf1); the oxidative stress AP-1 response elements recognized by Yap1/Skn7; and the stress response elements (STREs) which are controlled by Msn2/4 transcription factors (21). Gene expression after a diverse range of stressful conditions is commonly regulated by those transactivators (Msn2/Msn4p) that bind to STREs that occur in a large number of gene promoters. However, some genes regulated by Msn2/Msn4p, like *HSP12* and *HSP26,* are expressed under pressure, while others, like those implicated in trehalose metabolism, are not. Also, pressure is a particular condition in which *HSP30* is up-regulated when *HSP12* and *HSP26* are too. *HSP30* activation is not related to Msn2/Msn4p or other classical stress transcription factors (63).

Recent studies with an *msn2/4* strain (*MAT*a *ade2 can1 his3 leu2 trp1 ura3 msn2-D3::HIS3 msn4-1::TRP1* [22]) show that this strain is more sensitive to HHP than the wild type, and as observed for heat treatment (9, 45), the sensitivity is enhanced with increasing pressure or exposure time. Also, an interesting feature of this strain is that it does not acquire piezotolerance. Logarithmically growing *msn2/4* cells submitted to a low hydrostatic pressure treatment prior to a high pressure show no enhancement in their survival, even after a 15-min recovery period at atmospheric pressure. On the other hand, heat-treated *msn2/4* cells acquire pressure tolerance (18). Moreover, all genes dependent upon Msn2/4 for their induction are repressed by elevated intracellular levels of cyclic AMP (cAMP), indicating that cAMP mediates Msn2/4 regulation of these genes (10). Also, high cAMP levels prevent stress-induced activation by hyperphosphorylation of Msn2/4 (27). Furthermore, artificial increases in cAMP levels in a strain carrying an *rca1* mutation in the *PDE2* gene, which allows an increase in the intracellular level of cAMP by adding cAMP to the medium, and a *lacZ* reporter gene under the control of four STRE motifs (10, 75) abolished the induction of β-galactosidase by hydrostatic pressure, which increased 12-fold after pressure treatment in the absence of cAMP (18). These observations indicate that Msn2/Msn4p factors might be the main transcriptional controllers for HHP stress.

Nitric oxide (NO) is a widespread signaling molecule involved in the regulation of a large number of cellular functions (48), and a possible role for NO as a mediator of stress response in *S. cerevisiae* has been examined (14, 19). A cytoprotection is clearly observed when yeast cells are treated with a 1 mM concentration of NO donors sodium nitroprusside and *S*-nitroso-*N*-acetylpenicillamine. Furthermore, *S. cerevisiae* cells at early exponential phase submitted to pressure at 50 MPa showed an induction in the expression level of the nitric oxide synthase-inducible isoform (NOS_2). In contrast, regarding NO-induced thermotolerance, a heat pretreatment does not lead to an induction of any NOS isoform. The intracellular concentration of NOS_2 remains the same after a piezotreatment during first exponential growth, but decreases during heat pretreatment (19). This observation not only confirms that hydrostatic pressure induces a different stress response than heat but also suggests that NO may play a role as a signal molecule in the hydrostatic pressure stress-induced pathway. It has been observed that NO interacts with and enhances the activity of the transcription factor Ace1p (14), but no other results, to our knowledge, have been reported for other transcription factors involved in the NO signaling stress response in *S. cerevisiae*.

CONCLUDING REMARKS

Yeasts are single-celled organisms subjected to environmental changes and must have a way to rapidly adjust their metabolism to cope with that. In addition, *S. cerevisiae* is economically important in the baking and brewing industries. So, when "at work" they have to tolerate an environment going from high sugar to high ethanol concentrations, besides high temperature and moderate pressure in big fermentation tanks. Moreover, yeasts used for making bread, wine, or beer are often supplied commercially as a dry powder, facing dehydration stress.

There are numerous advantages in using yeasts, particularly *Saccharomyces cerevisiae,* to study the effects of HHP in cells. As a start, *S. cerevisiae* is the best-characterized eukaryotic cell, also at the molecular level. Furthermore, it is quite impressive how the basic structures and processes have been well conserved throughout the eukaryotic life. Finally, most food spoilage fungi belong to the ascomycetes and are thus closely related to *S. cerevisiae*.

Research on the effects of pressure as a fundamental thermodynamic parameter on the yeast cell constitutes a very exciting field of the biological sciences, and in this chapter I discussed some remarkable and recent results in this area.

Returning to the three axes that sustain life, I first considered the source of energy, and it is clear that pressure stress, as most stresses studied in *S. cerevisiae* so far, necessitates extra energy generation to handle the new situation, as seen by the induction of genes related to glycolysis. Also, considering the need of stressed cells for circular flux of carbon to rapidly buffer and manage energy and osmotic stability, genes involved in gluconeogenesis and glycogen metabolism are up-regulated too. In addition, genes involved in amino acid and nucleotide anabolism are down-regulated.

Water is the second axis, and as extensively reviewed by Mentré and Hui Bon Hoa (47), the effect of HHP on living cells is a direct consequence of the specific properties of structured water associated with macromolecules. Under pressure the proportion of structured water generally increases, therefore inducing cell matrices, like cell membranes, to be more ordered and rigid. The plasma membrane appears to play a key role in the defense against pressure shock, and phase transition induced by HHP could act as a signal to trigger transduction cascades that lead to the induction of many pressure shock or stress proteins to sustain cellular integrity. Thus, one is brought to the third axis: the range of conditions an organism can tolerate. This is a new field and much work has to be done, such as to correlate the transcripts already seen with the active proteins effectively synthesized in pressure-treated yeast cells. Generally, only a fraction of the transcriptionally regulated genes shows a parallel response at protein level.

It should be noted that in nature, organisms are usually faced with a combination of physical, chemical, and biological challenges, and for that reason studies on the mechanisms of cellular stress responses need to consider the interaction between different stresses. On the other hand, biotechnological research aimed at the use of HHP as a novel unit operation in the food and pharmaceutical industry should focus on efficient killing of microorganisms, avoiding the emergence of pressure-resistant populations.

Recently, Smits and Brul (67) published a review considering the importance of the knowledge of stress resistance and adaptation in *S. cerevisiae* to the application of treatments to inhibit hazardous fungi in food manufacture. The authors emphasized the importance of understanding the physiology of yeast cells under conditions of stress and at different stages of development. Combining physiological studies with an up-to-date molecular approach (the "-omics" era) can give us a detailed understanding of the strengths and weaknesses of eukaryotic organisms.

Acknowledgments. Thanks are due to George Lindsey from Cape Town University, Cape Town, South Africa, for so many fruitful discussions and, moreover, for letting me use his yet-to-be-published results.

I am grateful to the Brazilian agencies CAPES, CNPq, FAPES, FINEP, and FUNDECI-BNB for support.

REFERENCES

1. **Abbott, A.** 1999. A post-genomic challenge: learning to read patterns of protein synthesis. *Nature* **402:**715–720.
2. **Abe, F.** 2004. Piezophysiology of yeast: occurrence and significance. *Cell. Mol. Biol.* **50:**437–445.
3. **Alexandre, H., V. Ansanay-Galeote, S. Dequin, and B. Blondin.** 2001. Global gene expression during short-term ethanol stress in *Saccharomyces cerevisiae*. *FEBS Lett.* **498:**98–103.
4. **Ayscough, K., N. M. Hajibagheri, R. Watson, and G. Warren.** 1993. Stacking of Golgi cisternae in *Schizosaccharomyces pombe* requires intact microtubules. *J. Cell Sci.* **106:**1227–1237.

5. **Bard, M., D. A. Bruner, C. A. Pierson, N. D. Lees, B. Biermann, L. Frye, C. Koegel, and R. Barbuch.** 1996. Cloning and characterization of ERG25, the *Saccharomyces cerevisiae* gene encoding C-4 sterol methyl oxidase. *Proc. Natl. Acad. Sci. USA* **93:**186–190.

6. **Bartlett, D., M. Glaser, and R. Welti.** 1997. Membrane penetration depth and lipid phase preference of acyl-labeled dansyl phosphatidylcholines in phosphatidylcholine vesicles. *Biochim. Biophys. Acta* **1328:**48–54.

7. **Bartlett, D. H.** 2002. Pressure effects on in vivo microbial processes. *Biochim. Biophys. Acta* **1595:**367–381.

8. **Botstein, D., D. Amberg, J. Mulholland, T. Huffaker, A. Adams, D. Dubrim, and T. Stearns.** 1997. The yeast cytoskeleton, p. 1–90. *In* E. W. Jones, J. R. Pringle, and J. R. Broach (ed.), *The Molecular and Cellular Biology of the Yeast Saccharomyces,* vol. 3. *Cell Cycle and Cell Biology.* Cold Spring Harbor Laboratory Press, Cold Spring Harbor, NY.

9. **Boy-Marcotte, E., G. Lagniel, M. Perrot, F. Bussereau, A. Boudsocq, M. Jacquet, and J. Labarre.** 1999. The heat shock response in yeast: differential regulations and contributions of the Msn2p/Msn4p and Hsf1p regulons. *Mol. Microbiol.* **33:**274–283.

10. **Boy-Marcotte, E., D. Tadi, M. Perrot, H. Boucherie, and M. Jacquet.** 1996. High cAMP levels antagonize the reprogramming of gene expression that occurs at the diauxic shift in *Saccharomyces cerevisiae*. *Microbiology* **142:**459–467.

11. **Cartwright, C. P., F. J. Veazey, and A. H. Rose.** 1987. Effect of ethanol on activity of the plasma-membrane ATPase and accumulation of glycine by *Saccharomyces cerevisiae*. *J. Gen. Microbiol.* **133:**857–865.

12. **Chen, C., and C. W. Tseng.** 1997. Effect of high hydrostatic pressure on the temperature dependence of *Saccharomyces cerevisiae* and *Zygosaccharomyces rouxii*. *Process Biochem.* **32:**337–343.

13. **Chi, Z., and N. Arneborg.** 1999. Relationship between lipid composition, frequency of ethanol-induced respiratory deficient mutants, and ethanol tolerance in *Saccharomyces cerevisiae*. *J. Appl. Microbiol.* **86:**1047–1052.

14. **Chiang, K. T., M. Shinyashiki, C. H. Switzer, J. S. Valentine, E. B. Gralla, D. J. Thiele, and J. M. Fukudo.** 2000. Effects of nitric oxide on the copper-responsive transcription factor Ace1 in *Saccharomyces cerevisiae*: cytotoxic and cytoprotective actions of nitric oxide. *Arch. Biochem. Biophys.* **377:**296–303.

15. **Costa, V., and P. Moradas-Ferreira.** 2001. Oxidative stress and signal transduction in *Saccharomyces cerevisiae*: insights into ageing, apoptosis and diseases. *Mol. Asp. Med.* **22:**217–246.

16. **Costa, V., E. Reis, A. Quintanilha, and P. Moradas-Ferreira.** 1993. Acquisition of ethanol tolerance in *Saccharomyces cerevisiae*: the key role of the mitochondrial superoxide dismutase. *Arch. Biochem. Biophys.* **300:**608–614.

17. **Craig, E.A.** 1992. The heat shock response of *Saccharomyces cerevisiae,* p. 501–537. *In* E. W. Jones, J. R. Pringle, and J. R. Broach (ed.), *The Molecular and Cellular Biology of the Yeast Saccharomyces,* vol. 2. *Gene Expression.* Cold Spring Harbor Laboratory Press, Cold Spring Harbor, NY.

18. **Domitrovic, T., P. M. B. Fernandes, and E. Kurtenbach.** 2005. Hydrostatic pressure induces transcription via the stress response element (STRE) of *Saccharomyces cerevisiae*. Proceedings of the XXXIV Annual Meeting of the Brazilian Society of Biochemistry and Molecular Biology. CD-ROM.

19. **Domitrovic, T., F. L. Palhano, C. Barja-Fidalgo, M. DeFreitas, M. T. D. Orlando, and P. M. B. Fernandes.** 2003. Role of nitric oxide in the response of *Saccharomyces cerevisiae* cells to heat shock and high hydrostatic pressure. *FEMS Yeast Res.* **3:**341–346.

20. **Dubrin, D. G., H. D. Jones, and K. F. Wertman.** 1993. Actin structure and function: roles in mitochondrial organization and morphogenesis in budding yeast and identification of the phalloidin-binding site. *Mol. Biol. Cell* **4:**1277–1294.

21. **Estruch, F.** 2000. Stress-controlled transcription factors, stress-induced genes and stress tolerance in budding yeast. *FEMS Microbiol Rev.* **24:**469–486.

22. **Estruch, F., and M. Carlson.** 1993. Two homologous zinc finger genes identified by multicopy suppression in a SNF1 protein kinase mutant of *Saccharomyces cerevisiae*. *Mol. Cell. Biol.* **13:**3872–3881.

23. **Fernandes, P. M. B., T. Domitrovic, C. M. Kao, and E. Kurtenbach.** 2004. Genome expression pattern in *Saccharomyces cerevisiae* cells in response to high hydrostatic pressure. *FEBS Lett.* **556:**153–160.

24. **Fernandes, P. M. B., M. Farina, and E. Kurtenbach.** 2001. Effect of hydrostatic pressure on the morphology and ultrastructure of wild-type and trehalose synthase mutant cells of *Saccharomyces cerevisiae*. *Lett. Appl. Microbiol.* **32:**42–46.

25. **Fernandes, P. M. B., A. D. Panek, and E. Kurtenbach.** 1997. Effect of hydrostatic pressure on a mutant of *Saccharomyces cerevisiae* deleted in the trehalose-6-phosphate synthase gene. *FEMS Microbiol. Lett.* **152:**17–21.

26. **Gabriel, M., and M. Kopecká.** 1995. Disruption of the actin cytoskeleton in budding yeast results in formation of an aberrant cell wall. *Microbiology* **141:**891–899.

27. **Garreau, H., R. N. Hasan, G. Renault, F. Estruch, E. Boy-Marcotte, and M. Jacquet.** 2000. Hyperphosphorylation of Msn2p and Msn4p in response to heat shock and the diauxic shift is inhibited by cAMP in *Saccharomyces cerevisiae*. *Microbiology* **146:**2113–2120.

28. **Gasch, A. P., P. T. Spellman, C. M. Kao, O. Carmel-Harel, M. B. Eisen, G. Storz, D. Botstein, and P. O. Brown.** 2000. Genomic expression programs in the response of yeast cells to environmental changes. *Mol. Biol. Cell* **11:**4241–4257.

29. **Gross, M., and R. Jaenicke.** 1994. Proteins under pressure. The influence of high hydrostatic pressure on structure, function and assembly of proteins and protein complexes. *Eur. J. Biochem.* **221:**617–630.

30. **Gygi, S. P., Y. Rochon, B. R. Franza, and R. Aebersold.** 1999. Correlation between protein and mRNA abundance in yeast. *Mol. Cell. Biol.* **19:**1720–1730.

31. **Hartmann, C., and A. Delgado.** 2004. Numerical simulation of the mechanics of a yeast cell under high hydrostatic pressure. *J. Biomech.* **37:**977–987.

32. **Heremans, K. A. H.** 1982. High pressure effects upon proteins and other biomolecules. *Annu. Rev. Biophys. Bioeng.* **11:**1–21.

33. **Ideker, T., V. Thorsson, J. A. Ranish, R. Christmas, J. Buhler, J. K. Eng, R. Bumgarner, D. R. Goodlett, R. Aebersold, and L. Hood.** 2001. Integrated genomic and proteomic analyses of a systematically perturbed metabolic network. *Science* **292:**929–934.

34. **Iwahashi, H., S. C. Kaul, K. Obuch, and Y. Komatsu.** 1991. Induction of barotolerance by heat shock treatment in yeast. *FEMS Microbiol. Lett.* **80:**325–328.

35. **Iwahashi, H., S. Nwaka, and K. Obuchi.** 2000. Evidence for contribution of neutral trehalase in barotolerance of *Saccharomyces cerevisiae*. *Appl. Environ. Microbiol.* **66:**5182–5185.

36. **Iwahashi, H., H. Shimizu, M. Odani, and Y. Komatsu.** 2003. Piezophysiology of genome wide gene expression levels in the yeast *Saccharomyces cerevisiae*. *Extremophiles* **7:**291–298.

37. **Jamieson, D. J.** 1998. Oxidative stress responses of the *Saccharomyces cerevisiae*. *Yeast* **14:**1511–1527.

38. **Jones, R. P.** 1989. Biological principles for effects of ethanol. *Enzyme Microb. Technol.* **11:**130–153.

39. **Kajiwara, S., A. Shirai, T. Fujii, T. Toguri, K. Nakamura, and K. Ohtaguchi.** 1996. Polyunsaturated fatty acid biosynthesis in *Saccharomyces cerevisiae:* expression of ethanol tolerance and the *FAD2* gene from *Arabidopsis thaliana*. *Appl. Environ. Microbiol.* **62:**4309–4313.

40. **Karreman, R. J., W. F. Brandt, and G. G. Lindsey.** 2005. The yeast *Saccharomyces cerevisiae* stress response protein HSP 12 decreases the gel strength of agarose used as a model system for the β-glucan layer of the cell wall. *Carbohydr. Polym.* **60:**193–198.

41. **Karreman, R. J., E. Dague, F. Gaboriaud, F. Quilès, J. F. Duval, and G. G. Lindsey.** 2007. The stress response protein Hsp12p increases the flexibility of the yeast *Saccharomyces cerevisiae* cell wall. *Biochim. Biophys. Acta* **1774:**131–137.

42. **Kobori, H., M. Sato, A. Tameike, K. Hamada, S. Shimada, and M. Osumi.** 1996. Changes in microfilaments and microtubules of yeasts induced by pressure stress, p. 83–94. *In* R. Hayashi and C. Balny (ed.), *High Pressure Bioscience and Biotechnology.* Elsevier Science, B. V. Amsterdam, The Netherlands.

43. **Lammi, M. J., M. A. Elo, R. K. Sironen, H. M. Karjalainen, K. Kaarniranta, and H. J. Helminen.** 2004. Hydrostatic pressure-induced changes in cellular protein synthesis. *Biorheology* **41:**309–313.

44. **Macdonald, A. G.** 1984. The effects of pressure on the molecular structure and physiological functions of cell membranes. *Philos. Trans. R. Soc. Lond. B* **304:**47–68.

45. **Martinez-Pastor, M. T., G. Marchler, C. Schuller, A. Marchler-Bauer, H. Ruis, and F. Estruch.** 1996. The *Saccharomyces cerevisiae* zinc finger proteins Msn2p and Msn4p are required for transcriptional induction through the stress response element (STRE). *EMBO J.* **15:**2227–2235.

46. **Mentre, P., L. Hamraoui, G. Hui Bon Hoa, and P. Debey.** 1999. Pressure-sensitivity of endoplasmic reticulum membrane and nucleolus as revealed by electron microscopy. *Cell. Mol. Biol.* **45:**353–362.

47. **Mentré, P., and G. Hui Bon Hoa.** 2001. Effects of high hydrostatic pressures on living cells: a consequence of the properties of macromolecules and macromolecule-associated water. *Int. Rev. Cytol.* **201:**1–84.

48. **Monkada, S., R. M. J. Palmer, and E. A. Higgs.** 1991. Nitric oxide: physiology, pathophysiology and pharmacology. *Pharmacol. Rev.* **43:**109–141.

49. **Motshwene, P., R. Karreman, G. Kgari, W. Brandt, and G. Lindsey.** 2004. LEA (late embryonic abundant)-like protein Hsp 12 (heat-shock protein 12) is present in the cell wall and enhances the barotolerance of the yeast *Saccharomyces cerevisiae*. *Biochem. J.* **377:**769–774.

50. **Osumi, M., M. Sato, H. Kobori, Z. H. Feng, S. A. Ishijima, K. Hamada, and S. Shimada.** 1996. Morphological effects of pressure stress on yeasts, p. 37–46. *In* R. Hayashi and C. Balny (ed.), *High Pressure Bioscience and Biotechnology.* Elsevier Science, B. V. Amsterdam, The Netherlands.

51. **Palhano, F. L., H. L. Gomes, M. T. D. Orlando, E. Kurtenbach, and P. M. B. Fernandes.** 2004. Pressure response in the yeast *Saccharomyces cerevisiae:* from cellular to molecular approaches. *Cell. Mol. Biol.* **50:**447–457.

52. **Palhano, F. L., M. T. D. Orlando, and P. M. B. Fernandes.** 2004. Induction of baroresistance by hydrogen peroxide, ethanol and cold-shock in *Saccharomyces cerevisiae. FEMS Microbiol. Lett.* **233:**139–145.

53. **Penninckx, M. J.** 2002. An overview on glutathione in *Saccharomyces* versus non-conventional yeasts. *FEMS Yeast Res.* **2:**295–305.

54. **Perrier-Cornet, J. M., M. Hayert, and P. Gervais.** 1999. Yeast cell mortality related to a high-pressure shift: occurrence of cell membrane permeabilization. *J. Appl. Microbiol.* **87:**1–7.

55. **Perrier-Cornet, J. M., S. Tapin, S. Gaeta, and P. Gervais.** 2005. High-pressure inactivation of *Saccharomyces cerevisiae* and *Lactobacillus plantarum* at subzero temperatures. *J. Biotechnol.* **115:**405–412.

56. **Piper, P. W.** 1997. The yeast heat shock response, p. 75–89. *In* W. H. Mager and S. Hohmann (ed.), *Yeast Stress Response.* R. G. Landes Co., Austin, TX.

57. **Quinn, P. J., F. Joo, and L. Vigh.** 1989. The role of unsaturated lipids in membrane structure and stability. *Prog. Biophys. Mol. Biol.* **53:**71–103.

58. **Riezman, H.** 2004. Why do cells require heat shock proteins to survive heat stress? *Cell Cycle* **3:**61–63.

59. **Sahara, T., T. Goda, and S. Ohgiya.** 2002. Comprehensive expression analysis of time-dependent genetic responses in yeast cells to low temperature. *J. Biol. Chem.* **277:**50015–50021.

60. **Sajbidor, J., and J. Grego.** 1992. Fatty acid alterations in *Saccharomyces cerevisiae* exposed to ethanol stress. *FEMS Microbiol. Lett.* **93:**13–16.

61. **Sales, K., W. Brandt, E. Rumbak, and G. Lindsey.** 2000. The LEA-like protein HSP12 in *Saccharomyces cerevisiae* has a plasma membrane location and protects membranes against desiccation and ethanol-induced stress. *Biochim. Biophys. Acta* **1463:**267–278.

62. **Sanchez, Y., J. Taulin, K. A. Borkovich, and S. Lindquist.** 1992. Hsp104 is required for tolerance to many forms of stress. *EMBO J.* **11:**2357–2364.

63. **Seymour, I. J., and P. W. Piper.** 1999. Stress induction of HSP30, the plasma membrane heat shock protein gene of *Saccharomyces cerevisiae,* appears not to use known stress-regulated transcription factors. *Microbiology* **145:**231–239.

64. Reference deleted.

65. **Siderius, M., and W. H. Mager.** 1997. General stress response: in search of a common denominator, p. 213–230. *In* W. H. Mager and S. Hohmann (ed.), *Yeast Stress Response.* R. G. Landes Co., Austin, TX.

66. **Singer, M. A., and S. Lindquist.** 1998. Thermotolerance in *Saccharomyces cerevisiae:* the Yin and Yang of trehalose. *Trends Biotechnol.* **16:**460–468.

67. **Smits, G. J., and S. Brul.** 2005. Stress tolerance in fungi—to kill a spoilage yeast. *Curr. Opin. Biotechnol.* **16:**225–230.

68. **Suzuki, K., Y. Miyosaka, and Y. Taniguchi.** 1971. The effect of pressure on deoxyribonucleic acid. *J. Biochem.* **69:**595–598.

69. **Swan, T. M., and K. Watson.** 1998. Stress tolerance in a yeast sterol auxotroph: role of ergosterol, heat shock proteins and trehalose. *FEMS Microbiol. Lett.* **169:**191–197.

70. **Takahashi, K., T. Kubo, K. Kobayashi, J. Imanishi, M. Takigawa, Y. Arai, and Y. Hirasawa.** 1997. Hydrostatic pressure influences mRNA expression of transforming growth factor-beta 1 and heat shock protein 70 in chondrocyte-like cell line. *J. Orthop. Res.* **15:**150–158.

71. **Tamura, K., M. Miyashita, and H. Iwahashi.** 1998. Stress tolerance of pressure-shocked *Saccharomyces cerevisiae. Biotechnol. Lett.* **20:**1167–1169.

72. **Varela, J. C. S., and W. H. Mager.** 1996. Response of *Saccharomyces cerevisiae* to changes in external osmolarity. *Microbiology* **142:**721–731.

73. **Weber, G., and H. G. Drickamer.** 1983. The effect of high pressure upon proteins and other biomolecules. *Q. Rev. Biophys.* **16:**89–112.

74. **Wharton, D. A.** 2002. *Life at the Limits. Organisms in Extreme Environments.* University Press, Cambridge, United Kingdom.

75. **Wilson, B. A., M. Khalil, F. Tamanoi, and J. F. Cannon.** 1993. New activated Ras2 mutations identified in *Saccharomyces cerevisiae. Oncogene* **8:**3441–3445.

76. **Yale, J., and H. J. Bohnert.** 2001. Transcript expression in *Saccharomyces cerevisiae* at high salinity. *J. Biol. Chem.* **276:**15996–16007.

77. **Yancey, P. H.** 2005. Organic osmolytes as compatibles, metabolic and counteracting cytoprotectants in high osmolarity and other stresses. *J. Exp. Biol.* **208:**2819–2830.

78. **You, K. M., C. Rosenfield, and D. C. Knipple.** 2003. Ethanol tolerance in the yeast *Saccharomyces cerevisiae* is dependent on cellular oleic acid content. *Appl. Environ. Microbiol.* **69:**1499–1503.

79. **Zara, S., G. Antonio Farris, M. Budroni, and A. T. Bakalinsky.** 2002. Hsp12 is essential for biofilm formation by a Sardinian wine strain of *S. cerevisiae. Yeast* **19:**269–276.

80. **Zimmerman, A. M.** 1971. High pressure studies in cell biology. *Int. Rev. Cytol.* **30:**1–47.

81. **Zimmerman, A. M.** 1970. *High Pressure Effects on Cellular Processes.* Academic Press, New York, NY.

High-Pressure Microbiology
Edited by C. Michiels, D. H. Bartlett, and A. Aertsen
© 2008 ASM Press, Washington, DC

Chapter 9

Effects of Growth-Permissive Pressures on the Physiology of *Saccharomyces cerevisiae*

Fumiyoshi Abe

The discovery of piezophiles that prefer pressures greater than atmospheric pressure for growth prompted researchers to investigate the survival strategies they employ in high-pressure environments (36, 57). While molecular adaptation to high-pressure environments has been extensively analyzed in various marine prokaryotes (10, 12, 29), a number of pioneering studies have investigated eukaryotic cell division, protein synthesis, and cellular enzyme activities under high-pressure conditions, yielding knowledge on the fundamental aspects of the effects of pressure on eukaryotes (16, 34, 51, 58). There have been renewed high-pressure studies on biological processes using the modern techniques of genetics and molecular biology (2, 8, 10, 12, 38). This chapter focuses on the effects of growth-permissive pressures of less than 50 MPa on the growth and physiology of the yeast *Saccharomyces cerevisiae*.

Hydrostatic pressure affects a variety of biological processes in living cells depending on the magnitude and duration of applied pressure in combination with temperature, pH, oxygen supply, and composition of culture media. Accordingly, the effects are very complex, sometimes making interpretations difficult. One of the limitations in propagating microorganisms under high-pressure conditions is the supply of oxygen when aerobes are placed in closed hydrostatic chambers. In addition, it is generally quite difficult to manipulate the genome of microbes isolated from natural environments to introduce or disrupt the desired genes. The yeast *S. cerevisiae* is a facultative anaerobe and is one of the best-characterized eukaryotes. The complete genome sequence has been released (*Saccharomyces* Genome Database [http://www.yeastgenome.org/]), and powerful genetic tools are readily available for analyses. The yeast genome contains approximately 6,000 genes, including more than 4,800 nonessential ones. Recent large-scale phenotypic screens of the *S. cerevisiae* deletion mutant collection (Yeast Deletion Clones; Invitrogen, Carlsbad, CA) have revealed numerous unexpected genes and metabolic pathways that are involved in the tolerance to environmental stresses (14, 37). Accordingly, systematic analyses using this organism make a great contribution to obtaining a more complete picture of the effects of

Fumiyoshi Abe • Extremobiosphere Research Center, Japan Agency for Marine-Earth Science and Technology (JAMSTEC), 2-15 Natsushima-cho, Yokosuka 237-0061, Japan.

high pressure and establishing the molecular basis of pressure effects. Here I review recent advances in explorations of the effects of growth-permissive pressures on the growth and physiology of *S. cerevisiae*. Investigations of the effects of lethal levels of hydrostatic pressure, generally greater than 100 MPa, are reviewed in chapter 8. To date, few reports have been published regarding yeast physiology at growth-permissive pressures, and many of the results in this chapter were derived from my laboratories. Table 1 summarizes the phenomena occurring in yeast cells at pressures equal or less than 50 MPa.

HIGH-PRESSURE CULTIVATION AND MICROSCOPIC OBSERVATIONS OF *S. CEREVISIAE*

The simplest and most convenient system for high-pressure cultivation of *S. cerevisiae* is a pressure syringe, generally made of stainless steel or titanium, with a diameter of approximately 10 cm, length of 30 cm, and internal volume of about 500 ml, which can typically be used at pressures up to 200 MPa. Individual syringes should have a pressure gauge to check for pressure leaks during pressurization. In general, to examine the effects of high pressure on growth, exponentially growing yeast cells (optical density at 600 < 1.0) in rich or minimal medium are placed in sterilized polypropylene tubes in a volume of 1.5 to 50 ml. The tubes are sealed with Parafilm to transmit low pressures, but the caps should not be screwed on tightly. A hand pump can be used for pressurization. The adiabatic temperature increase due to rapid compression is usually negligible (2 to 3°C increase/100 MPa). When pressures up to 100 MPa are applied, rapid compression and decompression by approximately 10 MPa/s do not cause a marked loss of yeast cell viability. Some labo-

Table 1. Effects of growth-permissive pressure on the growth and physiology of *Saccharomyces cerevisiae*[a]

Pressure	Effect	Note	Reference(s)
25 MPa	Arrest of growth	Wild-type Trp$^-$ strains	6
	HPG	Wild-type Trp$^+$ strains	6
	HPG	HPG1, HPG2, HPG3, and HPG4 strains (Trp$^-$)	6, 9, 41
	HPG	doa4Δ, ubp6Δ, and ubp14Δ strains (Trp$^-$)	39
	Alteration of glycolytic intermediate levels	G6P[b] increases.	7
	Metabolic change	Calorimetry	53
	Inhibition of amino acid uptake	Severity, Trp > Lys > His > Leu	6, 9
	Pressure-inducible gene expression	Growth at 25 or 30 MPa	27, 40
	Enhancement of esterase activity	Nonspecific esterases; application to flow cytometry	1
50 MPa	Arrest of growth	All *S. cerevisiae* strains tested	6, 9
	Alteration of glycolytic intermediate levels	FBP[c] decreases.	7
	Inhibition of ethanol fermentation	Internal ATP level is unchanged.	4, 5
	Vacuolar and cytoplasmic acidification	Fluorescence analysis; internal pH decreases by 0.3–0.5 units	3–5

[a]It should be noted that most effects are dependent on time, strain, growth phase, and analytical procedure.
[b]G6P, glucose-6-phosphate.
[c]FBP, fructose-1,6-bisphosphate.

ratories have built specialized reactors for continuous high-pressure cultivation of microorganisms (42, 55).

An optical device was developed to allow the microscopic observation of *S. cerevisiae* during high-pressure treatment (44). Cell division of *S. cerevisiae* is retarded, with a lag phase of 2 h even at atmospheric pressure due to the stringent experimental conditions required for visualization (e.g., immobilization of the cells, oxygen deficiency, and accumulation of carbon dioxide). The lag phase is increased at 10 MPa, and cell division is no longer observed after incubation for 16 h. Whereas no morphological changes in the cells are observed at 10 MPa, the cell volume is decreased to 85 to 90% of the original volume at the higher pressure of 250 MPa (44). The decrease in cell volume is attributable to the leakage of internal solutes. Analysis of individual cells using flow cytometry indicates that cells are slightly enlarged when cultured at pressures of 30 and 40 MPa (27). Thus, the cell size of *S. cerevisiae* may be controlled at 30 to 40 MPa in a manner different from that at atmospheric pressure.

The fission yeast *Schizosaccharomyces pombe* is more sensitive to high pressure than *S. cerevisiae*. Even at 50 MPa, the cell cycle-specific actin distribution is lost in *S. pombe*, whereas the distribution is not affected in *S. cerevisiae* at pressures of up to 100 MPa (46). In higher eukaryotic cells, oligomerization of F-actin is a typical pressure-sensitive process that is accompanied by a large positive volume change (52). Comparison of pressure-induced dissociations of F-actin in vitro between *S. pombe* and *S. cerevisiae* could contribute to the identification of critical amino acid residue(s) and/or regulatory proteins in the stability of F-actin.

TRYPTOPHAN AVAILABILITY IS DIRECTLY INVOLVED IN THE EFFECTS OF PRESSURE ON YEAST CELL GROWTH

Many mesophilic bacteria are known to become filamentous, with multiple nuclei, when growing cells are exposed to pressures of 40 to 50 MPa (36; see also chapter 5). Thus, biosynthesis of DNA, RNA, and proteins in bacteria occurs at high pressure in the absence of cell division. There is no report showing filamentous growth of *S. cerevisiae* at high pressure. In yeast cells as well as other eukaryotic cells, the entry into the DNA-synthetic phase of the cell cycle is strictly regulated in the G_1 phase. When cells are exposed to moderately stressful conditions such as nutrient deprivation, heat, cold, or the presence of toxic chemicals, the cell cycle is arrested in the G_1 phase. The effects of high pressure on the cell growth of *S. cerevisiae* are closely related to tryptophan availability. Laboratory wild-type yeast strains (e.g., YPH499 or W303-1A) usually have some nutrient auxotrophies such as *trp1* (tryptophan), *leu2* (leucine), *lys2* (lysine), *his3* (histidine), *ade2* (adenine), or *ura3* (uracil) for the selection of plasmid-bearing transformants. Such strains must take up the corresponding nutrients from the medium. The uptake of tryptophan is readily impaired by increasing pressure (6). Consequently, pressures of 15 to 25 MPa severely impair the growth of tryptophan-auxotrophic (Trp$^-$) strains, leading to arrest of the cell cycle in the G_1 phase (6). In contrast, tryptophan-prototrophic (Trp$^+$) strains (e.g., X2180-1A, BY4742, and most industrial strains) are capable of growth at a pressure of 25 MPa, even though the doubling time is extended. Increasing hydrostatic pressure diminishes the uptake of amino acids in the order of tryptophan > lysine > histidine > leucine with respect to the rate of uptake in a unit of time (6). The basal rate of tryptophan uptake measured at 0.1 MPa and 24°C is much lower than that of the other three amino acids (tryptophan, 8.3 pmol/10^7 cells per min; lysine, 90.3 pmol/10^7 cells per min; hisitidine, 21.0 pmol/10^7 cells per min; and leucine, 55.4 pmol/10^7 cells per min) (9). Accordingly, the cells are likely to be starved of tryptophan during growth at high pressure. The uptake of adenine and uracil by yeast cells has not been examined at high pressure.

The addition of excess amounts of tryptophan to the medium (e.g., 1 g/liter) enables the Trp$^-$ cells to grow at 25 MPa, whereas the addition of other amino acids has no effect (6). Tryptophan uptake in *S. cerevisiae* is mediated by high-affinity-type tryptophan permease Tat2 and low-affinity-type tryptophan permease Tat1 (47). Overexpression of either Tat2 or Tat1 confers high-pressure growth on Trp$^-$ cells (6, 9). Overexpression of Tat2 or Tat1 also confers the ability to grow at the low temperature of 15°C. Upon incubation of the wild-type cells at 25 MPa, both Tat2 and Tat1 are degraded, causing a reduction in tryptophan uptake activity (6, 9). This process is actively regulated by the cellular ubiquitin system (see below).

The structure of lipid bilayers is particularly sensitive to changes in hydrostatic pressure and temperature (54). In artificial lipid bilayers such as dipalmitoylphosphatidylcholine, the temperature for the main transition (T_m) from the ripple gel phase to the liquid crystalline phase is 41.6°C at atmospheric pressure, but it is increased to 66°C at 100 MPa (24). Increasing pressure as well as decreasing temperature enhances the order of hydrocarbon chains and decreases membrane fluidity. In this sense, tryptophan uptake by yeast cells is sensitive to decreases in membrane fluidity caused by either high pressure or low temperature.

ISOLATION OF *HPG* MUTANTS

Mutants capable of growth at high pressure have been isolated from the tryptophan-auxotrophic wild-type strain YPH499 in the hope that phenotypic characterization and cloning of the corresponding mutant genes would contribute to an understanding of the regulation of tryptophan permease under high-pressure conditions (9). Cells were treated or not with 2% ethylmethanesulfonate for 10 to 30 min, so that the surviving fraction of the cells decreased to 10 to 60%. After being washed, the cells were spread on yeast-peptone-dextrose (YPD) agar and incubated at 4 to 10°C and 0.1 MPa for 1 to 3 months or subjected to pressures of 30 to 50 MPa in yeast-peptone-dextrose medium for 1 to 3 months at 24°C using hydrostatic pressure vessels. The resulting mutants are designated high-pressure-growth (*HPG*) mutants. All *HPG* mutants have acquired the ability to grow at the low temperatures of 8 to 15°C as well as to grow at pressures of 25 to 35 MPa, although the parental strain cannot grow at pressures greater than 15 MPa or temperatures lower than 15°C (9). The *HPG* mutants are classified into four semidominant linkage groups designated *HPG1*, *HPG2*, *HPG3*, and *HPG4*. The *HPG1* and *HPG2* genes were successfully identified (9, 41).

REGULATION OF TRYPTOPHAN PERMEASES BY THE UBIQUITIN SYSTEM IN RESPONSE TO HIGH PRESSURE

HPG1 appears to be allelic to *RSP5,* which encodes Rsp5 ubiquitin ligase (9). Ubiquitination is a selective degradation system of cellular proteins in eukaryotes. After ubiquitin molecules are covalently bound to target proteins, the ubiquitinated proteins are delivered to the proteasome or the vacuoles for degradation (23, 31). The four *HPG1* mutation sites (*HPG1-1* [Rsp5$^{Pro514 > Thr}$], *HPG1-2* [Rsp5$^{Cys517 > Tyr}$], *HPG1-3* [Rsp5$^{Cys517 > Phe}$], and *HPG1-4*

[Rsp5$^{Ala799 > Thr}$]) are located in the catalytic HECT (<u>h</u>omologous to <u>E</u>6-AP <u>C</u> terminus) domain of Rsp5 (9). Ubiquitination deficiency causes a remarkable stabilization of Tat1 and Tat2 in the *HPG1* mutants at a pressure of 25 MPa, whereas both permeases are degraded in a ubiquitination-dependent manner in the wild-type cells (9). Consequently, the mutant is capable of growth at high pressure. *BUL1* is a gene encoding an Rsp5-binding protein. Disruption of *BUL1* also causes a remarkable stabilization of Tat2 at high pressure, indicating that Bul1 contributes to pressure-induced degradation of tryptophan permeases (9).

Another HPG gene, *HPG2,* is allelic to *TAT2* (41). When yeast cells are treated with the immunosuppressive drug rapamycin or starved of nutrients, degradation of Tat2 is initiated by covalent binding of ubiquitin molecules to the 29th and/or 31st lysine at the N-terminal tail of the Tat2 protein (13). The *HPG2* mutation sites are located in the regulatory domains of the N-terminal (*HPG2-1* [Tat2$^{Glu27 > Phe}$]) and C-terminal (*HPG2-2* [Tat2$^{Asp563 > Asn}$] and *HPG2-3* [Tat2$^{Glu570 > Lys}$]) tails of the Tat2 protein (41). The amino acid substitutions are likely to interfere with ubiquitination on the 29th and/or 31st lysine by Rsp5 ubiquitin ligase. Consequently, the mutant forms of Tat2 are stabilized, leading to *HPG* of *HPG2* cells.

Ubiquitinated proteins undergo deubiquitination for recycling of ubiquitin prior to breakdown by the proteasomes or in the vacuoles. Deubiquitination is catalyzed by ubiquitin-specific proteases. Of the 17 potential ubiquitin-specific proteases encoded by the yeast genome, Doa4 (Ubp4), Ubp6, and Ubp14 appear to be involved in degradation of Tat2. Doa4 is known to act by facilitating ubiquitin recycling from soluble ubiquitinated substrates targeted to the proteasome and also on the late endosome/multivesicular body in deubiquitination and trafficking of plasma membrane proteins. Deletion of *DOA4* stabilizes general amino acid permease Gap1 (43) and uracil permease Fur4 (19). Ubp6 is a component of the proteasome which recognizes the proteasome base and its subunit Rpn1 (22). Ubp14 is known to catalyze disassembly of free polyubiquitin chains, which correlates with defects in ubiquitin-dependent proteolysis in the proteasome (11). Disruption of one of the three *UBP* genes stabilizes Tat2 at high pressure, and consequently the disruptants exhibit *HPG* at 25 MPa (39).

Taking all the reports together, regulation of tryptophan permease Tat2 by the ubiquitin system in response to high pressure can be illustrated as follows (Fig. 1). Upon incubation of the wild-type cells at high pressure, Tat2 is assumed to be partially denatured. Then, denatured forms of Tat2 are recognized by the Rsp5-Bul1-Bul2 ubiquitin ligase complex, followed by ubiquitination on the lysine residue(s). The ubiquitinated Tat2 is targeted mainly to the vacuoles and partly to the proteasomes for degradation. Prior to degradation, the ubiquitin-specific proteases Doa4 and Ubp6 act to remove polyubiquitin chains from the ubiquitinated Tat2. Upon the loss of any processes through the degradation pathway, Tat2 proteins are stabilized and in some mutants accumulated in the plasma membrane. Consequently, the mutant cells become endowed with the ability to grow at high pressure. In this model, it is assumed that the partially denatured form of Tat2 is still active in the *HPG* mutants. Indeed, the basal level of Tat2 protein is increased 2.2- or 3.8-fold in the *HPG1-1* mutant or the double deletion mutant for *BUL1* and *BUL2*, respectively, with an increase in tryptophan uptake of 1.4- or 1.5-fold, respectively (9). To validate this model, it is necessary to confirm the partially denatured forms of Tat2 by some physicochemical techniques and to elucidate how Rps5 recognizes the partially denatured Tat2 proteins. Tat1 is likely to be degraded in a similar manner, although the roles of Doa4, Ubp6, and Ubp14 have not been examined for the regulation of Tat1.

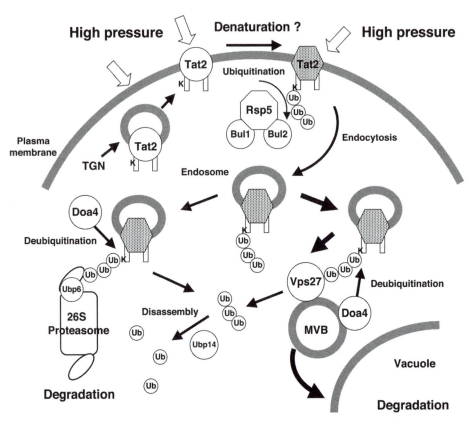

Figure 1. Model depicting the regulation of the high-affinity tryptophan permease Tat2 in response to high pressure. Upon incubation of the wild-type cells at high pressure, Tat2 is assumed to be partially denatured with retention of some activity. Then, the denatured Tat2 is recognized by the ubiquitin system, followed by degradation in the vacuoles or by the proteasomes. Upon the loss of any factors involved in the degradation pathway, Tat2 is stabilized in the plasma membrane. Consequently, the mutant cells become endowed with the ability to grow at high pressure. Ub, ubiquitin; K, lysine residue(s); Rsp5, ubiquitin ligase Rsp5; Bul1/2, binding proteins of Rsp5; Doa4, Ubp6, Ubp14, ubiquitin-specific proteases; MVB, multivesicular body; Vps27, an endosomal protein that functions at the MVB.

In this manner, if tryptophan-auxotrophic strains are used to isolate *HPG* mutants, the corresponding mutations will frequently occur on proteins involved in the degradation of Tat2. Using this approach, any factors affecting the stabilization of Tat2 through ubiquitination can be evaluated by examining *HPG* against the genetic background of tryptophan auxotrophy. It is worthwhile to examine the isolation of *HPG* mutants from nutrient-prototrophic strains.

ROLE OF HEAT SHOCK PROTEINS IN *HPG*

S. cerevisiae cells are killed by higher pressures, in the range of 100 to 200 MPa, which disrupt ultrastructures such as microtubules, actin filaments, and nuclear membranes (25, 32). The effects of short exposure of *S. cerevisiae* cells to lethal pressures are reviewed elsewhere in this volume (chapter 8) and are not discussed in detail here. Briefly, among many heat shock proteins, a molecular chaperone, Hsp104, plays an essential role in tolerance to the high pressures of 150 to 200 MPa (i.e., piezotolerance) in an ATP-dependent manner (26). Thus, Hsp104 contributes to the refolding of denatured proteins caused by high pressure.

A subset of *HSP* genes was systematically analyzed to understand their roles in *HPG* (40). Of the 17 *HSP*s and related genes, *HSP104, HSP10,* and *HSP78* are up-regulated 3- to 4-fold at 25 MPa compared with cells cultured at 0.1 MPa. *HSP30, HSP42,* and *HSP82* are moderately up-regulated, 2- to 2.6-fold, at 25 MPa, whereas *HSP26* and *HSP31* are down-regulated about 2-fold. Hsp104 is known to cooperate with Ydj1 (Hsp40) and Ssa1 (Hsp70) to unfold and reactivate denatured, aggregated proteins (21). Hsp10 is a mitochondrial matrix cochaperonin that inhibits the ATPase activity of Hsp60 and is involved in protein folding and sorting in the mitochondria (18). Hsp78 is an oligomeric mitochondrial matrix chaperone that cooperates with Ssc1 in mitochondrial thermotolerance after heat shock (33). Accordingly, the loss of mitochondrial functions at high pressure might be compensated for by up-regulation of *HSP60* and *HSP78*. Hsp30 is a hydrophobic plasma membrane protein that negatively regulates the H^+-ATPase Pma1 (45, 49). It is also induced by heat shock, ethanol treatment, weak organic acid, glucose limitation, and entry into the stationary phase. Hydrostatic pressure causes intracellular acidification in a manner analogous to that of weak acid treatment (3–5). Therefore, intracellular acidification may cause *HSP30* induction with hydrostatic pressure. The precise mechanism by which increasing pressure activates the transcription of these genes is still unclear.

Genome-wide expression profiles upon growth at 30 MPa were characterized using DNA microarray hybridization (27). Of 5,721 open reading frames analyzed, 366 genes are up-regulated more than twofold and 253 genes are down-regulated more than twofold. According to the functional categories of the Munich Information Center for Protein Sequences (http://mips.gsf.de/), the ratio of induced genes (induced genes per number of genes in each category) was high in the categories of the stress response and metabolism of carbon, lipids, and amino acids. Thus, up-regulation of genes involved in such categories contributes to establishing the cellular defense against high pressure.

Functional analysis was performed to identify *HSP* genes responsible for *HPG*. Upon the loss of *HSP31*, the doubling time for growth was prolonged 1.7-fold (from 5.9 to 10.1 h) at a pressure of 25 MPa, while it was prolonged 1.3-fold (from 2.4 to 3.1 h) at 0.1 MPa (40). No marked retardation was observed in the growth of other *HSP* deletion mutants at 25 MPa (40). This result suggests that Hsp31 plays a partial role in growth under moderate-pressure conditions. Hsp31 is a 25.5-kDa protein and a possible chaperone and cysteine protease with similarity to *Escherichia coli* Hsp31 (35). In general, a pressure of 25 MPa is less severe and does not affect the conformation of soluble proteins as far as investigated in vitro. Therefore, some proteins might be misfolded during de novo protein synthesis in the endoplasmic reticulum at 25 MPa. Such misfolded proteins may be processed by the Hsp31 chaperone.

INTRACELLULAR ACIDIFICATION CAUSED BY HIGH PRESSURE

Intracellular pH is usually maintained at around 7.0 in a majority of organisms, and neutral pH is essential for their growth and viability. The cytoplasmic pH and vacuolar pH of yeast cells were individually analyzed using pH-sensitive fluorescent dyes, 5-carboxy (and 6-carboxy) SNARF-1 and 6-carboxyfluorescein, respectively, in a hydrostatic pressure chamber with transparent windows (4, 5). In *S. cerevisiae,* neutral cytoplasmic pH is maintained by the plasma membrane H^+-ATPase Pma1 (48), and an acidic vacuolar pH of around 6.0 is maintained by the vacuolar H^+-ATPase (V-H^+-ATPase) on the vacuolar membrane (28). Increasing the pressure to 50 MPa decreases the cytoplasmic pH by about 0.3 unit (5). A decrease in the cytoplasmic pH significantly impairs the activity of phosphofructokinase, a key enzyme in glycolysis, which is sensitive to pH changes. Pressure also decreases vacuolar pH by 0.3 to 0.5 units (3–5). Internal acidification only occurs in the presence of fermentable sugars such as glucose, fructose, or mannose, and not when ethanol or glycerol is supplied as the carbon source (4). Taking the results of analysis using glycolytic mutants together, the production of carbon dioxide (CO_2) through ethanol fermentation appears to be connected to pressure-induced internal acidification: CO_2 is easily soluble in water; at atmospheric pressure, more than 99% of aqueous CO_2 exists as the dissolved gas and less than 1% exists as carbonic acid, H_2CO_3, which partly dissociates to yield H^+, HCO_3^-, and CO_3^{2-}. The reaction volume (ΔV) of the reaction $H_2CO_3 \rightarrow H^+ + HCO_3^-$ is negative (-26.0 ml/mol), which means that the dissociation of the weak acid is enhanced by increasing hydrostatic pressure. Therefore, large numbers of protons are likely to accumulate in the cytoplasm when growing yeast cells are exposed to high pressure. To maintain a favorable cytoplasmic pH, the yeast vacuole is assumed to serve as a proton sequestrant by pumping in protons at high pressure (5). The findings so far suggest that chemical reactions involving intracellular low-molecular-weight compounds should be taken into account to elucidate the physiological responses of living organisms to increasing hydrostatic pressure.

DYNAMICS OF TRYPTOPHAN IMPORT ANALYZED USING HYDROSTATIC PRESSURE

The effects of hydrostatic pressure on enzymatic reactions are interpreted within the framework of the simplest kinetic mechanism in which the transition state presents the highest energy barrier, and the chemical transformation of substrate to product is considered to be a singular, rate-limiting step. The following equation represents the quantitative estimation of the effect of pressure:

$$(\partial \ln k/\partial p)_T = -\Delta V^{\neq}/RT$$

where k is the rate constant, p is pressure (megapascals), T is absolute temperature (kelvin), and R is the gas constant (milliliters · megapascals/kelvins · mole). $\Delta V^{\neq}$ is the apparent volume change of activation (activation volume in milliliters per mole) that represents the difference in the volume between the initial state and the activated state of the reaction, and it can only be obtained by measuring the rate constants of the reaction as a function of hydrostatic pressure. It is likely that the most significant effect of pressure will be produced on conformational changes associated with reactions. The activation volumes associated with tryptophan import through the tryptophan permeases Tat1 and Tat2 were determined (Fig. 2). Notably, there is a significant difference in the activation volumes for tryptophan import between Tat1 and Tat2, which are 89.3 and 50.8 ml/mol, respectively (9). The inhibition of tryptophan uptake is attributable to positive activation volumes asso-

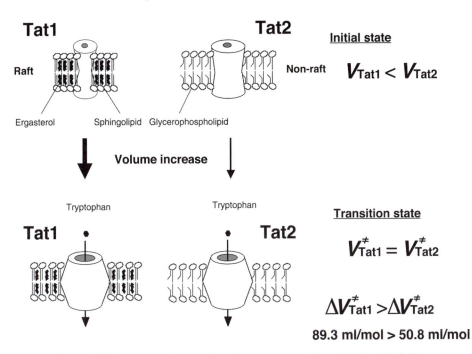

Figure 2. Model depicting the dynamics of tryptophan import through Tat1 and Tat2. Tat1 is associated with lipid rafts, whereas Tat2 is localized in nonrafts. The large activation volumes ($\Delta V^{\neq}$) for Tat1- and Tat2-mediated tryptophan import are accounted for mainly by volume changes associated with protein conformational changes. The initial volume of Tat1 is smaller than that of Tat2 because Tat1 is localized in the highly ordered lipid microdomain of lipid rafts. V, the volume of the permease in the initial state; $V^{\neq}$, the volume of the permease in the activated state; $\Delta V^{\neq}$, the activation volume accompanied by tryptophan import through the permease.

ciated with catalysis and the reduced probability of attaining these volumes in a more ordered lipid environment. This indicates that Tat1 undergoes a more dramatic conformational change during the activation associated with tryptophan import than Tat2 does.

The results of biochemical analyses indicate that Tat1 exists in tight and ordered lipid microdomains, so-called rafts, whereas Tat2 exists in disordered fluid domains, so-called nonrafts (9). The rafts consist of sphingolipid and ergosterol (cholesterol, in the case of animal cells) and are enriched in the plasma membrane (50). The remarkable difference in the activation volume (89.3 versus 50.8 ml/mol) is accounted for by the difference in the volume of the initial states; i.e., the initial volume of Tat1 is likely to be smaller than that of Tat2. This explanation is consistent with the findings that Tat1 is localized in the highly ordered lipid phase, in which the volume is small, whereas Tat2 is localized in the disordered phase, in which the volume is large. In this sense, hydrostatic pressure is a useful tool for probing lipid rafts and the proteins resident in membranes by measuring activation volumes to elucidate the dynamics of membrane protein function in living cells.

According to the results obtained with mammalian Na^{+},K^{+}-ATPase (17), ATPase activity is accompanied by a smaller activation volume ($\Delta V^{\neq} = 53$ ml/mol) when the membrane maintains fluidity under lower-pressure conditions (0.1 to 24 MPa) (17), which is a situation similar to that of Tat2 ($\Delta V^{\neq}$ = 50.8 ml/mol) (9). The activity is accompanied by a higher $\Delta V^{\neq}$ (83 ml/mol) when the membrane is ordered under higher-pressure conditions (pressure > 24 MPa) (17), which is a situation similar to that of Tat1 ($\Delta V^{\neq}$ = 89.3 ml/mol) (9). This indicates that the Na^{+},K^{+}-ATPase undergoes larger conformational changes at higher pressures because of an increase in the order parameter by shifting the melting transitions in phospholipids and aliphatic chains. Similar results have been reported in terms of the activation volume of the ATPase under different pressure conditions (15, 30).

POSSIBLE ROLE OF ORGANIC OSMOLYTES IN *S. CEREVISIAE* PHYSIOLOGY AT HIGH PRESSURE

Organic osmolytes such as amino acids and their derivatives, polyols, sugars, and methylamines are used by the cells of water-stressed organisms to maintain cell volume. In deep-sea animals, methylamines are known to enhance protein folding and ligand binding and counteract perturbations due to high pressure. Interestingly, the addition of 100 to 150 mM trimethylamine *N*-oxide to the culture medium has a protective effect against the high-pressure treatment of *S. cerevisiae* (56). This compound is assumed to counteract pressure effects simply by opposing the tendency of pressure to facilitate hydration in some situations. Trehalose, a nonreducing disaccharide, is known to serve as a cryoprotectant to prevent the formation of ice crystals in cells. In *S. cerevisiae*, trehalose is assumed to act to protect intracellular molecules from damage by lethal pressures (20, 26), although its contribution to *HPG* has yet to be examined. How trehalose influences the status of intracellular water molecules at high hydrostatic pressure should be examined in future studies with respect to the hydration of macromolecules and water geometry.

FUTURE PERSPECTIVES

By exploiting genomic information and powerful tools for genetic manipulation, the effects of hydrostatic pressure on *S. cerevisiae* have been analyzed by investigators in a broad range of experimental fields, including physiology, biochemistry, molecular biology, and food sciences. A more complete understanding of the effects of high pressure will be achieved by introducing high-throughput techniques to traditional yeast cell biology. In addition, a more mechanistic understanding will be achieved by a combinination with biophysical techniques. Using hydrostatic pressure as a parameter, piezophysiology will uncover novel biological phenomena that are accompanied by large volume changes, not only in *S. cerevisiae* but also in many other organisms (2).

REFERENCES

1. **Abe, F.** 1998. Hydrostatic pressure enhances vital staining with carboxyfluorescein or carboxydichlorofluorescein in *Saccharomyces cerevisiae:* efficient detection of labeled yeasts by flow cytometry. *Appl. Environ. Microbiol.* **64:**1139–1142.
2. **Abe, F.** 2004. Piezophysiology of yeast: occurrence and significance. *Cell. Mol. Biol.* **50:**437–445.
3. **Abe, F., and K. Horikoshi.** 1995. Hydrostatic pressure promotes the acidification of vacuoles in *Saccharomyces cerevisiae.* *FEMS Microbiol. Lett.* **130:**307–312.
4. **Abe, F., and K. Horikoshi.** 1997. Vacuolar acidification in *Saccharomyces cerevisiae* induced by elevated hydrostatic pressure is transient and is mediated by vacuolar H^{+}-ATPase. *Extremophiles* **1:**89–93.

5. **Abe, F., and K. Horikoshi.** 1998. Analysis of intracellular pH in the yeast *Saccharomyces cerevisiae* under elevated hydrostatic pressure: a study in baro-(piezo-) physiology. *Extremophiles* **2:**223–228.

6. **Abe, F., and K. Horikoshi.** 2000. Tryptophan permease gene *TAT2* confers high-pressure growth in *Saccharomyces cerevisiae. Mol. Cell. Biol.* **20:**8093–8102.

7. **Abe, F., and K. Horikoshi.** 2000. Metabolic changes in glycolysis in yeast induced by elevated hydrostatic pressure. A study in baro-(piezo-)physiology, p. 335–337. *In* M. H. Manghnani, W. J. Nellis, and M. F. Nicole (ed.), *Science and Technology of High-Pressure Research.* University Press, Hyderabad, India.

8. **Abe, F., and K. Horikoshi.** 2001. The biotechnological potential of piezophiles. *Trends Biotechnol.* **19:**102–108.

9. **Abe, F., and H. Iida.** 2003. Pressure-induced differential regulation of the two tryptophan permeases Tat1 and Tat2 by ubiquitin ligase Rsp5 and its binding proteins, Bul1 and Bul2. *Mol. Cell. Biol.* **23:**7566–7584.

10. **Abe, F., C. Kato, and K. Horikoshi.** 1999. Pressure-regulated metabolism in microorganisms. *Trends Microbiol.* **7:**447–452.

11. **Amerik, A. Y., S. Swaminathan, B. A. Krantz, K. D. Wilkinson, and M. Hochstrasser.** 1997. In vivo disassembly of free polyubiquitin chains by yeast Ubp14 modulates rates of protein degradation by the proteasome. *EMBO J.* **16:**4826–4838.

12. **Bartlett, D. H.** 2002. Pressure effects on in vivo microbial processes. *Biochim. Biophys. Acta* **1595:**367–381.

13. **Beck, T., A. Schmidt, and M. N. Hall.** 1999. Starvation induces vacuolar targeting and degradation of the tryptophan permease in yeast. *J. Cell Biol.* **146:**1227–1237.

14. **Chasse, S. A., and H. G. Dohlman.** 2004. Identification of yeast pheromone pathway modulators by high-throughput agonist response profiling of a yeast gene knockout strain collection. *Methods Enzymol.* **389:**399–409.

15. **Chong, P. L., P. A. Fortes, and D. M. Jameson.** 1985. Mechanisms of inhibition of (Na,K)-ATPase by hydrostatic pressure studied with fluorescent probes. *J. Biol. Chem.* **260:**14484–14490.

16. **Cossins, A. R., and A. G. Macdonald.** 1989. The adaptation of biological membranes to temperature and pressure: fish from the deep and cold. *J. Bioenerg. Biomembr.* **21:**115–135.

17. **de Smedt, H., R. Borghgraef, F. Ceuterick, and K. Heremans.** 1979. Pressure effects on lipid-protein interactions in (Na^{+}+K^{+})-ATPase. *Biochim. Biophys. Acta* **556:**479–489.

18. **Dubaquie, Y., R. Looser, and S. Rospert.** 1997. Significance of chaperonin 10-mediated inhibition of ATP hydrolysis by chaperonin 60. *Proc. Natl. Acad. Sci. USA* **94:**9011–9016.

19. **Dupré, S., and R. Haguenauer-Tsapis.** 2001. Deubiquitination step in the endocytic pathway of yeast plasma membrane proteins: crucial role of Doa4p ubiquitin isopeptidase. *Mol. Cell. Biol.* **21:**4482–4494.

20. **Fujii, S., H. Iwahashi, K. Obuchi, T. Fujii, and Y. Komatsu.** 1996. Characterization of a barotolerant mutant of the yeast *Saccharomyces cerevisiae:* importance of trehalose content and membrane fluidity. *FEMS Microbiol. Lett.* **141:**97–101.

21. **Glover, J. R., and S. Lindquist.** 1998. Hsp104, Hsp70, and Hsp40: a novel chaperone system that rescues previously aggregated proteins. *Cell* **94:**73–82.

22. **Guterman, A., and M. H. Glickman.** 2004. Complementary roles for Rpn11 and Ubp6 in deubiquitination and proteolysis by the proteasome. *J. Biol. Chem.* **279:**1729–1738.

23. **Hicke, L.** 1999. Gettin' down with ubiquitin: turning off cell-surface receptors, transporters and channels. *Trends Cell Biol.* **9:**107–112.

24. **Ichimori, H., T. Hata, H. Matsuki, and S. Kaneshina.** 1998. Barotropic phase transitions and pressure-induced interdigitation on bilayer membranes of phospholipids with varying acyl chain lengths. *Biochim. Biophys. Acta* **1414:**165–174.

25. **Iwahashi, H., S. C. Kaul, K. Obuchi, and Y. Komatsu.** 1991. Induction of barotolerance by heat shock treatment in yeast. *FEMS Microbiol. Lett.* **80:**325–328.

26. **Iwahashi, H., K. Obuchi, S. Fujii, and Y. Komatsu.** 1997. Effect of temperature on the role of Hsp104 and trehalose in barotolerance of *Saccharomyces cerevisiae. FEBS Lett.* **416:**1–5.

27. **Iwahashi, H., M. Odani, E. Ishidou, and E. Kitagawa.** 2005. Adaptation of *Saccharomyces cerevisiae* to high hydrostatic pressure causing growth inhibition. *FEBS Lett.* **579:**2847–2852.

28. **Kakinuma, Y., Y. Ohsumi, and Y. Anraku.** 1981. Properties of H^{+}-translocating adenosine triphosphatase in vacuolar membranes of *Saccharomyces cerevisiae. J. Biol. Chem.* **256:**10859–10863.

29. **Kato, C., and D. H. Bartlett.** 1997. The molecular biology of barophilic bacteria. *Extremophiles* **1:**111–116.

30. **Kato, M., R. Hayashi, T. Tsuda, and K. Taniguchi.** 2002. High pressure-induced changes of biological membrane. Study on the membrane-bound Na^{+}/K^{+}-ATPase as a model system. *Eur. J. Biochem.* **269:**110–118.

31. **Katzmann, D. J., M. Babst, and S. D. Emr.** 2001. Ubiqutin-dependent sorting into the multivesicular body pathway requires the function of a conserved endosomal protein sorting complex, ESCRT-1. *Cell* **106:**145–155.

32. **Kobori, H., M. Sato, K. Tameike, K. Hamada, S. Shimada, and M. Osumi.** 1995. Ultrastructural effects of pressure stress to the nucleus in *Saccharomyces cerevisiae:* a study by immunoelectron microscopy using frozen thin sections. *FEMS Microbiol. Lett.* **132:**253–258.

33. **Leonhardt, S. A., K. Fearson, P. N. Danese, and T. L. Mason.** 1993. *HSP78* encodes a yeast mitochondrial heat shock protein in the Clp family of ATP-dependent proteases. *Mol. Cell. Biol.* **13:**6304–6313.

34. **Macdonald, A. G.** 1967. The effect of high hydrostatic pressure on the cell division and growth of *Tetrahymena pyriformis. Exp. Cell Res.* **47:**569–580.

35. **Malki, A., T. Caldas, J. Abdallah, R. Kern, V. Eckey, S. J. Kim, S. S. Cha, H. Mori, and G. Richarme.** 2005. Peptidase activity of the *Escherichia coli* Hsp31 chaperone. *J. Biol. Chem.* **280:**14420–14426.

36. **Marquis, R. E.** 1976. High-pressure microbial physiology. *Adv. Microb. Physiol.* **14:**159–241.

37. **Martin, A. C., and D. G. Drubin.** 2003. Impact of genome-wide functional analyses on cell biology research. *Curr. Opin. Cell Biol.* **15:**6–13.

38. **Mentre, P., and G. Hui Bon Hoa.** 2001. Effects of high hydrostatic pressure on living cells: a consequence of the properties of macromolecules and macromolecule-associated water. *Int. Rev. Cytol.* **201:**1–84.

39. **Miura, T., and F. Abe.** 2004. Multiple ubiquitin-specific protease genes are involved in degradation of yeast tryptophan permease Tat2 at high pressure. *FEMS Microbiol. Lett.* **239:**171–179.

40. **Miura, T., H. Minegishi, R. Usami, and F. Abe.** 2006. Systematic analysis of *HSP* gene expression and effects on cell growth and survival at high hydrostatic pressure in *Saccharomyces cerevisiae. Extremophiles* **10:**279–284.

41. **Nagayama, A., C. Kato, and F. Abe.** 2004. The C-terminal mutation stabilizes the yeast tryptophan permease Tat2 under high-pressure and low-temperature conditions. *Extremophiles* **8:**143–149.

42. **Nelson, C. M., M. R. Schuppenhauer, and D. S. Clark.** 1992. High-pressure, high-temperature bioreactor for comparing effects of hyperbaric and hydrostatic pressure on bacterial growth. *Appl. Environ. Microbiol.* **58:**1789–1793.

43. **Nikko, E., A. M. Marini, and B. André.** 2003. Permease recycling and ubiquitination status reveal a particular role for Bro1 in the multivesicular body pathway. *J. Biol. Chem.* **278:**50732–50743.

44. **Perrier-Cornet, J. M., P. A. Marechal, and P. Gervais.** 1995. A new design intended to relate high pressure treatment to yeast cell mass transfer. *J. Biotechnol.* **41:**49–58.

45. **Piper, P. W., C. Ortiz-Calderon, C. Holyoak, P. Coote, and M. Cole.** 1997. Hsp30, the integral plasma membrane heat shock protein of *Saccharomyces cerevisiae,* is a stress-inducible regulator of plasma membrane H(+)-ATPase. *Cell Stress Chaperones* **2:**12–24.

46. **Sato, M., H. Kobori, S. A. Ishijima, Z. H. Feng, K. Hamada, S. Shimada, and M. Osumi.** 1996. *Schizosaccharomyces pombe* is more sensitive to pressure stress than *Saccharomyces cerevisiae. Cell Struct. Funct.* **21:**167–174.

47. **Schmidt, A., M. N. Hall, and A. Koller.** 1994. Two FK506 resistance-conferring genes in *Saccharomyces cerevisiae, TAT1* and *TAT2,* encode amino acid permeases mediating tyrosine and tryptophan uptake. *Mol. Cell. Biol.* **14:**6597–6606.

48. **Serrano, R.** 1993. Structure, function and regulation of plasma membrane H^+-ATPase. *FEBS Lett.* **325:**108–111.

49. **Seymour, I. J., and P. W. Piper.** 1999. Stress induction of *HSP30,* the plasma membrane heat shock protein gene of *Saccharomyces cerevisiae,* appears not to use known stress-regulated transcription factors. *Microbiology* **145:**231–239.

50. **Simons, K., and E. Ikonen.** 1997. Functional rafts in cell membranes. *Nature* **387:**569–572.

51. **Somero, G. N.** 1992. Adaptations to high hydrostatic pressure. *Annu. Rev. Physiol.* **54:**557–577.

52. **Swezey, R. R., and G. N. Somero.** 1985. Pressure effects on actin self-assembly: interspecific differences in equilibrium kinetics of the G to F transformation. *Biochemistry* **24:**852–860.

53. **Tamura, K., Y. Kamiki, and M. Miyashita.** 1999. Measurement of microbial activities under high pressure by calorimetry, p. 47–50. *In* H. Ludwig (ed.), *Advances in High Pressure Bioscience and Biotechnology.* Springer, Berlin, Germany.

54. **Winter, R., and W. Dzwolak.** 2004. Temperature-pressure configurational landscape of lipid bilayers and proteins. *Cell. Mol. Biol.* **50:**397–417.

55. **Yanagibayashi, M., Y. Nogi, L. Li, and C. Kato.** 1999. Changes in the microbial community in Japan Trench sediment from a depth of 6292 m during cultivation without decompression. *FEMS Microbiol. Lett.* **170:**271–279.

56. **Yancey, P. H., W. R. Blake, and J. Conley.** 2002. Unusual organic osmolytes in deep-sea animals: adaptation to hydrostatic pressure and other perturbations. *Comp. Biochem. Physiol. A* **133:**667–676.
57. **Yayanos, A. A.** 1995. Microbiology to 10,500 meters in the deep sea. *Annu. Rev. Microbiol.* **49:**777–805.
58. **Zimmerman, A. M.** 1971. High-pressure studies in cell biology. *Int. Rev. Cytol.* **30:**1–47.

High-Pressure Microbiology
Edited by C. Michiels, D. H. Bartlett, and A. Aertsen
© 2008 ASM Press, Washington, DC

Chapter 10

Factors Affecting Inactivation of Food-Borne Bacteria by High Pressure

Margaret F. Patterson and Mark Linton

High-pressure processing (HPP) is one of a number of technologies that are being considered commercially as alternatives to traditional methods such as thermal treatment, drying, or freezing of foods. All these new technologies need to be underpinned with robust data to ensure the safety and quality of the foods produced. This information is necessary to obtain regulatory approval as well as to ensure consumer confidence.

The potential for HPP to treat foods to extend shelf life and improve microbiological safety was first reported over 100 years ago (21). There has been much research carried out since then on the response of microorganisms to pressure. However, much of the early work was carried out in buffer systems rather than in real foods. It is now well established that microbial inactivation achieved in buffers may not always be the same as in foods and that many factors can influence the response of microorganisms to pressure. The aim of this chapter is to give an overview of the key factors that influence microbial behavior. Table 1 has been designed to give examples of the inactivation of bacteria by pressure in support of the text. The data have been calculated from results presented in individual papers, and unless stated otherwise, it is assumed that the cells were in the stationary phase of growth.

INACTIVATION KINETICS

It would be expected that increasing the magnitude of pressure applied and the time it is applied for will increase the lethal effect on microorganisms. If the inactivation follows first-order kinetics, a plot of treatment time versus $\log_{10}$ of survivors will result in a straight line. This has been reported for some organisms, and the data can be used to calculate pressure D_{10} values (time required at a particular pressure to achieve a 1-log reduction in numbers). D_{10} values are a useful way of comparing pressure resistance, for example, between

Margaret F. Patterson • Agriculture, Food and Environmental Science Division (Food Microbiology Branch), Agri-food and Biosciences Institute, and Department of Food Science, Queen's University, Belfast, Newforge Lane, Belfast BT9 5PX, Northern Ireland, United Kingdom.   **Mark Linton** • Agriculture, Food and Environmental Science Division (Food Microbiology Branch), Agri-food and Biosciences Institute, Newforge Lane, Belfast BT9 5PX, Northern Ireland, United Kingdom.

Table 1. Resistance of bacteria to high-pressure treatment

Microorganism(s)	Substrate	Treatment conditions	Inactivation	Reference
Vibrio spp.				
V. parahaemolyticus				
ATCC 17802	Alkaline peptone water	310 MPa/21°C/3 min	6.9-$\log_{10}$ reduction	9
	Oysters	310 MPa/21°C/3 min	5.1-$\log_{10}$ reduction	
Environmental isolate AST	Alkaline peptone water	310 MPa/21°C/3 min	8.4-$\log_{10}$ reduction	
	Oysters	310 MPa/21°C/3 min	4.8-$\log_{10}$ reduction	
V. parahaemolyticus				
Non-serotype O3:K6 (5 strains)	PBS	250 MPa/8–10°C	Mean D = 21 ± 9 s	14
Serotype O3:K6 (7 strains)	PBS	250 MPa/8–10°C	Mean D = 50 ± 4 s	
V. cholerae (10 strains)	PBS	250 MPa/8–10°C	Mean D = 41 ± 9 s	
V. vulnificus (22 strains)	PBS	200 MPa/8–10°C	Mean D = 26 ± 7 s	
Pseudomonas fluorescens grown at 8°C				31
Cells in exponential phase	UHT milk	250 MPa/8°C/5 min	~5.3-$\log_{10}$ reduction	
Cells in stationary phase			~2-$\log_{10}$ reduction	
Campylobacter spp.				
C. jejuni				
NCTC 11351	BBFBP[a] broth	300 MPa/20°C/10 min	2.5-$\log_{10}$ reduction	30
	UHT milk	300 MPa/20°C/10 min	~0.5-$\log_{10}$ reduction	
NCTC 11322	BBFBP broth	300 MPa/20°C/10 min	>7.1-$\log_{10}$ reduction	
	UHT milk	300 MPa/20°C/10 min	~1-$\log_{10}$ reduction	
C. coli				
NCTC 11350	BBFBP broth	300 MPa/20°C/10 min	5.6-$\log_{10}$ reduction	
NCTC 11366	BBFBP broth	300 MPa/20°C/10 min	>7.2-$\log_{10}$ reduction	
C. lari				
NCTC 11457	BBFBP broth	300 MPa/20°C/10 min	1.2-$\log_{10}$ reduction	
NCTC 11937	BBFBP broth	300 MPa/20°C/10 min	>7.4-$\log_{10}$ reduction	
C. fetus subsp. *fetus*				
NCTC 10348	BBFBP broth	300 MPa/20°C/10 min	5.8-$\log_{10}$ reduction	
NCTC 5850	BBFBP broth	300 MPa/20°C/10 min	>7.1-$\log_{10}$ reduction	
Yersinia enterocolitica				
CECT 559 serotype O:1	Model bovine milk cheese, pH 5.4	300 MPa/20°C/10 min	≥3.36-$\log_{10}$ reduction	16
CECT 4055 serotype O:3			1.95-$\log_{10}$ reduction	
CECT 4054 serotype O:8			3.48-$\log_{10}$ reduction	

Organism/strain	Medium	Treatment condition	Reduction	Reference
Escherichia coli O157				
C9490 NCTC 12079 H1071	PBS, pH 7.0	500 MPa/20°C/15 min	<0.5-log$_{10}$ reduction ~5-log$_{10}$ reduction ~5.6-log$_{10}$ reduction	5
NCTC 12079	Poultry meat	500 MPa/20°C/15 min 500 MPa/40°C/15 min 500 MPa/50°C/15 min	1-log$_{10}$ reduction 4.1-log$_{10}$ reduction >8.0-log$_{10}$ reduction	38
	UHT milk	500 MPa/20°C/15 min 500 MPa/40°C/15 min 500 MPa/50°C/15 min	<0.5-log$_{10}$ reduction ~0.6-log$_{10}$ reduction >8.0-log$_{10}$ reduction	
Cocktail of 3 strains (ATCC 43895, SEA13B88, and 932)	Grapefruit juice, pH 3.0 Apple juice, pH 3.7 Orange juice, pH 3.7 Carrot juice, pH 6.2	615 MPa/15°C/1min	2.4-log$_{10}$ reduction 0.2-log$_{10}$ reduction 1.07-log$_{10}$ reduction 4.51-log$_{10}$ reduction	51
Salmonella spp.				
Serovar Enteritidis FDA	Apricot juice, pH 3.8 Orange juice, pH 3.76 Cherry juice, pH 3.30	250 MPa/30°C/10 min	4.78-log$_{10}$ reduction 5.53-log$_{10}$ reduction 6.67-log$_{10}$ reduction	4
Serovar Enteritidis VL Serovar Typhimurium E21274 Serovar Typhimurium ATCC 14028	Tryptic soy broth	345 MPa/25°C/5 min	5.45-log$_{10}$ reduction 7.48-log$_{10}$ reduction 5.71-log$_{10}$ reduction 7.30-log$_{10}$ reduction	2
L. monocytogenes				
Scott A strain 35091 SLR1	Tryptic soy broth	345 MPa/25°C/5 min	0.96-log$_{10}$ reduction 2.10-log$_{10}$ reduction 3.53-log$_{10}$ reduction	2
Lactobacillus spp.				
L. helveticus *L. lactis* subsp. *cremoris*	Ringers solution	400 MPa/15 min/room temp	~2.5-log$_{10}$ reduction >8.0-log$_{10}$ reduction	23
Staphylococcus	Apricot juice, pH 3.8 Orange juice, pH 3.76 Cherry juice, pH 3.30	250 MPa/30°C/10 min	4.3-log$_{10}$ reduction 4.52-log$_{10}$ reduction 4.7-log$_{10}$ reduction	4

[a]BBFBP, Bolton broth supplemented with ferrous sulfate, sodium metabisulfite, and sodium pyruvate.

different microorganisms, when all other treatment conditions are identical. For example, in Table 1, D_{10} values are quoted for different *Vibrio* species, and it can be clearly seen that they have significantly different sensitivities to pressure.

However, there have been many reports showing that the inactivation is not always first order. It is relatively common to find that the curve showing treatment time versus $\log_{10}$ of survivors is concave (second-order kinetics), with an initial rapid decrease in $\log_{10}$ of survivors followed by a tailing effect, where there is little further inactivation as treatment time increases. Such inactivation curves have been found with different species, including strains of *Listeria monocytogenes* (Fig. 1), *Escherichia coli, Salmonella enterica* serovar Typhimurium, and *Yersinia enterocolitica.* These tailing effects have also been reported in thermal resistance studies, but the tails appear to be more common and more pronounced with pressure treatment.

It is known that experimental conditions, such as the substrate and growth conditions, may influence the shape of the curve. It is also believed that the curve shape may reflect an inherent phenotypic variation in pressure resistance in the microbial population, rather than the presence of persistent resistant cells. For example, Metrick et al. (32) pressure treated *Salmonella* serovar Typhimurium and *S. enterica* serovar Senftenberg and isolated cells from the resistant "tail." When these cells were grown and again exposed to the pressure treatment, there was no significant difference in pressure resistance compared with that of the original culture.

Tailing phenomena cannot be ignored, as the production of safe, pressure-treated foods requires the effectiveness of the high-pressure treatment to be predicted accurately. Therefore, accurate information on inactivation kinetics is essential (12). A number of alternative models, such as log-logistic and Weibull models, have been used successfully to describe nonlinear inactivation curves (8). Models such as these can be useful in identifying the optimum treatment conditions to ensure microbiological safety while at the same time maximizing the eating quality of food products.

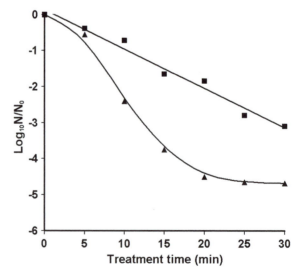

Figure 1. Inactivation of *Listeria monocytogenes* in PBS using a pressure treatment of 375 MPa at 20°C. N, count after pressure treatment; N_0, initial count. ■, NCTC 11994; ▲, poultry isolate.

SPECIES VARIATION

In general terms, gram-positive bacteria are more pressure resistant than gram-negative bacteria. This may be due to the more complex nature of the cell envelope in gram-negative bacteria (44). The envelopes of gram-positive bacteria are less complex than those of gram-negative bacteria, with only a plasma membrane and a thick peptidoglycan outer layer, which can constitute up to 90% of the cell envelope. In contrast, the envelopes of gram-negative bacteria are relatively complex, consisting of an inner and outer membrane with a thin layer of peptidoglycan sandwiched between. It is thought that the cell membrane is a key site for pressure damage in microorganisms (11, 54), which may partly explain why gram-negative bacteria tend to be more sensitive to pressure. However, there are exceptions to this general trend. For example, there are published examples of gram-negative bacteria, such as certain strains of *E. coli,* which are extremely pressure resistant (Table 1). These isolates were naturally occurring, and it is not known why they should have developed such an extreme pressure resistance.

Pressure-resistant isolates can also be selected by repeating cycles of lethal pressure treatment followed by growth of the surviving cell fraction. Hauben et al. (19) used this approach to isolate pressure-resistant mutants. It was noted that the pressure resistance of the mutants diminished when the treatment temperature increased up to 50°C. The authors suggested that the development of barotolerance may have important practical implications when using pressure as a food preservation technique, especially at ambient temperatures.

Vibrio spp. are one of the most pressure-sensitive groups of bacteria reported to date. A population of around 10^6 CFU of *Vibrio parahaemolyticus* per ml was killed by exposure to 170 MPa within 10 min at 25°C in clam juice and 30 min in phosphate buffer (49). Berlin et al. (6) also treated various *Vibrio* spp. *(V. parahaemolyticus, V. vulnificus, V. cholerae* O:1, *V. cholerae* non-O:1, *V. hollisae,* and *V. mimicus)* to pressures of 200 to 300 MPa for 5 to 15 min at 25°C in artificial seawater containing 2.34% NaCl. A treatment with 250 MPa for 10 min was sufficient to reduce all strains by at least 10^6 CFU/ml. The pressure treatment did not induce a viable-but-nonculturable (VBNC) state; however, cells already existing in a VBNC state appeared to be more pressure resistant.

Staphylococcus aureus tends to be relatively pressure resistant compared to other vegetative bacteria. Treatment in 10 mM phosphate-buffered saline (PBS) (pH 7.0) for 15 min at an initial temperature of 20°C showed that no significant inactivation occurred until the pressure exceeded 400 MPa (39).

The fact that there is a variation in pressure resistance between microbial species is not unexpected and is not unique to this technology. *Salmonella* serovar Senftenberg 775W, for example, is recognized as being significantly more heat resistant than other salmonella serovars (35). It is of interest that in some cases a positive correlation between pressure resistance and heat resistance has been found. This has been reported for *L. monocytogenes* (20) and certain strains of *E. coli* O157:H7 (5). However, no such correlation was found between heat resistance and pressure resistance for *Salmonella* (20). This confirms earlier observations that serovar Senftenberg 775W, which is five times more heat resistant than serovar Typhimurium, is actually more sensitive to pressure (32). Sherry et al. (43) compared the resistances of 40 *Salmonella* serovars to heat, irradiation, and high-pressure stresses and also found no correlation in resistance among the different stresses.

Cells in the stationary phase of growth tend to be more pressure resistant than those in the exponential phase (5). An example of this is shown in Table 1 with *Pseudomonas fluorescens*. It has been proposed that exponential-phase cells are inactivated under high pressure by irreversible damage to the cell membrane, while stationary-phase cells have a more robust cytoplasmic membrane that can better withstand pressure treatment (28). This assumption was based on the fact that exponential-phase cells showed changes in their cell envelopes that were not seen in stationary-phase cells. Other factors, such as growth temperature, also can influence the response to pressure. It has been noted that cells in the stationary phase become more resistant as growth temperature increases, while those in the exponential phase become less resistant as growth temperature increases (31).

STRAIN VARIATION

As noted above, significant variation in pressure resistance has been reported among different strains of the same species. It has been suggested that differences in pressure resistance among strains may be related to differences in susceptibility to membrane damage, although the exact nature of the lethal effect and the role of the membrane structure in determining resistance to pressure still have to be clarified (5). For example, Alpas et al. (2) reported significant variability in pressure resistance among strains of *L. monocytogenes, Salmonella, S. aureus,* and *E. coli* O157:H7. However, they also found that the range of pressure differences within a species decreased when the temperature during the pressure treatment was increased from 25 to 50°C. This may be helpful in a commercial situation, where the combination of pressure and mild heat could be used to enhance the lethal effect of the treatment. These results emphasized the importance of using strains with appropriate resistance when modeling different food preservation processes. It has been recommended that single-strain studies with above-average, but not with the most extreme, resistance be used to develop kinetic inactivation models for pressure treatment of a particular food. However, in recognition of natural variability, a cocktail of strains should be used to validate these kinetic models in the food (3).

TREATMENT TEMPERATURE

In many cases, HPP is promoted as a nonthermal preservation technology that is particularly useful for foods that suffer quality losses if heated. However, for certain applications pressure treatment at ambient temperature does not give adequate microbial kill. This is especially true for inactivation of bacterial spores, which can be extremely pressure resistant (see chapter 3). For example, Meyer et al. (33) recommended a temperature of 80°C and a pressure of 828 MPa (or 90°C and 690 MPa) to pressure sterilize vegetables and other low-acid foods. Also, the more resistant vegetative bacterial species, or some resistant strains of otherwise relatively sensitive species, may survive at moderate pressures. Microbial inactivation can be increased under these circumstances by combining pressure with mild heat. This approach has been used successfully by a number of authors. As shown in Table 1, the inactivation of *E. coli* O157 by 500 MPa in ultrahigh-temperature (UHT) milk was increased from < 0.5 to 8.0 log cycles by increasing the treatment temperature from 20 to 50°C (38). Similarly, the inactivation of pressure-resistant strains of *L. monocytogenes, E. coli, S. aureus,* and *S. enterica* serovar Enteritidis was enhanced significantly when pres-

sure was applied at higher temperatures. In most cases, there was more than an 8.0-log reduction when the cultures were pressurized at 345 MPa at 50°C for 5 min, compared to
less than a 4.1-log reduction when pressurized at 25°C (1). When the pressure treatment
was carried out in the presence of either citric or lactic acid, the inactivation increased by
an additional 1.2 to 3.9 log cycles at pH 4.5.

In other studies, when *E. coli* O157:H7 was pressure treated in orange juice, the inactivation levels were also increased by increasing the temperature from 20 to 30°C (24).
Simpson and Gilmour (45) showed that increasing the temperature of pressure treatment
(200 MPa for 15 min) from 25 to 45°C had no effect on the levels of inactivation of *L.
monocytogenes* in UHT milk, but increasing the temperature to 55°C increased the inactivation by approximately 6 log cycles. With a higher-pressure treatment (375 MPa for 15
min), inactivation increased with increasing temperature up to 35°C. At 45°C complete inactivation (>7-log reduction) was achieved. Kalchayanand et al. (22) found that pressure
treatment of *L. monocytogenes* (207 MPa for 5 min in 0.1% peptone water) at 35°C gave
about a 2.5-log inactivation, but when the temperature was increased to 45°C, cell death increased to approximately 9 log cycles. Capellas et al. (10) also found that pressure treating
Staphylococcus carnosus in fresh cheese at 500 MPa for 5 min at 50°C gave a 7-log inactivation, whereas treatment at 500 MPa for 30 min at 10 or 25°C gave no substantial decrease in numbers.

There is evidence that the combination of pressure with low (subzero) treatment temperature can be more effective at killing microorganisms than treatment at 20°C. Takahashi et al.
(50) treated *V. parahaemolyticus* and *S. aureus* suspended in 2 mM sodium phosphate buffer,
pH 7.0, with pressures ranging from 100 to 400 MPa at either −20 or 20°C. In most cases the
microbial kill was greater at the lower temperature. López-Caballero et al. (27) also found
greater bacterial inactivation when the pressure treatment was carried out at refrigeration
temperatures rather than at ambient temperature. They reported 1- to 2-log-greater inactivation of *Pseudomonas fluorescens* at 5°C than with pressure treatment at 20 or 35°C. When *L.
innocua* was pressure treated at 300 MPa for 15 min in liquid whole egg, a 1.5- to 2-log inactivation was achieved at −15 or 2°C, compared with virtually no inactivation at 20°C.
When a higher-pressure treatment was used, this effect was not significant (40).

Pressure treatment at subzero temperatures is being considered as a method of rapidly
freezing or thawing products to maintain quality attributes. The fact that the treatment may
also give enhanced microbial kill could be an additional bonus.

SUBSTRATE COMPOSITION

Many studies have shown that the high-pressure resistance of microorganisms increases
in more complex substrates. Patterson et al. (39) found that *E. coli* O157:H7 and *L. monocytogenes* were more resistant to pressure in UHT milk than in buffer or poultry meat.
Campylobacter jejuni was more pressure resistant in UHT whole milk than in laboratory
broth (Table 1). Similarly, the pathogen was much more resistant in UHT whole milk,
UHT skim milk, soya milk, and chicken puree than in buffer or broth (48). Pressure treatment at 300 to 325 MPa for 10 min was sufficient to reduce *C. jejuni* to undetectable levels
in buffer or broth, but an additional 50 to 75 MPa was required to achieve a similar level of
inactivation in these foods. Simpson and Gilmour (45) found that survival of *L. monocytogenes* was significantly greater in UHT milk than in cooked chicken mince or cooked

minced beef, which conferred only slightly more resistance than PBS. Further investigation (46) revealed that protein (bovine serum albumin), carbohydrate (glucose), and lipid (olive oil) added to PBS all increased the pressure resistance of three strains of *L. monocytogenes*. The concentration of bovine serum albumin required to affect pressure resistance varied depending on the strain and duration of treatment. Survival in 1 to 5% glucose was variable, with some strains more sensitive to pressure and some more resistant, but in 10% glucose all strains had increased pressure resistance. All strains were more resistant to pressure in 30% olive oil emulsion.

The pH of the substrate also significantly affects piezotolerance. Bacterial cells are most resistant at neutral pH, with piezotolerance decreasing when pH is either increased or decreased (53). Garcia-Graells et al. (18) noted that the sensitivity of pressure-resistant mutants of *E. coli* was increased in fruit juices, at least partly due to the reduced pH. Pressure sensitivity increased more in apple juice (pH 3.3) than in orange juice (pH 3.8). Similar results were obtained by Linton et al. (24), who showed that pressure inactivation of *E. coli* O157:H7 in orange juice was inversely correlated with pH in the range of 5.0 to 3.4. Erkman and Dogan (17) found that *L. monocytogenes* was more sensitive to pressure in orange juice than in peach juice, which, in turn, gave increased pressure sensitivity compared to raw milk. These changes were attributed to differences in pH.

Water activity (a_w) also affects high-pressure resistance; a decrease in a_w generally increases pressure resistance. Oxen and Knorr (36) described a_w-dependent barotolerance in the yeast *Rhodotorula rubra*. The protective effect occurred at a_w values less than 0.92 and was independent of solute. However, the protective effect could be counteracted by increasing the temperature during pressure treatment. At 200 MPa and 45°C there was a 7-log inactivation of *R. rubra* in 30 g of sucrose/100 g, compared with virtually no inactivation at 25°C. Van Opstal et al. (55) found that *E. coli* was more resistant to pressure in the presence of sucrose (10 to 50%) and that above 30% sucrose there was little or no inactivation. Smiddy et al. (47) found that pressure inactivation of *E. coli*, *V. mimicus*, *Listeria innocua*, and *L. monocytogenes* was greater in PBS than in oysters. When *E. coli* was pressure treated at 400 MPa for 5 min, there was a reduction of approximately 1 log cycle in oysters, compared with an approximately 3-log reduction in PBS. The difference between the two substrates increased with increasing pressure. The increase in resistance was attributed, at least in part, to the increase in salt concentration in oysters, as pressure resistance was increased in broth containing 3.5% sodium chloride. Molina-Hoppner et al. (34) explained the increased barotolerance of *Lactococcus lactis* in the presence of sodium chloride and sucrose as the result of accumulation of compatible solutes in the cells. A comparable level of protection was achieved by the addition of 3 M NaCl (a_w, 0.917) or 0.5 M sucrose (a_w, 0.985). Different compatible solutes were accumulated depending on whether the reduction in a_w was achieved by ionic or nonionic solutes. Therefore, care should be taken when attributing changes in pressure resistance to changes in a_w, as these may be the result of changes in solute concentration and not simply the result of lowering a_w.

SUBLETHAL INJURY AND RECOVERY OF DAMAGED CELLS

High pressure not only inactivates bacterial cells but also causes sublethal injury to a proportion of the population. Kalchayanand et al. (22) found that the degree of injury increased with pressure and with the temperature during pressure treatment, but this was dependent on the species. When *S. aureus* and *Salmonella* serovar Typhimurium were treated at pressures greater than 345 MPa, all the survivors were found to be injured cells. Increasing the temperature increased the number of sublethally injured cells, and this effect was more pronounced for *S. aureus* and serovar Typhimurium than for *E. coli* O157:H7. When *L. monocytogenes* was treated at 35 or 45°C, cells were killed either outright or not at all; no injured cells were detected.

The effect of time on pressure-induced injury depends on the species and the severity of the treatment. Simpson and Gilmour (45) reported that increasing the duration of pressure (375 MPa at 18°C) increased the degree of injury of *L. monocytogenes* in all of the substrates tested. However, Kalchayanand et al. (22) found that increasing the duration of pressure significantly affected injury only in *S. aureus* and not the other three species tested, including *L. monocytogenes*. The pressure used in this case was 207 MPa at 25°C, lower than that used by Simpson and Gilmour, and the substrate was 0.1% peptone water. Therefore, the effect of duration of pressure on the degree of injury may depend on the magnitude of pressure. The substrate may also be responsible for changes in the degree of injury, as Metrick et al. (32) found that serovar Typhimurium cells in buffer were more susceptible to injury than those in strained chicken.

Pressure-induced injury is normally associated with changes in membrane permeability and may be indicated by comparison of counts on selective and nonselective agars. For example, Ramaswamy et al. (41) used violet red bile agar to determine injury in *E. coli,* Sherry et al. (43) used modified brilliant green agar to measure injury in salmonellae, and Kalchayanand et al. (22) used modified Oxford agar for *L. monocytogenes* and TSAYE plus 7.5% NaCl to measure injury in *S. aureus.*

Changes in membrane permeability may also be indicated by the uptake of fluorescent dyes, such as propidium iodide, that are normally membrane impermeative or by measuring the release of ATP. Pagán and Mackey (37) used these techniques to show that the phase of growth also has a bearing on the degree of injury. They reported that the loss of viability in exponential-phase cells was associated with a permanent loss of membrane integrity but that stationary-phase cells repaired to a greater or lesser extent after pressure release.

The fate of pressure-injured cells depends on the conditions after pressure treatment, as pressure-injured cells can repair in a medium containing the necessary nutrients, given appropriate conditions (7). This is a problem in low-acid foods, as the recovery of injured cells during storage may result in food-borne disease or spoilage. For example, Metrick et al. (32) reported complete recovery of injured cells in strained chicken, but not in phosphate buffer, after 4 h at 37°C. However, injury may be an advantage in acid foods, as injured cells are not able to repair and die off during storage. Garcia-Graells et al. (18) studied the survival of a pressure-resistant mutant of *E. coli* during storage in fruit juices after treatments of up to 500 MPa for 15 min at 20°C. The results for different juices at the same pressure (300 MPa) or the same juice (mango) at different pressures indicated that the rate of inactivation during storage was inversely correlated with the pH of the juice and positively correlated with the pressure applied. Similarly, Linton et al. (25) found that the application of high pressure prior to refrigerated storage significantly increased the susceptibility of *E. coli* O157:H7 to acidity. The number of survivors in orange juice (pH 3.4) decreased from approximately 10^6 CFU/ml immediately after pressure treatment (400 MPa for 1 min at 10°C) to undetectable levels after 6 days at 3°C. It is likely that this effect could be enhanced by storage at higher temperatures (18).

Using the right combination of parameters, microbiological safety may be ensured in pressure-processed, high-acid foods by holding back the product for several days to guarantee complete inactivation of pathogens such as *E. coli* O157:H7. Pressure-injured cells may be killed by other stresses, not just by low pH. For example, there are many reported cases of high pressure sensitizing cells to antimicrobials (10, 13). Pressure-injured cells may also be killed by other conditions during pressure treatment or storage, such as reduced a_w (55).

PRESSURE IN COMBINATION WITH OTHER PRESERVATION TECHNIQUES

As discussed above, the use of high pressure in combination with mild heat or subzero temperatures has proved successful at enhancing microbial kill. Other hurdles such as redox potential and preservatives may be used in combination with high pressure to help reduce microbial numbers or to retard the growth of survivors.

For example, high-pressure treatment of foods is normally carried out in vacuum packs in order to minimize the presence of air. The use of high pressure combined with storage in vacuum packs has a significant impact on the development of the spoilage microflora and tends to favor those bacteria that are both piezotolerant and able to grow under reduced-oxygen conditions, such as lactic acid bacteria, some *Bacillus* spp., and other gram-positive bacteria (26). These may be present in large numbers before they cause unacceptable changes in food compared with typical spoilage microorganisms such as *Pseudomonas* spp.

Low a_w may also be used to enhance the effects of high-pressure treatment. When *E. coli* was pressure treated in sucrose and stored for 24 h at 20°C the number of survivors was further reduced, indicating that sublethal injury had occurred and that the injured cells were subsequently killed by the low a_w. The level of pressure-induced sublethal injury was dependent on the magnitude of pressure treatment and inversely correlated with sucrose concentration (55).

There are many reports of the advantages of combining high pressure with antimicrobials. For example, ter Steeg et al. (52) reported that HPP, nisin, and lowered temperature acted synergistically to reduce the severity of pressure treatment required to inactivate *L. plantarum*, *E. coli*, and *Saccharomyces cerevisiae*. The tailing of survival curves, often seen for high-pressure inactivation, could be eliminated using this combination treatment.

A combination of high pressure with multiple hurdles can also be useful to inactivate vegetative bacteria and bacterial endospores. High pressure, heat, acidity, and nisin were used to inactivate *Bacillus coagulans* spores. A 6-log inactivation was achieved using 400 MPa in pH 4.0 buffer at 70°C for 30 min in combination with 0.8 IU of nisin/ml (42). Marcos et al. (29) showed that a combination of reduced pH, reduced a_w, antimicrobial agents (nitrite), and high pressure inactivated a three-strain cocktail of *Salmonella* in fermented sausage. The high-pressure treatment led to a greater decrease in the numbers of *Salmonella* during ripening, compared with untreated controls.

High pressure may also be combined with other food preservation techniques to enhance microbial inactivation. This would allow the severity of one or all of the treatments to be reduced. For instance, the radiation dose required to yield a 10-fold reduction in *Clostridium sporogenes* spores in chicken meat was halved by combining irradiation with subsequent pressure treatment at 680 MPa for 20 min at 80°C (15).

In conclusion, it is clear that high hydrostatic pressure has the potential to kill bacteria and other microorganisms. However, many interacting factors will affect the response of individual microorganisms to high pressure. We need to be aware of these interactions so that optimum treatment conditions can be used to produce high-quality foods that are microbiologically safe and have an extended shelf life.

REFERENCES

1. **Alpas, H., N. Kalchayanand, F. Bozoglu, and B. Ray.** 2000. Interactions of high hydrostatic pressure, pressurization temperatures and pH on death and injury of pressure-resistant and pressure-sensitive strains of foodborne pathogens. *Int. J. Food Microbiol.* **60:**33–42.

2. **Alpas, H., N. Kalchayanand, F. Bozoglu, A. Sikes, C. P. Dunne, and B. Ray.** 1999. Variation in resistance to hydrostatic pressure among strains of food-borne pathogens. *Appl. Environ. Microbiol.* **65:**4248–4251.

3. **Balasubramaniam, V. M., E. Y. Ting, C. M. Stewart, and J. A. Robbins.** 2004. Recommended laboratory practices for conducting high-pressure microbial inactivation experiments. *Innovat. Food Sci. Emerg. Technol.* **5:**299–306.

4. **Bayındırlı, A., H. Alpas, F. Bozoğlu, and M. Hızal.** 2006. Efficiency of high pressure treatment on inactivation of pathogenic microorganisms and enzymes in apple, orange, apricot and sour cherry juices. *Food Control* **17:**52–58.

5. **Benito, A., G. Ventoura, M. Casadei, T. Robinson, and B. Mackey.** 1999. Variation in resistance of natural isolates of *Escherichia coli* O157 to high hydrostatic pressure, mild heat, and other stresses. *Appl. Environ. Microbiol.* **65:**1564–1569.

6. **Berlin, D. L., D. S. Herson, D. T. Hicks, and D. G. Hoover.** 1999. Response of pathogenic *Vibrio* species to high hydrostatic pressure. *Appl. Environ. Microbiol.* **65:**2776–2780.

7. **Bozoglu, F., H. Alpas, and G. Kalentunç.** 2004. Injury recovery of foodborne pathogens in high hydrostatic pressure treated milk during storage. *FEMS Immunol. Med. Microbiol.* **40:**243–247.

8. **Buzrul, S., and H. Alpas.** 2004. Modeling the synergistic effect of high pressure and heat on the inactivation kinetics of *Listeria innocua:* a preliminary study. *FEMS Microbiol. Lett.* **238:**29–36.

9. **Calik, H., M. T. Morrissey, P. W. Reno, and H. An.** 2002. Effect of high-pressure processing on *Vibrio parahaemolyticus* strains in pure culture and Pacific oysters. *J. Food Sci.* **67:**1506–1510.

10. **Capellas, M., M. Mor-Mur, R. Gervilla, J. Yuste, and B. Guamis.** 2000. Effect of high pressure combined with mild heat or nisin on inoculated bacteria and mesophiles of goat's milk fresh cheese. *Food Microbiol.* **17:**633–641.

11. **Casadei, M. A., P. Mañas, G. W. Niven, E. Needs, and B. M. Mackey.** 2002. Role of membrane fluidity in pressure resistance of *Escherichia coli* NCTC 8164. *Appl. Environ. Microbiol.* **68:**5965–5972.

12. **Chen, H., and D. G. Hoover.** 2003. Modeling the combined effect of high hydrostatic pressure and mild heat on the inactivation kinetics of *Listeria monocytogenes* Scott A in whole milk. *Innovat. Food Sci. Emerg. Technol.* **4:**25–34.

13. **Chung, Y.-K., M. Vurma, E. J. Turek, G. W. Chism, and A. E. Yousef.** 2005. Inactivation of barotolerant *Listeria monocytogenes* in sausage by combination of high-pressure processing and food-grade additives. *J. Food Prot.* **68:**744–750.

14. **Cook, D. W.** 2003. Sensitivity of *Vibrio* species in phosphate-buffered saline and oysters to high-pressure processing. *J. Food Prot.* **66:**2276–2282.

15. **Crawford, Y. J., E. A. Murano, D. G. Olson, and K. Shenoy.** 1996. Use of high hydrostatic pressure and irradiation to eliminate *Clostridium sporogenes* spores in chicken breast. *J. Food Prot.* **59:**711–715.

16. **De Lamo-Castellvi, S., M. Capellas, T. López-Pedemonte, M. M. Hernandez-Herrero, B. Guamis, and A. X. Roig-Sagués.** 2005. Behavior of *Yersinia enterocolitica* strains inoculated in model cheese treated with high hydrostatic pressure. *J. Food Prot.* **68:**528–533.

17. **Erkman, O., and C. Dogan.** 2004. Effects of ultra high hydrostatic pressure on *Listeria monocytogenes* and natural flora in broth, milk and fruit juices. *Int. J. Food Sci. Technol.* **39:**91–97.

18. **Garcia-Graells, C., K. J. A. Hauben, and C. W. Michiels.** 1998. High-pressure inactivation and sublethal injury of pressure-resistant *Escherichia coli* mutants in fruit juices. *Appl. Environ. Microbiol.* **64:**1566–1568.

19. **Hauben, K. J. A., D. H. Bartlett, C. C. F. Soontjens, K. Cornellis, E. Y. Wuytack, and C. W. Michiels.** 1997. *Escherichia coli* mutants resistant to inactivation by high hydrostatic pressure. *Appl. Environ. Microbiol.* **63:**945–950.

20. **Hellemons, J. C., and J. P. M. Smelt.** 1997. Annual report for EU project CT96-1175. *Cited by* J. P. Smelt, J. C. Hellemons, and M. Patterson, 2001, Effects of high pressure on vegetative microorganisms, p. 55–76, *in* M. E. G. Hendrickx and D. Knorr, *Ultra High Pressure Treatments of Foods.* Kluwer Academic/Plenum Publishers, New York, NY.

21. **Hite, B. H.** 1899. The effect of pressure on the preservation of milk. *W. Va. Agric. Exp. Stn. Bull.* **58:**15–35.

22. **Kalchayanand, N., A. Sikes, C. P. Dunne, and B. Ray.** 1998. Factors influencing death and injury of foodborne pathogens by hydrostatic pressure-pasteurization. *Food Microbiol.* **15:**207–214.

23. **Krasowska, M., A. Reps, and A. Jankowska.** 2005. Effect of high pressures on the activity of selected strains of lactic acid bacteria. *Milchwissenschaft* **60:**382–385.

24. **Linton, M., J. M. J. McClements, and M. F. Patterson.** 1999. Inactivation of *Escherichia coli* O157:H7 in orange juice using a combination of high pressure and mild heat. *J. Food Prot.* **62:**277–279.

25. **Linton, M., J. M. J. McClements, and M. F. Patterson.** 1999. Survival of *Escherichia coli* O157:H7 during storage in pressure-treated orange juice. *J. Food Prot.* **62:**1038–1040.

26. **Linton, M., J. M. J. McClements, and M. F. Patterson.** 2000. Changes in the microbiological quality of vacuum-packaged, minced chicken treated with high hydrostatic pressure. *Innovat. Food Sci. Emerg. Technol.* **5:**151–159.

27. **López-Caballero, M. E., J. Carballo, M. T. Solas, and F. Jiménez-Colmenero.** 2002. Responses of *Pseudomonas fluorescens* to combined high pressure/temperature treatments. *Eur. Food Res. Technol.* **214:**511–515.

28. **Mañas, P., and B. M. Mackey.** 2004. Morphological and physiological changes induced by high hydrostatic pressure in exponential- and stationary-phase cells of *Escherichia coli:* relationship with cell death. *Appl. Environ. Microbiol.* **70:**1545–1554.

29. **Marcos, B., T. Aymerich, and M. Garriga.** 2005. Evaluation of high pressure processing as an additional hurdle to control *Listeria monocytogenes* and *Salmonella enterica* in low-acid fermented sausages. *Food Microbiol. Safety* **70:**339–344.

30. **Martínez-Rodriguez, and B. M. Mackey.** 2005. Factors affecting the pressure resistance of some *Campylobacter* species. *Lett. Appl. Microbiol.* **41:**321–326.

31. **McClements, J. M. J., M. F. Patterson, and M. Linton.** 2001. The effect of growth stage and growth temperature on high hydrostatic pressure inactivation of some psychrotrophic bacteria in milk. *J. Food Prot.* **64:**514–522.

32. **Metrick, C., D. G. Hoover, and D. F. Farkas.** 1989. Effect of high hydrostatic pressure on heat resistant and heat sensitive strains of *Salmonella. J. Food Sci.* **54:**1547–1549.

33. **Meyer, R. S., K. L. Cooper, D. Knorr, and H. L. M. Lelieveld.** 2000. High pressure sterilization of foods. *Food Technol.* **54:**67–72.

34. **Molina-Hoppner, A., W. Doster, R. F. Vogel, and M. G. Gänzle.** 2004. Protective effect of sucrose and sodium chloride for *Lactococcus lactis* during sublethal and lethal high-pressure treatments. *Appl. Environ. Microbiol.* **70:**2013–2020.

35. **Ng, H., H. G. Bayne, and J. A. Garibaldi.** 1969. Heat resistance of *Salmonella:* the uniqueness of *Salmonella* Senftenberg 775W. *Appl. Microbiol.* **17:**78–82.

36. **Oxen, P., and D. Knorr.** 1993. Baroprotective effects of high solute concentrations against inactivation of *Rhodotorula rubra. Lebensm.-Wiss. Technol.* **26:**220–223.

37. **Pagán, R., and B. Mackey.** 2000. Relationship between membrane damage and cell death in pressure-treated *Escherichia coli* cells: differences between exponential- and stationary-phase cells and variation among strains. *Appl. Environ. Microbiol.* **66:**2829–2834.

38. **Patterson, M. F., and D. J. Kilpatrick.** 1998. The combined effect of high hydrostatic pressure and mild heat on inactivation of pathogens in milk and poultry. *J. Food Prot.* **61:**432–436.

39. **Patterson, M. F., M. Quinn, R. Simpson, and A. Gilmour.** 1995. Sensitivity of vegetative pathogens to high hydrostatic pressure treatment in phosphate-buffered saline and foods. *J. Food Prot.* **58:**524–529.

40. **Ponce, E., R. Pla, M. Mor-Mur, R. Gervilla, and B. Guamis.** 1998. Inactivation of *Listeria innocua* inoculated in liquid whole egg by high hydrostatic pressure. *J. Food Prot.* **61:**119–122.

41. **Ramaswamy, H. S., E. Riahi, and E. Idziak.** 2003. High-pressure destruction kinetics of *E. coli* (29055) in apple juice. *J. Food Sci.* **68:**1750–1756.

42. **Roberts, C. M., and D. G. Hoover.** 1996. Sensitivity of *Bacillus coagulans* spores to combinations of high hydrostatic pressure, heat, acidity and nisin. *J. Appl. Bacteriol.* **81:**363–368.

43. **Sherry, A. E., M. F. Patterson, and R. H. Madden.** 2004. Comparison of 40 *Salmonella enterica* serovars injured by thermal, high-pressure and irradiation stress. *J. Appl. Microbiol.* **96:**887–893.

44. **Shigehisa, T., T. Ohmori, A. Saito, S. Taji, and R. Hayashi.** 1991. Effects of high pressure on the characteristics of pork slurries and inactivation of micro-organisms associated with meat and meat products. *Int. J. Food Microbiol.* **12:**207–216.

45. **Simpson, R. K., and A. Gilmour.** 1997. The resistance of *Listeria monocytogenes* to high hydrostatic pressure in foods. *Food Microbiol.* **14:**567–573.

46. **Simpson, R. K., and A. Gilmour.** 1997. The effect of high hydrostatic pressure on *Listeria monocytogenes* in phosphate-buffered saline and model food systems. *J. Appl. Microbiol.* **83:**181–188.

47. **Smiddy, M., L. O'Gorman, R. D. Sleator, J. P. Kerry, M. F. Patterson, A. L. Kelly, and C. Hill.** 2005. Greater high pressure resistance of bacteria in oysters than in buffer. *Innovat. Food Sci. Emerg. Technol.* **6:**83–90.

48. **Solomon, E. B., and D. G. Hoover.** 2004. Inactivation of *Campylobacter jejuni* by high hydrostatic pressure. *Lett. Appl. Microbiol.* **38:**505–509.

49. **Styles, M. F., D. G. Hoover, and D. F. Farkas.** 1991. Response of *Listeria monocytogenes* and *Vibrio parahaemolyticus* to high hydrostatic pressure. *J. Food Sci.* **56:**1404–1407.

50. **Takahashi, K., H. Ishii, and H. Ishikawa.** 1991. Sterilisation of microorganisms by hydrostatic pressure at low temperatures, p. 225–232. *In* R. Hayashi (ed.), *High Pressure Science of Food.* Sa-Ei Publishing, Kyoto, Japan.

51. **Teo, A., Y.-L. S. Ravishankar, and C. E. Sizer.** 2001. Effect of low-temperature, high pressure treatment on the survival of *Escherichia coli* O157:H7 and *Salmonella* in unpasteurized fruit juices. *J. Food Prot.* **64:**1122–1127.

52. **ter Steeg, P. F., J. C. Hellemons, and A. E. Kok.** 1999. Synergistic actions of nisin, sublethal ultrahigh pressure, and reduced temperature on bacteria and yeast. *Appl. Environ. Microbiol.* **65:**4148–4154.

53. **Tewari, G., D. S., Jayas, and R. A. Holley.** 1999. High pressure processing of foods: an overview. *Sci. Aliments* **19:**619–661.

54. **Ulmer, H. M., M. G. Gänzle, and R. F. Vogel.** 2000. Effects of high pressure on survival and metabolic activity of *Lactobacillus plantarum* TMW1.460. *Appl. Environ. Microbiol.* **66:**3966–3973.

55. **Van Opstal, I., S. C. M. Vanmuysen, and C. W. Michiels.** 2003. High sucrose concentration protects *E. coli* against high pressure inactivation but not against pressure sensitization to the lactoperoxidase system. *Int. J. Food Microbiol.* **88:**1–9.

High-Pressure Microbiology
Edited by C. Michiels, D. H. Bartlett, and A. Aertsen
© 2008 ASM Press, Washington, DC

Chapter 11

Introduction to Deep-Sea Microbiology

Douglas H. Bartlett

The following chapters in this volume describe the influence of pressure on deep-sea microbes, those members of the domains *Bacteria, Archaea,* and *Eukarya* which experience elevated pressure as a standard feature of their surroundings. Most of the chapters describe features of piezophiles, microbes which reproduce preferentially or exclusively at pressures above those present on the surface of our planet.

The science of deep-sea life has come a long way since the naturalist Edward Forbes formulated his azoic hypothesis in the mid-19th century. Based on dredging studies in the Aegean Sea, he postulated that no marine life could exist in the deep sea below a depth of about 550 m (3). Today it is clear that the species richness at depth could be many times greater than that present in the better-understood and -appreciated tropical rainforests and shallow-coral reef environments, both of which teem with biodiversity (14).

Many of the advances in the study of deep-sea life have followed on the heels of great accomplishments in engineering and their use in the exploration of the unknown (4). This was true in 1934 when William Beebe and Otis Barton descended in their tethered sphere to the unheard-of depth of 932 m. It was also the case in 1960 when Piccard and Walsh descended in the bathyscaphe *Trieste* to approximately 11 km in the Challenger Deep, the deepest spot on Earth.

As previously noted by Robert Marquis (22), microbial piezobiology has its origins at the turn of the 19th century in France. Since that time, deep-sea microbiology has experienced bursts of research intensity followed by periods of lulls in activity. Thanks to the availability of hydraulic pumps and tanks used for gas liquefaction, A. Certes, working in the laboratory of none other than Louis Pasteur, and P. Regnard, working with Paul Bert, investigated the effects of pressure on a wide variety of organisms and biological extracts. The German scientist Fisher was also active in deep-sea microbiology during the same general period (13). Certes obtained seawater and sediments from depths as great as 5 km from the *Travailleur* and *Talisman* expeditions of 1882–1883. He came to believe that it was possible for bacteria to become adapted to elevated pressure (although with few data) (8). This work and that of the previous Challenger expedition of 1873–1876 dispelled Forbes's azoic-deep-sea postulate (20).

Douglas H. Bartlett • Mail code 0202, Marine Biology Research Division, Center for Marine Biotechnology and Biomedicine, Scripps Institution of Oceanography, University of California, San Diego, La Jolla, CA 92093-0202.

ZoBELL AND KRISS

The second birth of the field took place in the 1940s at Scripps Institute of Oceanography with the work of Claude ZoBell, his student Richard Morita, and other associates (Fig. 1A). Both ZoBell and Morita participated in the Danish *Galathea* "Round the World" expedition of 1950–1952 (39). Their account of this expedition is inspirational reading to all deep-sea microbiologists who have since followed in their wake. The authors obtained evidence of microbial growth at high pressure (and low temperature) by comparing most-probable-number (MPN) estimates of various physiological groups of microbes incubated at atmospheric pressure and elevated pressure using pumps, vessels, and pressurizable "culture tubes" they had previously optimized in earlier studies with other sources of microbes. Microbes were obtained from sediment subcores recovered from water depths of ~6.6 to 10.2 km at various locations in the Kermadec-Tonga Trench, Indian Ocean, and Philippine Trench (37, 39). Some of the pressurized sediment subsamples were shipped back to California via air express at high pressure in pressure vessels packed in dry ice. Clear evidence of piezophilic growth was noted with many samples, and in some cases the differences in MPN counts between incubations at atmospheric pressure and high pressure were over 100-fold. It should be noted that this was not the first evidence obtained for the existence of piezophilic microbes. ZoBell first introduced the concept of piezophily (then termed barophily) based on MPN measurements of mud samples obtained from 5.8 km off the coast of Bermuda (38, 39).

A contemporary of ZoBell and Morita was the Soviet marine microbiologist Kriss. He obtained mud samples from a water column depth up to 9.6 km within the Kuril-Kamchatka Trench and from a water column depth up to 4 km within the Arctic Ocean. The results of these experiments are described in his well-read textbook on marine microbiology (thanks to a broadly circulated English language translation) (18). In his case only 1 bacterial isolate out of 146 tested showed any evidence of piezophily. However, much of this work seems to have been performed at 26°C, which would have certainly killed all known non-hydrothermal-vent deep-sea piezophiles. In this light Kriss's results may not appear to be particularly surprising. However, it is still a mystery as to how ZoBell and Morita succeeded in cultivating their piezophiles. ZoBell emphasized the fact that the temperature of most of the Galathea expedition samples ranged from 5 to 10°C when they were hauled aboard and that great haste was taken in sample processing. Still, many of the inocula were only about 10 µl, and the ambient temperature during these manipulations was often in excess of 40°C in their shipboard microbiological laboratory. As they pointed out, "many of the more sensitive organisms were probably killed by the heat" (39).

Kriss also performed some fascinating high-pressure microbiology experiments with mud and soil samples. Atmospheric-pressure plate count results indicated that incubation of either mud or soil samples at pressures as high as 186 MPa eventually gave rise to increasing CFU counts on the meat peptone agar used (18). Microbial growth at these incredible pressures has not since been reported, except for an unsubstantiated report of terrestrial bacterial growth at gigapscal pressures in a diamond anvil cell (27). Both of these types of high-pressure incubations need to be reinvestigated.

Even after the reports of ZoBell, the issue of whether piezophiles existed was still openly debated, largely because of his inability to subculture the apparent piezophiles (17, 23). While there was universal consensus that piezotolerant deep-sea microbes must exist,

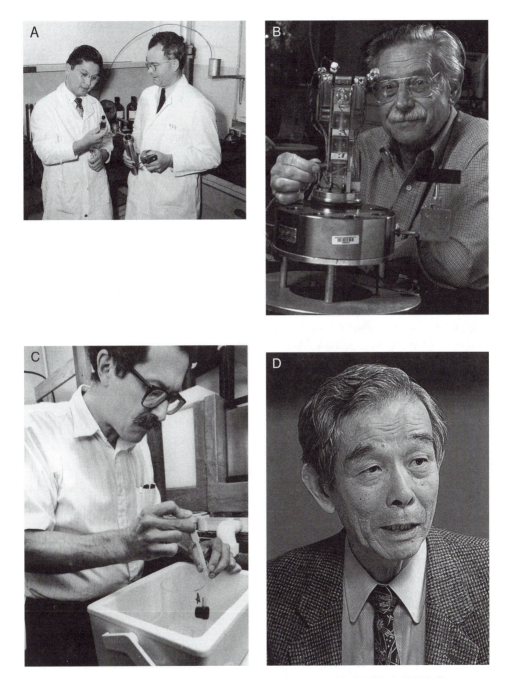

Figure 1. Some notable deep thinkers. (A) Richard Y. Morita (left) and Claude E. ZoBell (right); (B) Holger W. Jannasch; (C) A. Aristides Yayanos; (D) Koki Horikoshi.

Kriss and Mitskevich had previously noted that depending on nutrient conditions, many non-deep-sea microbes grow quite well, even better in some cases, at pressures of 10 to as high as 35 MPa (19). So, there might not be a need for adaptation to elevated pressure if pressure exerted little selective influence.

JANNASCH

Additional evidence arguing against the existence of piezophiles followed the 1968 Alvin debacle in which the deep-sea submersible was lost at sea (fortunately without human loss) for 11 months at a depth of 1.5 km. After retrieval it was noted that a box lunch present in the sub was largely preserved until recovery and accompanying decompression and refrigeration. Decompression had apparently led to increased activity of the associated deep-sea microbes. This led to a series of well-conceived and -engineered experiments by Holger Jannasch (Fig. 1B) and colleagues exploring the metabolic rates of deep-sea microbial populations in situ, at depths extending to 5.3 km (16). Regardless of the nutrient employed or the strategy followed for measuring its utilization, microbial uptake values were always higher when the measurements were performed at atmospheric pressure rather than the in situ condition, even though the other physical and chemical conditions were identical. While these results ran contrary to the overall message from Kriss's work, both lines of investigation were consistent with the view that piezophiles did not exist. Caveats to this interpretation of the data did exist. Jannasch himself noted that such a conclusion would be incorrect (i) if the nutrients he chose for his metabolic-rate measurements were not well used by the deep-sea microbial populations or (ii) if the activity of the deep-sea microbes was being swamped out by large numbers of surface forms that had reached the deep sea by sedimentation (17).

History has proven that both of these factors are indeed important in the assessment of pressure sensitivity of microbial consortia. Subsequent deep-sea microbial activity measurements have uncovered piezophilic community responses when the samples interrogated have a large proportion of autochthonous members. For example, rate measurements in deep-sea sediments can be quite pressure insensitive, and those associated with the gut flora of holothurians and fish can be piezophilic (29). Under conditions of water column stratification in the Mediterranean, piezophilic community responses are seen even at 11 MPa (6).

YAYANOS

The debate over the existence of piezophiles came to a halt in 1979 when Art Yayanos (Fig. 1C) and his team became the first of what would eventually be numerous groups from around the world to obtain and maintain a pure culture of a piezophile (35). It was obtained from decomposing amphipods collected from the central north Pacific at a depth of 5,700 m and maintained at in situ temperature and pressure. The resulting turbid suspension was inoculated into nutrient medium with silica gel and incubated at high pressure. The growth properties of one of the isolated colonies were further examined and its piezophilic nature was clarified. As a result of this and subsequent work, the existence of piezophiles moved from hypothesis to fact, and for the first time, they became available for studies of physiology, biochemistry, genetics, and molecular biology, as detailed in some of the chapters of this book. Yayanos and colleagues went on to isolate the first obligately piezophilic microbes, examine piezophile extreme sensitivity to UV light, correlate the rates of piezophile reproduction with capture depth, and explore some facets of piezophile lipid and protein biochemical adaptation to high pressure (10–12, 33, 34, 36). Along the way Yayanos also promoted the name change from barophile to piezophile (34). *Piezo* is Greek for pressure, whereas *baro* is Greek for weight. *Baro* is the prefix which had previously been coined by ZoBell and Oppenheimer (40). The *piezo* descriptor is used throughout the following chapters.

HORIKOSHI

A dramatic infusion of talent and resources in high-pressure studies of deep-sea microbes occurred in 1990 with the launching of the DEEPSTAR program at the Japan Agency for Marine-Earth Science and Technology (JAMSTEC) (15). Scientists associated with DEEPSTAR and other subsequent JAMSTEC programs have since utilized deep-diving manned and unmanned submersibles, a deep-sea piezophile/thermophile collecting and cultivating system, and a variety of other pressurized devices to isolate and characterize thousands of deep-sea microbes and their products. It would be difficult to overemphasize the impact that JAMSTEC contributions have made since 1990 to studies of high-pressure influences on biological systems, at levels ranging from ecosystems to the molecular and biophysical. A guiding light for many of these efforts has been the current JAMSTEC Extremobiosphere Research Center Director General, Koki Horikoshi (Fig. 1D). More recently a major effort in high-pressure, deep-sea microbiology has also been initiated at the Third Oceanographic Institute in Xiamen, China (32).

LIFE IN A PRESSURE COOKER

One of the great discoveries of the 20th century was the discovery of abundant life in and around hydrothermal vents. Pressure plays critical roles in deep-sea vents, not the least of which is to keep many gaseous substrates for metabolism in solution at high temperature. It also provides an opportunity for chemical reactions to occur under supercritical water conditions (2). In these environs animal development may require elevated pressure (24). High pressure can also extend the temperatures limits tolerated by microbes (26). A number of piezophilic thermophiles and hyperthermophiles have been isolated (5). Remarkably, only a few studies have actually enriched for hydrothermal-vent-associated microbes at elevated pressure (1, 25). The deep-sea vent microbe for which the most information is available regarding the influence of pressure on its growth and metabolism is *Methanocaldococcus jannaschii,* which is described in chapter 16.

PIEZOPHILE BLUEPRINTS

Two piezophile genome sequences have already been published (7, 30), and more are in the queue. Future advances in the understanding of piezophiles, as with many branches of biology, will benefit from advances in genome technology and its innovative application. Two striking examples of this are the use of environmental genomics and fosmid end sequencing by DeLong and colleagues using microbial samples collected at depths up to 4 km at station ALOHA in the North Pacific Subtropical Gyre (9), and the massive sequencing of a portion of the 16S rRNA gene using pyrosequencing technology by Sogin et al. (28) using water samples collected in the North Atlantic Ocean at depths also up to 4 km. Both of these reports have been extensively reviewed (21, 31). These and other molecular adventures suggest that there is much to learn about adaptations to pressure by many microbial groups as yet uncultured from depth, including the bacterial phyla *Acidobacteria, Actinobacteria, Planctomycetes,* and *Verrucomicrobia* as well as members of both the *Crenarchaeota* and *Euryarchaeota* kingdoms within the domain *Archaea.*

This would seem to be an opportune time for another period of intensive inquiry into the microbial inhabitants of the largest portion of the biosphere.

REFERENCES

1. **Alain, K., V. G. Marteinsson, M. L. Miroshnichenko, E. A. Bonch-Osmolovskaya, D. Prieur, and J.-L. Birrien.** 2002. *Marinitoga piezophila* sp. nov., a rod-shaped, thermo-piezophilic bacterium isolated under high hydrostatic pressure from a deep-sea hydrothermal vent. *Int. J. Syst. Evol. Microbiol.* **52:**1331–1339.

2. **Alargov, D. K., S. Deguchi, K. Tsujii, and K. Horikoshi.** 2002. Reaction behaviors of glycine under super- and subcritical water conditions. *Origins Life Evol. Biosph.* **32:**1–11.

3. **Anderson, T. R., and T. Rice.** 2006. Deserts on the sea floor: Edward Forbes and his azoic hypothesis for a lifeless deep ocean. *Endeavor* **30:**131–137.

4. **Ballard, R. D.** 2000. *The Eternal Darkness. A Personal History of Deep-Sea Exploration.* Princeton University Press, Princeton, NJ.

5. **Bartlett, D. H., F. M. Lauro, and E. A. Eloe.** 2007. Microbial adaptation to high pressure, p. 333–348. *In* C. Gerday and N. Glandsdorf (ed.), *Physiology and Biochemistry of Extremophiles.* ASM Press, Washington, DC.

6. **Bianchi, A., and J. Garcin.** 1993. In stratified waters the metabolic rate of deep-sea bacteria decreases with decompression. *Deep-Sea Res. I* **40:**1703–1710.

7. **Bult, C. J., O. White, G. J. Olsen, L. Zhou, R. D. Fleischmann, G. G. Sutton, J. A. Blake, L. M. Fitz-Gerald, R. A. Clayton, J. D. Gocayn, A. R. Kerlavage, B. A. Dougherty, J.-F. Tomb, M. D. Adams, C. I. Reich, R. Overbeek, E. F. Kirkness, K. G. Weinstock, J. M. Merrick, A. Glodek, J. L. Scott, N. S. M. Geoghagen, J. F. Weidman, J. L. Fuhrmann, D. Nguyen, T. R. Utterback, J. M. Kelley, J. D. Peterson, P. W. Sadow, M. C. Hanna, M. D. Cotton, K. M. Roberts, M. A. Hurst, B. P. Kaine, M. Borodovsky, H.-P. Klenk, C. M. Fraser, H. O. Smith, C. R. Woese, and J. C. Venter.** 1996. Complete genome sequence of the methanogenic archaeon, *Methanococcus jannaschii. Science* **273:**1058–1073.

8. **Certes, A.** 1884. Sur la culture, a l'abri des germes atmospherique, des eaux et des sediments rapportes par les expeditions du Travailleur et du Talisman. *C. R. Acad. Sci. Paris* **11:**690–693.

9. **DeLong, E. F., C. M. Preston, T. Mincer, V. Rich, S. J. Hallam, N. U. Frigaard, A. Martinez, M. B. Sullivan, R. Edwards, B. R. Brito, S. W. Chisholm, and D. M. Karl.** 2006. Community genomics among stratified microbial assemblages in the ocean's interior. *Science* **311:**496–503.

10. **DeLong, E. F., and A. A. Yayanos.** 1985. Adaptation of the membrane lipids of a deep-sea bacterium to changes in hydrostatic pressure. *Science* **228:**1101–1103.

11. **DeLong, E. F., and A. A. Yayanos.** 1986. Biochemical function and ecological significance of novel bacterial lipids in deep-sea prokaryotes. *Appl. Environ. Microbiol.* **51:**730–737.

12. **DeLong, E. F., and A. A. Yayanos.** 1987. Properties of the glucose transport system in some deep-sea bacteria. *Appl. Environ. Microbiol.* **53:**527–532.

13. **Fisher, B.** 1894. Die Baktrien des Meeres nach den Untersuchungen der Plankton-Expedition unter gleichzeitiger Berücksichtigung einiger älterer und neuerer Untersuchungen. *Ergebnisse der Plankton-Expedition der Hu,boldt-Siftung* **4:**1–83.

14. **Grassle, J. F., and N. J. Maciolek.** 1992. Deep-sea species richness—regional and local diversity estimates from quantitative bottom samples. *Am. Nat.* **139:**313–341.

15. **Hamamoto, T., and K. Horikoshi.** 1993. Deepsea microbiology research within the Deepstar program. *J. Mar. Biotechnol.* **1:**119–124.

16. **Jannasch, H. W., and C. D. Taylor.** 1984. Deep-sea microbiology. *Annu. Rev. Microbiol.* **38:**487–514.

17. **Jannasch, H. W., and C. O. Wirsen.** 1977. Microbial life in the deep sea. *Sci. Am.* **236:**42–52.
18. **Kriss, A. E.** 1963. *Marine Microbiology.* John Wiley & Sons, Inc., New York, NY.
19. **Kriss, A. E., and I. N. Mitskevich.** 1967. Effect of nutrient medium on the tolerance of barotolerant bacteria to high pressure. *Mikrobiologiya* **36:**203–206.
20. **Kunzig, R.** 2003. Deep-sea biology: living with the endless frontier. *Science* **302:**991.
21. **Lauro, F. M., and D. H. Bartlett.** 17 January 2007. Prokaryotic lifestyles in deep-sea habitats. *Extremophiles* doi:10.1007/s00792-006-0059-5.
22. **Marquis, R. E.** 1982. Microbial barobiology. *BioScience* **32:**267–271.
23. **Marquis, R. E., and P. Matsumura.** 1978. Microbial life under pressure, p. 105–158. *In* D. J. Kushner (ed.), *Microbial Life in Extreme Environments.* Academic Press, London, United Kingdom.
24. **Marsh, A. G., L. S. Mullineaux, C. M. Young, and D. T. Manahan.** 2001. Larval dispersal potential of the tubeworm *Riftia pachyptila* at deep-sea hydrothermal vents. *Nature* **411:**77–80.
25. **Marteinsson, V. T., J.-L. Birrien, A.-L. Reysenbach, M. Vernet, D. Marie, A. Gambacorta, P. Messner, U. B. Sleytr, and D. Prieur.** 1999. *Thermococcus barophilus* sp. nov., a new barophilic and hyperthermophilic archaeon isolated under high hydrostatic pressure from a deep-sea hydrothermal vent. *Int. J. Syst. Bacteriol.* **49:**351–359.
26. **Marteinsson, V. T., P. Moulin, J.-L. Birrien, A. Gambacorta, M. Vernet, and D. Prieur.** 1997. Physiological responses to stress conditions and barophilic behavior of the hyperthermophilic vent archaeon *Pyrococcus abyssi. Appl. Environ. Microbiol.* **63:**1230–1236.
27. **Sharma, A., J. H. Scott, G. D. Cody, M. L. Fogel, R. M. Hazen, R. J. Hemley, and W. T. Huntress.** 2002. Microbial activity at gigapascal pressures. *Science* **295:**1514–1516.
28. **Sogin, M. L., H. G. Morrison, J. A. Huber, D. M. Welch, S. M. Huse, P. R. Neal, J. M. Arrieta, and G. J. Herndl.** 2006. Microbial diversity in the deep sea and the underexplored "rare biosphere." *Proc. Natl. Acad. Sci. USA* **103:**12115–12120.
29. **Tabor, P. S., J. W. Deming, K. Ohwada, and R. R. Colwell.** 1982. Activity and growth of microbial populations in pressurized deep-sea sediment and animal gut samples. *Appl. Environ. Microbiol.* **44:**413–422.
30. **Vezzi, A., S. Campanaro, M. D'Angelo, F. Simonato, N. Vitulo, F. M. Lauro, A. Cestaro, G. Malacrida, B. Simionati, C. Cannata, C. Romualdi, D. H. Bartlett, and G. Valle.** 2005. Life at depth: *Photobacterium profundum* genome sequence and expression analysis. *Science* **307:**1459–1461.
31. **Worden, A. Z., M. L. Cuvelier, and D. H. Bartlett.** 2006. In-depth analyses of marine microbial community genomics. *Trends Microbiol.* **14:**331–336.
32. **Xiao, X., P. Wang, X. Zeng, D. H. Bartlett, and F. Wang.** 2007. *Shewanella psychrophila* sp. nov. and *Shewanella piezotolerans* sp. nov., isolated from west Pacific deep-sea sediment. *Int. J. Syst. Evol. Microbiol.* **57:**60–65.
33. **Yayanos, A. A.** 1986. Evolutional and ecological implications of the properties of deep-sea barophilic bacteria. *Proc. Natl. Acad. Sci. USA* **83:**9542–9546.
34. **Yayanos, A. A.** 1995. Microbiology to 10,500 meters in the deep sea. *Annu. Rev. Microbiol.* **49:**777–805.
35. **Yayanos, A. A., A. S. Dietz, and R. Van Boxtel.** 1979. Isolation of a deep-sea barophilic bacterium and some of its growth characteristics. *Science* **205:**808–810.
36. **Yayanos, A. A., A. S. Dietz, and R. Van Boxtel.** 1981. Obligately barophilic bacterium from the Mariana trench. *Proc. Natl. Acad. Sci. USA* **78:**5212–5215.
37. **ZoBell, C. E.** 1952. Bacterial life at the bottom of the Philippine Trench. *Science* **115:**507–508.
38. **ZoBell, C. E., and F. H. Johnson.** 1949. The influence of hydrostatic pressure on the growth and viability of terrestrial and marine bacteria. *J. Bacteriol.* **57:**179–189.
39. **ZoBell, C. E., and R. Y. Morita.** 1959. Deep-sea bacteria. Scientific resuts of the Danish deep-sea expedition round the world 1950–2. *Galathea Rep.* **1:**139–154.
40. **ZoBell, C. E., and C. H. Oppenheimer.** 1950. Some effects of hydrostatic pressure on the multiplication and morphology of marine bacteria. *J. Bacteriol.* **60:**771–781.

High-Pressure Microbiology
Edited by C. Michiels, D. H. Bartlett, and A. Aertsen
© 2008 ASM Press, Washington, DC

Chapter 12

Isolation, Cultivation, and Diversity of Deep-Sea Piezophiles

Chiaki Kato, Yuichi Nogi, and Shizuka Arakawa

We have isolated numerous cold deep-sea adapted microorganisms using deep-sea research submersibles. In keeping with the nomenclature described in the preceding chapter, these microbes are described here as piezophiles (formerly referred to as barophiles; 52). The general growth characteristics of piezophiles are presented in Fig. 1.

Many of the isolates we have obtained are novel psychrophilic ("cold-loving") bacteria, and we have identified several new piezophilic species, i.e., *Photobacterium profundum, Shewanella violacea, Moritella japonica, Moritella yayanosii, Psychromonas kaikoae,* and *Colwellia piezophila.* These piezophiles belong to five genera within the *Gammaproteobacteria* subgroup and produce significant amounts of unsaturated fatty acids in their cell membranes, presumably to maintain appropriate fluidity levels in cold and high-pressure environments. Piezophilic microorganisms have been identified in many deep-sea sediments obtained from many of the world's oceans. Therefore, these microbes are well distributed on our planet. In this chapter, we focus on the isolation, taxonomy, and diversity of piezophilic microorganisms and their habitats.

PIEZOPHILES ARE ADAPTED TO THE DEEP-SEA ENVIRONMENT

It has been suggested that life may have originated in the deep sea some 3.5 to 4 billion years ago, where it was protected from the damaging effects of UV light. The deep sea is a particularly high-pressure environment, and hydrostatic pressure could have been a very important stimulus for early forms of life. Scientists have proposed that life might have originated in deep-sea hydrothermal vents (46), and thus it appears to be possible that high-pressure-adapted mechanisms of gene expression could represent a feature present in early forms of life (21). It has been reported that the primary chemical reactions involved in the polymerization of organic materials (i.e., amino acids) could have occurred in such high-pressure and high-temperature environments (16). Thus, the study of deep-sea microorganisms not only may enhance our understanding of specific adaptations to abyssal

Chiaki Kato, Yuichi Nogi, and Shizuka Arakawa • Extremobiosphere Research Center, Japan Agency for Marine-Earth Science and Technology, Yokosuka 237-0061, Japan.

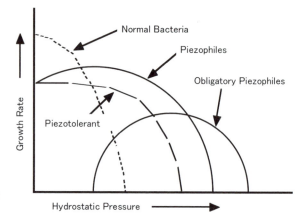

Figure 1. Characterization of piezophilic growth properties.

and hadal ocean realms but also may provide valuable insights into the origin and evolution of life on our planet.

TAXONOMY OF THE PIEZOPHILES

Bacteria living in the deep sea have several unusual features which allow them to thrive in their extreme environments. We have isolated and characterized several piezophilic and piezotolerant bacteria from cold deep-sea sediments at depths ranging from 2,500 to 11,000 m using sterilized sediment samplers by means of the submersibles *SHINKAI 6500* and *KAIKO,* systems operated by Japan Agency for Marine-Earth Science and Technology (JAMSTEC) (19, 25, 26). Most isolated strains are not only piezophilic but also psychrophilic and cannot be cultured at temperatures higher than 20°C.

The isolated deep-sea piezophilic bacterial strains have been characterized in an effort to understand the interaction between the deep-sea environment and its microbial inhabitants (22, 31, 53, 54). Thus far, all piezophilic bacterial isolates fall into the *Gammaproteobacteria* according to phylogenetic classifications based on 5S and 16S rRNA gene sequence information (8, 19, 31). DeLong et al. (8) reported that the cultivated psychrophilic and piezophilic deep-sea bacteria were affiliated with one of five genera within the *Gammaproteobacteria* subgroup: *Shewanella, Photobacterium, Colwellia, Moritella,* and an unidentified genus. The only deep-sea piezophilic bacterial species of these genera were named to be *Shewanella benthica* (11, 29) and *Colwellia hadaliensis* (12) prior to the reports by the JAMSTEC group. We have identified several novel piezophilic species within these genera based on the results of chromosomal DNA-DNA hybridization studies and several other taxonomic properties. Both previously described and novel species have been identified among the piezophilic bacterial isolates. Based upon these studies we have indicated that cultivated psychrophilic and piezophilic deep-sea bacteria could be affiliated with one of five genera within the *Gammaproteobacteria* subgroup: *Shewanella, Photobacterium, Colwellia, Moritella,* and *Psychromonas,* which was formally classified as "an unidentified genus" (41). Figure 2 shows the phylogenetic relations between the taxonomically identified piezophilic species and other bacteria within the *Gammaproteobacteria* subgroup. The taxonomic features of the piezophilic genera were determined as described below.

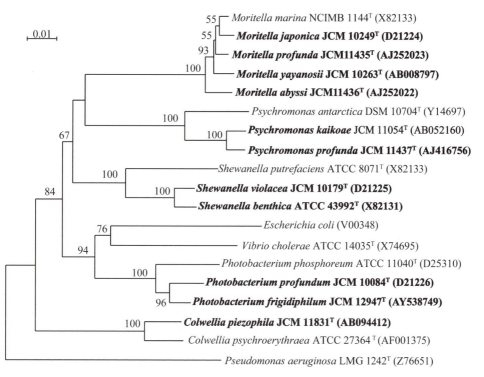

Figure 2. Phylogenetic tree showing the relationships between isolated deep-sea piezophilic bacteria (in bold) within the *Gammaproteobacteria* subgroup determined by comparing 16S rRNA gene sequences using the neighbor-joining method (references for species description are indicated in the text). The scale represents the average number of nucleotide substitutions per site. Bootstrap values (percent) are shown for frequencies above the threshold of 50%.

The Genus *Shewanella*

Members of the genus *Shewanella* are not unique to marine environments. They are gram-negative, aerobic, and facultatively anaerobic *Gammaproteobacteria* (29). The type strain of this genus is *Shewanella putrefaciens,* which is a bacterium formerly known as *Pseudomonas putrefaciens* (29, 43). Recently, however, several novel marine *Shewanella* species have been isolated and described. These isolates are not piezophilic species, and at this time *S. benthica* and *S. violacea* are the only known members of the genus *Shewanella* showing piezophilic growth properties (40). The piezophilic *Shewanella* strains PT99, DB5501, DB6101, DB6705, DB6906, DB172F, DB172R, and DB21MT-2 were all identified as members of the same species, *S. benthica* (24, 40). The psychrophilic and piezophilic *Shewanella* strains, including *S. violacea* and *S. benthica,* produce eicosapentaenoic acid (EPA). The production of a long-chain polyunsaturated fatty acid (PUFA) is a property shared by many deep-sea bacteria to maintain cell membrane fluidity under conditions of extreme cold and high hydrostatic pressure (14; see also chapter 14). *S. violacea* strain DSS12 has been studied extensively, particularly with respect to its molecular mechanisms of adaptation to high pressure (23, 34, 35). This strain is moderately piezophilic, with a fairly constant doubling time at pressures between 0.1 and 70 MPa, whereas the doubling times of most piezophilic *S. benthica* strains change substantially with increasing pressure. As there are few differences in the growth characteristics of strain DSS12 under different pressure conditions, this strain is a very convenient deep-sea bacterium for use in studies on the mechanisms of adaptation to high-pressure environments. The genome analysis of strain DSS12 has been performed because of its value as a model deep-sea piezophilic bacterium (36).

The Genus *Photobacterium*

The genus *Photobacterium* was one of the earliest known bacterial taxa and was first proposed by Beijerinck in 1889 (4). Phylogenetic analysis based on 16S rRNA gene sequences has shown that the genus *Photobacterium* falls within the *Gammaproteobacteria* and, in particular, is closely related to the genus *Vibrio* (42). *Photobacterium profundum*, a novel species, was identified through studies of the moderately piezophilic strains DSJ4 and SS9 (42). *P. profundum* strain SS9 has been extensively studied with regard to the molecular mechanisms of pressure regulation (3) and subsequently genome sequencing and expression analysis (48). Recently, *Photobacterium frigidiphilum* was reported to be slightly piezophilic: its optimal pressure for growth is 10 MPa (45). Thus, *P. profundum* and *P. frigidiphilum* are the only species within the genus *Photobacterium* known to display piezophily and the only two known to produce the long-chain PUFA EPA. No other known species of *Photobacterium* produces EPA (42).

The Genus *Colwellia*

Species of the genus *Colwellia* are defined as facultative anaerobic and psychrophilic bacteria (12) which belong to the *Gammaproteobacteria*. In the genus *Colwellia*, the only deep-sea piezophilic species reported was *C. hadaliensis* strain BNL-1 (12), although no public culture collections maintain this species and/or its 16S rRNA gene sequence information. Bowman et al. (5) reported that *Colwellia* species produce the long-chain PUFA docosahexaenoic acid (DHA). We have recently isolated the obligately piezophilic strain Y223G[T] from sediment collected from the bottom of a deep-sea fissure in the Japan Trench, and this strain was identified as *C. piezophila* (37). Regarding fatty acids, this strain does not produce EPA or DHA in the membrane layer, whereas high levels of monounsaturated fatty acids (16:1 fatty acids) are produced. This observation suggests that the possession of long-chain PUFA is not a requirement for piezophily; however, the production of unsaturated fatty acids could be a common property of piezophiles.

The Genus *Moritella*

The type strain of the genus *Moritella* is *Moritella marina*, previously known as *Vibrio marinus* (47). It is one of the most common psychrophilic organisms isolated from marine environments. *M. marina* is a piezosensitive bacterial species closely related to the genus *Shewanella* on the basis of 16S rRNA gene sequence data. Strain DSK1, a moderately piezophilic bacterium isolated from the Japan Trench, was identified as *Moritella japonica* (39). This was the first piezophilic species identified in the genus *Moritella*. Production of the long-chain PUFA DHA is a characteristic property of this genus. The extremely piezophilic bacterial strain DB21MT-5 isolated from the world's deepest sea bottom, the Mariana Trench Challenger Deep at a depth of 10,898 m, was also identified as a *Moritella* species and designated *M. yayanosii* (38). The optimal pressure for the growth of *M. yayanosii* strain DB21MT-5 is 80 MPa; this strain is unable to grow at pressures of less than 50 MPa but grows well at pressures as high as 100 MPa (22). The fatty acid composition of piezophilic strains changes as a function of pressure, and in general greater amounts of PUFAs are synthesized at higher growth pressures. Approximately 70% of the membrane lipids in *M. yayanosii* are unsaturated fatty acids, which is a finding consistent with its adaptation to very high pressures (13, 38). Two other species of the genus *Moritella, Moritella abyssi* and *Moritella profunda,* were isolated from a depth of 2,815 m off the West African coast (49); they are moderately piezophilic, with growth properties similar to those of *M. japonica.*

The Genus *Psychromonas*

The genus *Psychromonas* consists of psychrophilic bacteria which are closely related to the genera *Shewanella* and *Moritella.* The type species of the genus *Psychromonas, Psychromonas antarctica,* was isolated as an aerotolerant anaerobic bacterium from a high-salinity pond on the McMurdo ice shelf in Antarctica (33). This strain did not display piezophilic properties. *Psychromonas kaikoae,* isolated from sediment collected from the deepest cold-seep environment with chemosynthesis-based animal communities within the Japan Trench, at a depth of 7,434 m, is a novel obligatory piezophilic bacterium (41). The optimal temperature and pressure for growth of *P. kaikoae* are 10°C and 50 MPa, respectively, and both PUFAs EPA and DHA are produced in the membrane layer. *P. antarctica* does not produce either EPA or DHA in its membrane layer. Strain CNPT-3 was originally described as an unidentified genus of piezophiles by DeLong and colleagues (8), but it has since been shown to be closely related to *P. kaikoae.* Thus, the genus *Psychromonas* is the fifth genus reported to contain piezophilic species within the *Gammaproteobacteria.* In addition to *P. kaikoae, P. profunda* is another piezophilic member of the genus. It is a moderately piezophilic bacterium isolated from deep Atlantic sediments at a depth of 2,770 m (50). This strain is similar to the piezosensitive strain *Psychromonas marina,* which also produces small amounts of DHA. In the genus *Psychromonas,* only *P. kaikoae* produces both EPA and DHA.

PUFAs and Piezophily

The piezophilic and psychrophilic *Shewanella* and *Photobacterium* strains produce EPA (40, 42), *Moritella* strains produce DHA (38, 39), and *Psychromonas kaikoae* produces both EPA and DHA (41), but *Colwellia piezophila* does not produce either of these PUFAs (37). The fatty acid composition of piezophilic strains changes as a function of pressure, and in general greater amounts of PUFAs are synthesized under higher-pressure conditions for their growth (9, 10). Psychrophilic and piezophilic bacteria were believed to produce one of the long-chain PUFAs, either EPA or DHA, but this does not appear to be obligatory. For example, Allen et al. (1) reported that monounsaturated fatty acids, but not PUFAs, are required for the growth of the piezophilic bacterium *P. profundum* SS9 based on the analysis of pressure-sensitive mutants. In their experiment, 18:1 fatty acids proved to be necessary for growth under low-temperature and/or high-pressure conditions. *C. piezophila* had no 18:1 fatty acids but produced a large amount of 16:1 fatty acids in its cell membrane layer. All piezophilic and psychrophilic bacteria analyzed so far have 16:1 fatty

Table 1. Whole-cell fatty acid composition of piezophilic isolates (type strains)[a]

Fatty acid	% Fatty acid in:				
	Sh	Ph	Co	Mo	Ps
12:0	2	2	1		1
14:0	13	3	3	15	6
15:0		1	3	1	1
16:0	14	9	31	13	15
17:0					
18:0		1			
iso-13:0	5	2			
iso-14:0		4			
iso-15:0	11	2			
iso-16:0		15			
14:1		3	9	6	10
15:1			2		
16:1	31	31	50	53	55
17:1					
18:1	2	9		1	2
[EPA] 20:5	16	13			2
[DHA] 22:6				11	2
3OH-12:0	1	5	1		2
3OH-iso-13:0	5				
3OH-14:0					4
Unsaturated (%)	49	56	61	71	71
Saturated (%)	51	44	39	29	29
Ratio (unsaturated/saturated)	0.96	1.27	1.56	2.45	2.45

[a]Sh, *Shewanella benthica* ATCC 43992[T]; Ph, *Photobacterium profundum* JCM 10084[T]; Co, *Colwellia piezophila* Y223G[T]; Mo, *Moritella yayanosii* JCM 10263[T]; Ps, *Psychromonas kaikoae* JCM 11054[T].

acids (Table 1); thus, this fatty acid would appear to be one of the components required for high-pressure growth. The fatty acid compositions of those piezophilic strains are distinct depending on their genus, and commonly large amounts of unsaturated fatty acids (49 ~ 71%) are involved in their membrane layer as shown in Table 1.

HIGH-PRESSURE CULTIVATION EXPERIMENTS BY THE DEEPBATH SYSTEM

For handling piezophiles for further study, JAMSTEC developed a deep-sea baro-piezophile and thermophile isolation and cultivation system, referred to as the DEEPBATH system. The DEEPBATH system consists of four separate devices: (1) a pressure-retaining sampling device, (2) a dilution device under pressure conditions, (3) an isolation device, and (4) a cultivation device (20, 27). The system is controlled by central regulation systems, and the pressure and temperature ranges of the devices are from 0.1 to 65 MPa and from 0 to 150°C, respectively. The capacity of the cultivation devices (two sets) is 1.5 liters each, and therefore, cultures of up to 3 liters can be obtained. The construction of the system and the sample stream are shown in Fig. 3.

Using the DEEPBATH system, we have attempted to study the changes in the microbial community in the Japan Trench sediment from a depth of 6,292 m during cultivation with-

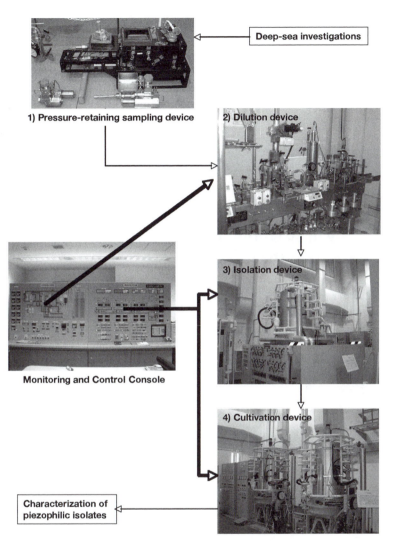

Figure 3. The DEEPBATH system. The system is composed of four devices: (1) a pressure-retaining sampling device, (2) a dilution device under pressure conditions, (3) an isolation device, and (4) a cultivation device. The system is controlled by the monitoring and control console.

out decompression (51). A deep-sea sediment sample (about 5 cm^3) was obtained using the pressure-retaining sampling device under conditions of 65 MPa and 2°C. The sample obtained was diluted almost 100-fold with sterilized seawater with the dilution device at a pressure of 65 MPa and a temperature of 2°C. Then 10 ml of the dilution solution was inoculated into the cultivation device (containing 1.5 liters of marine broth 2216 medium) with no change in pressure or temperature. Cultivation was repeated five times at 65 MPa and 10°C without decompression, and each time cultivation continued for approximately

4 days until the early stationary phase had been reached, as confirmed with a laser beam microbe density measurement system (27). Each cultivated bacterial suspension was centrifuged, DNA was purified, and total fatty acid composition was analyzed from the mixture to study microbial diversity. From the analyses of 16S rRNA gene sequences after cultivation at 65 MPa, two groups of the bacterial genera *Shewanella* and *Moritella* were identified (51). Changes in the fatty acid profiles during high-pressure cultivation were also indicative of these two genera, *Shewanella* producing EPA and *Moritella* producing DHA. The amount of EPA decreased while that of DHA increased during consecutive cultivation without decompression, which indicated that the genus *Moritella* became abundant (Fig. 4). These results suggested that the piezophilic *Moritella* species might be better adapted to growth under limiting oxygen conditions under high pressure than members of the piezophilic *Shewanella* (51).

WHERE DO THE PIEZOPHILES COME FROM: THE POLAR AND DEEP SEAS?

Most *Shewanella* spp. are isolated from ocean environments, and some are psychrophilic or psychrotolerant bacteria. The piezophilic *Shewanella* species *S. benthica* and *S. violacea* are also categorized as psychrophilic at atmospheric pressure (40). *Shewanella gelidimarina* and *Shewanella frigidimarina* isolated from Antarctic ice (6) and *Shewanella hanedai* isolated from the Arctic Ocean (18) are cold-adapted psychrotolerant bacteria that grow well at low temperature. A phylogenetic tree of these *Shewanella* species within the *Gammaproteobacteria* subgroup constructed based on 16S rRNA gene sequences is shown in Fig. 5.

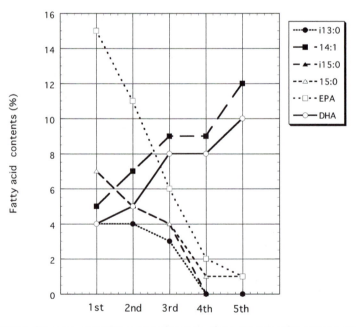

Figure 4. Changes in major fatty acid profiles during five consecutive high-pressure cultivations (65 MPa) of deep-sea sediment samples using the DEEPBATH system.

In this tree, two major branches are recognizable in the genus *Shewanella,* indicated by *Shewanella* group 1 and *Shewanella* group 2. Deep-sea *Shewanella* spp. forming the *Shewanella* piezophilic branch are categorized as members of group 1. Interestingly, most *Shewanella* species shown to be psychrophilic or psychrotolerant also belong to group 1. The other species in group 1, *Shewanella pealeana* and *Shewanella woodyi,* isolated from ocean squid and detritus, respectively, grow optimally at 25°C (28, 30), and thus these strains might also be included in the group of cold-adapted bacteria. Most *Shewanella* species in group 2 are not cold-adapted bacteria. They grow well under mesophilic conditions at 25 to 35°C. *S. frigidimarina,* which can grow optimally below 25°C, is the only exception in this category, although this species belongs to group 2 (24).

The growth of some of these *Shewanella* species under high-pressure conditions indicates that the members of *Shewanella* group 1 show piezophilic (*S. benthica* and *S. violacea*) or

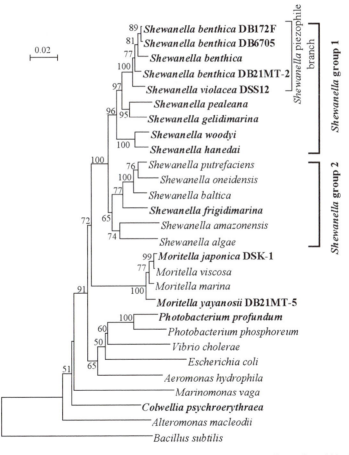

Figure 5. Phylogenetic tree showing the relationships of the *Shewanella* species within the *Gammaproteobacteria* subgroup constructed based on 16S rRNA gene sequences with the neighbor-joining method. The scale represents the average number of nucleotide substitutions per site. Bootstrap values (percent) were calculated from 1,000 trees. Psychrophilic and/or piezophilic bacteria are shown in bold.

piezotolerant (*S. gelidimarina* and *S. hanedai*) growth properties, although the members of *Shewanella* group 2 generally show piezosensitive growth, i.e., no growth at a pressure of 50 MPa (24). Only a limited number of experiments have been performed examining the growth of these bacteria under high-pressure conditions, but generally members of *Shewanella* group 1 are characterized as cold adapted and pressure tolerant, whereas the members of *Shewanella* group 2 are mostly mesophilic and pressure sensitive. Some *Shewanella* species are known to produce PUFAs, particularly EPA. It is clear that the members of *Shewanella* group 1 produce substantial amounts of EPA (11 to 16% of total fatty acids), whereas members of *Shewanella* group 2 produce no EPA or only limited amounts. In terms of other fatty acids, the membrane lipid profiles of members of the genus *Shewanella* are basically similar. This observation also supports the view described above (24).

On the basis of the properties of *Shewanella* species, we propose that two major branches of the genus *Shewanella* be recognized taxonomically, *Shewanella* group 1 and group 2 (Fig. 5). The two subgenus branches of *Shewanella* would be as follows: *Shewanella* group 1 is characterized as a group of high-pressure, cold-adapted species that produce substantial amounts of EPA, and *Shewanella* group 2 is characterized as a group of mostly mesophilic and pressure-sensitive species.

The deep-sea bottom and other cold-temperature environments are probably similar in terms of microbial diversity. Members of *Shewanella* group 1 live in such environments, and most of them show piezophilic or piezotolerant growth properties. In this regard, it is interesting to consider the influence of the ocean circulation, as deep ocean water is derived from polar waters (in the Arctic and/or Antarctic region) that sink to the deep-sea bottom (44), probably along with microbes (Fig. 6). It was reported that *Psychrobacter pacif-*

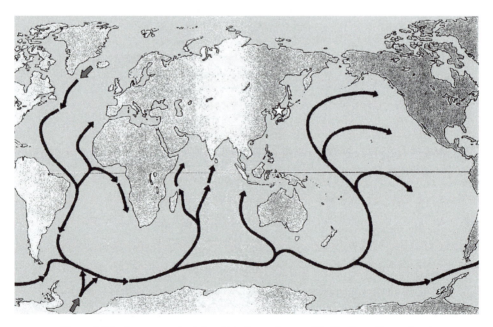

Figure 6. The deep ocean circulation (data from reference 44). The Japan Sea, a closed ocean, is indicated by the star.

icensis isolated from seawater of the Japan Trench at a depth of 5,000 to 6,000 m is taxonomically similar to the Antarctic isolates *Psychrobacter immobilis, Psychrobacter gracincola,* and *Psychrobacter fridigicola* (32). The occurrence of *Psychrobacter* in cold seawater deep in the Japan Trench and at the surface of the Southern Ocean suggests that bacterial habitation of the deep sea and bacterial evolution have been influenced by the global deep ocean circulation linked to the sinking of cooled seawater in polar regions. Thus, it is possible that the ocean circulation may be one of the major factors influencing microbial diversity on our planet.

PIEZOPHILES IN THE CLOSED JAPAN SEA OCEAN

The Japan Sea is a closed ocean that is separated by four straits (Mamiya, Sohya, Tsugaru, and Tsushima) from other oceans. These straits are very shallow, with depths of less than 200 m; thus, it might be difficult to mix deep-sea waters with other oceans, i.e., the Pacific Ocean. Therefore, since the Japan Sea bottom water could be isolated from other ocean bottoms, the former may serve as a model ocean for consideration of piezophilic diversity independently of ocean current. For such studies, we have identified piezophiles from the bottom of the Japan Sea at a depth of about 3,100 m (2).

We have analyzed the microbial community structures by the terminal restriction fragment length polymorphism for the bacterial 16S rRNA gene (17) and determined that the community is drastically changed at different pressure conditions of cultivation using the DEEPBATH system. In the original sediment, many bacteria appeared, but after cultivation at 0.1, 30, and 50 MPa, six, three, and two major peaks remained, respectively. The peaks corresponding to the bacterial genera *Psychromonas, Moritella,* and *Shewanella* were particularly apparent under high-pressure conditions. In addition, *Moritella* and *Shewanella* also grew well at 50 MPa. These genera contain piezophilic bacterial species (19, 25, 26), so they might include high-pressure-adapted microbes. From these cultivations, three pressure-adapted psychrophilic strains were isolated which were identified as *Psychromonas, Shewanella,* and *Moritella* strains based upon their 16S rRNA gene sequences. The isolated *Shewanella* strain (no. 302) was identical to piezophilic *Shewanella benthica* (40), with more than 99% similarity in its 16S rRNA gene sequence; thus, we named this strain *S. benthica* 302. The isolated *Psychromonas* strain (no. 503) and *Moritella* strain (no. 304) were also very closely related to the piezophilic species *Psychromonas profunda* (50) and *Moritella yayanosii* (38), respectively. Growth profiles under elevated pressure were examined for these isolates, and *Psychromonas* strain 503, *S. benthica* strain 302, and *Moritella* strain 304 grew very well up to 40 MPa (Fig. 7). *Moritella* strain 304 and *S. benthica* strain 302 could grow at 70 MPa, although their optimal pressure conditions were 20 and 30 MPa, respectively. These results indicated that *S. benthica* strain 302 and *Moritella* strain 304 are piezophilic and *Psychromonas* strain 503 is piezotolerant.

These isolates are closely related to the piezophilic species from Pacific Ocean trenches (19, 22, 25). How might such deep-sea-adapted microbes come from other oceans to the bottom of the Japan Sea? The answer might be that the isolates are not obligatory piezophiles, which would allow for growth and survival under atmospheric pressure conditions, and thus these organisms could have followed a path through one of the shallow straits from other oceans. Actually, Gamo et al. (15) reported from studies of geochemical tracer analyses that the surface seawater could be supplied to the bottom layers by the

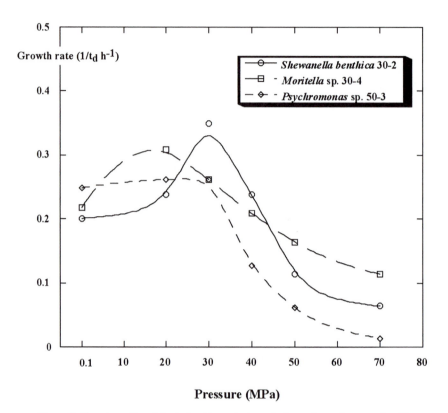

Figure 7. Growth profiles of the isolated bacteria from the Japan Sea sediment under different pressure conditions.

deep convection system in the Japan Sea. Our results suggested that the piezophilic microorganisms, with the exception of obligatory piezophiles, could be distributed to any of the oceans of our planet.

CONCLUSION

Cultured deep-sea piezophilic bacteria are affiliated with one of five genera within the *Gammaproteobacteria* subgroup: *Shewanella, Photobacterium, Colwellia, Moritella,* and *Psychromonas.* These piezophiles are characterized by high levels of unsaturated fatty acids in their cell membrane layers, but PUFAs like EPA and DHA are not necessarily required for high-pressure growth. The diversity of piezophilic bacteria is closely linked with the global deep-sea ocean circulation, but some of the closed oceans, like the Japan Sea, also contain piezophilic bacteria taxonomically similar to deep-sea microbes in the open oceans. These observations indicate that piezophilic bacteria could be present in any of the deep-sea cold and high-pressure environments.

Acknowledgments. We are very grateful to Koki Horikoshi for his direction in the study of extremophiles. We also thank the members of the ship and submersible operation division at JAMSTEC for their efforts in collecting samples from the deep-sea environment, and Howard K. Kuramitsu for assistance in editing the manuscript.

REFERENCES

1. **Allen, E. E., D. Facciotti, and D. H. Bartlett.** 1999. Monounsaturated but not polyunsaturated fatty acids are required for growth of the deep-sea bacterium *Photobacterium profundum* SS9 at high pressure and low temperature. *Appl. Environ. Microbiol.* **65:**1710–1720.
2. **Arakawa, S., Y. Nogi, T. Sato, Y. Yoshida, R. Usami, and C. Kato.** 2006. Diversity of piezophilic microorganisms in the closed ocean Japan Sea. *Biosci. Biotechnol. Biochem.* **70:**749–752.
3. **Bartlett, D. H.** 1999. Microbial adaptations to the psychrosphere/piezosphere. *J. Mol. Microbiol. Biotechnol.* **1:**93–100.
4. **Beijerinck, M. W.** 1889. Le *Photobacterium luminosum,* bactérie luminosum de la mer Nord. *Arch. Neerl. Sci.* **23:**401–427.
5. **Bowman, J. P., J. J. Gosink, S. A. McCammon, T. E. Lewis, D. S. Nichols, P. D. Nichols, J. H. Skerratt, J. T. Staley, and T. A. McMeekin.** 1998. *Colwellia demingiae* sp. nov., *Colwellia hornerae* sp. nov., *Colwellia rossensis* sp. nov., and *Colwellia psychrotropica* sp. nov.: psychrophilic Antarctic species with the ability to synthesize docosahexaenoic acid (22:6ω3). *Int. J. Syst. Bacteriol.* **48:**1171–1180.
6. **Bowman, J. P., S. A. McCammon, D. S. Nichols, J. H. Skerratt, S. M. Rea, P. D. Nichols, and T. A. McMeekin.** 1997. *Shewanella gelidimarina* sp. nov. and *Shewanella frigidimarina* sp. nov., novel Antarctic species with the ability to produce eicosapentaenoic acid (20:5ω3) and grow anaerobically by dissimilatory Fe(III) reduction. *Int. J. Syst. Bacteriol.* **47:**1040–1047.
7. **Colwell, R. R., and R. Y. Morita.** 1964. Reisolation and emendation of description of *Vibrio marinus* (Russell) Ford. *J. Bacteriol.* **88:**831–837.
8. **DeLong, E. F., D. G. Franks, and A. A. Yayanos.** 1997. Evolutionary relationship of cultivated psychrophilic and barophilic deep-sea bacteria. *Appl. Environ. Microbiol.* **63:**2105–2108.
9. **DeLong, E. F., and A. A. Yayanos.** 1985. Adaptation of the membrane lipids of a deep-sea bacterium to changes in hydrostatic pressure. *Science* **228:**1101–1103.
10. **DeLong, E. F., and A. A. Yayanos.** 1986. Biochemical function and ecological significance of novel bacterial lipids in deep-sea prokaryotes. *Appl. Environ. Microbiol.* **51:**730–737.
11. **Deming, J. W., H. Hada, R. R. Colwell, K. R. Luehrsen, and G. E. Fox.** 1984. The nucleotide sequence of 5S rRNA from two strains of deep-sea barophilic bacteria. *J. Gen. Microbiol.* **130:**1911–1920.
12. **Deming, J. W., L. K. Somers, W. L. Straube, D. G. Swartz, and M. T. Macdonell.** 1988. Isolation of an obligately barophilic bacterium and description of a new genus, *Colwellia* gen. nov. *Syst. Appl. Microbiol.* **10:**152–160.
13. **Fang, J., M. J. Barcelona, Y. Nogi, and C. Kato.** 2000. Biochemical function and geochemical significance of novel phospholipids of the extremely barophilic bacteria from the Mariana Trench at 11,000 meters. *Deep-Sea Res.* **147:**1173–1182.
14. **Fang, J., O. Chan, C. Kato, T. Sato, T. Peeples, and K. Niggemeyer.** 2003. Phospholipid FA of piezophilic bacteria from the deep sea. *Lipids* **38:**885–887.
15. **Gamo, T., N. Momoshima, and S. Tolmachyov.** 2001. Recent upward shift of the deep convection system in the Japan Sea, as inferred from the geochemical tracers tritium, oxygen, and nutrients. *Geophys. Res. Lett.* **28:**4143–4146.
16. **Imai, E., H. Honda, K. Hatori, A. Brack, and K. Matsuno.** 1999. Elongation of oligopeptides in a simulated submarine hydrothermal system. *Science* **283:**831–833.
17. **Inagaki, F., Y. Sakihama, A. Inoue, C. Kato, and K. Horikoshi.** 2002. Molecular phylogenetic analyses of reverse-transcribed bacterial rRNA obtained from deep-sea cold seep sediments. *Environ. Microbiol.* **4:**277–294.
18. **Jensen, M. J., B. M. Tebo, P. Baumann, M. Mandel, and K. H. Nealson.** 1980. Characterization of *Alteromonas hanedai* (sp. nov.), a non-fermentative luminous species of marine origin. *Curr. Microbiol.* **3:**311–315.
19. **Kato, C.** 1999. Barophiles (piezophiles), p. 91–111. *In* K. Horikoshi and K. Tsujii (ed.), *Extremophiles in Deep-Sea Environments.* Springer-Verlag, Tokyo, Japan.
20. **Kato, C.** 2006. The handling of piezophilic microorganisms. *Methods Microbiol.* **35:**733–741.
21. **Kato, C., and K. Horikoshi.** 1996. Gene expression under high pressure. *Prog. Biotechnol.* **13:**59–66.

22. **Kato, C., L. Li, Y. Nakamura, Y. Nogi, J. Tamaoka, and K. Horikoshi.** 1998. Extremely barophilic bacteria isolated from the Mariana Trench, Challenger Deep, at a depth of 11,000 meters. *Appl. Environ. Microbiol.* **64:**1510–1513.

23. **Kato, C., K. Nakasone, M. H. Qureshi, and K. Horikoshi.** 2000. How do deep-sea microorganisms respond to changes in environmental pressure?, p. 277–291. *In* K. B. Storey and J. M. Storey (ed.), *Cell and Molecular Response to Stress,* vol. 1. *Environmental Stressors and Gene Responses.* Elsevier Science B. V., Amsterdam, The Netherlands.

24. **Kato, C., and Y. Nogi.** 2001. Correlation between phylogenetic structure and function: examples from deep-sea *Shewanella. FEMS Microbiol. Ecol.* **35:**223–230.

25. **Kato, C., T. Sato, and K. Horikoshi.** 1995. Isolation and properties of barophilic and barotolerant bacteria from deep-sea mud samples. *Biodivers. Conserv.* **4:**1–9.

26. **Kato, C., T. Sato, Y. Nogi, and K. Nakasone.** 2004. Piezophiles: high pressure-adapted marine bacteria. *Mar. Biotechnol.* **6:**S195–S201.

27. **Kyo, M., T. Tuji, H. Usui, and T. Itoh.** 1991. Collection, isolation and cultivation system for deep-sea microbes study: concept and design. *Oceans* **1:**419–423.

28. **Leonardo, M. R., D. P. Moser, E. Barbieri, C. A. Brantner, B. J. MacGregor, B. J. Paster, E. Stackebrandt, and K. H. Nealson.** 1999. *Shewanella pealeana* sp. nov., a member of the microbial community associated with the accessory nidamental gland of the squid *Loligo pealei. Int. J. Syst. Bacteriol.* **49:**1341–1351.

29. **MacDonell, M. T., and R. R. Colwell.** 1985. Phylogeny of the Vibrionaceae, and recommendation for two new genera, *Listonella* and *Shewanella. Syst. Appl. Microbiol.* **6:**171–182.

30. **Makemson, J. C., N. R. Fulayfil, W. Landry, L. M. Van Ert, C. F. Wimpee, E. A. Widder, and J. F. Case.** 1997. *Shewanella woodyi* sp. nov., an exclusively respiratory luminous bacterium isolated from the Alboran Sea. *Int. J. Syst. Bacteriol.* **47:**1034–1039.

31. **Margesin, R., and Y. Nogi.** 2004. Psychropiezophilic microorganisms. *Cell. Mol. Biol.* **50:**429–436.

32. **Maruyama, A., D. Honda, H. Yamamoto, K. Kitamura, and T. Higashihara.** 2000. Phylogenetic analysis of psychrophilic bacteria isolated from the Japan Trench, including a description of the deep-sea species *Psychrobacter pacificensis* sp. nov. *Int. J. Syst. Evol. Microbiol.* **50:**835–846.

33. **Mountfort, D. O., F. A. Rainey, J. Burghardt, F. Kasper, and E. Stackebrant.** 1998. *Psychromonas antarcticus* gen. nov., sp. nov., a new aerotolerant anaerobic, halophilic psychrophile isolated from pond sediment of the McMurdo ice shelf, Antarctica. *Arch. Microbiol.* **169:**231–238.

34. **Nakasone, K., A. Ikegami, C. Kato, R. Usami, and K. Horikoshi.** 1998. Mechanisms of gene expression controlled by pressure in deep-sea microorganisms. *Extremophiles* **2:**149–154.

35. **Nakasone, K., A. Ikegami, H. Kawano, R. Usami, C. Kato, and K. Horikoshi.** 2002. Transcriptional regulation under pressure conditions by the RNA polymerase σ^{54} factor with a two component regulatory system in *Shewanella violacea. Extremophiles* **6:**89–95.

36. **Nakasone, K., H. Mori, T. Baba, and C. Kato.** 2003. Whole-genome analysis of piezophilic and psychrophilic microorganism. *Kagaku to Seibutu* **41:**32–39. (In Japanese.)

37. **Nogi, Y., S. Hosoya, C. Kato, and K. Horikoshi.** 2004. *Colwellia piezophila* sp. nov., a novel piezophilic species from deep-sea sediments of the Japan Trench. *Int. J. Syst. Evol. Microbiol.* **54:**1627–1631.

38. **Nogi, Y., and C. Kato.** 1999. Taxonomic studies of extremely barophilic bacteria isolated from the Mariana Trench, and *Moritella yayanosii* sp. nov., a new barophilic bacterial species. *Extremophiles* **3:**71–77.

39. **Nogi, Y., C. Kato, and K. Horikoshi.** 1998. *Moritella japonica* sp. nov., a novel barophilic bacterium isolated from a Japan Trench sediment. *J. Gen. Appl. Microbiol.* **44:**289–295.

40. **Nogi, Y., C. Kato, and K. Horikoshi.** 1998. Taxonomic studies of deep-sea barophilic *Shewanella* species, and *Shewanella violacea* sp. nov., a new barophilic bacterial species. *Arch. Microbiol.* **170:**331–338.

41. **Nogi, Y., C. Kato, and K. Horikoshi.** 2002. *Psychromonas kaikoae* sp. nov., a novel piezophilic bacterium from the deepest cold-seep sediments in the Japan Trench. *Int. J. Syst. Evol. Microbiol.* **52:**1527–1532.

42. **Nogi, Y., N. Masui, and C. Kato.** 1998. *Photobacterium profundum* sp. nov., a new, moderately barophilic bacterial species isolated from a deep-sea sediment. *Extremophiles* **2:**1–7.

43. **Owen, R., R. M. Legros, and S. P. Lapage.** 1978. Base composition, size and sequence similarities of genome deoxyribonucleic acids from clinical isolates of *Pseudomonas putrefaciens. J. Gen. Microbiol.* **104:**127–138.

44. **Schmitz, W. J., Jr.** 1995. On the interbasin-scale thermohaline circulation. *Rev. Geophys.* **33:**151–173.

45. **Seo, H. J., S. S. Bae, J.-H. Lee, and S.-J. Kim.** 2005. *Photobacterium frigidiphilum* sp. nov., a psychrophilic, lipolytic bacterium isolated from deep-sea sediments of Edison Seamount. *Int. J. Syst. Evol. Microbiol.* **55:**1661–1666.

46. **Stetter, K. O.** 1993. Life at the upper temperature border, p. 195–219. *In* J. Van Tran Than, K. Van Tran Than, J. C. Mounolou, J. Schneider, and C. McKay (ed.), *Frontiers of Life.* Frontières, Gif-sur-Yvette, France.

47. **Urakawa, H., K. Kita-Tsukamoto, S. E. Steven, K. Ohwada, and R. R. Colwell.** 1998. A proposal to transfer *Vibrio marinus* (Russell 1891) to a new genus *Moritella* gen. nov. as *Moritella marina* comb. nov. *FEMS Microbiol. Lett.* **165:**373–378.

48. **Vezzi, A., S. Campanaro, M. D'Angelo, F. Simonato, N. Vitulo, F. M. Laauro, A. Cestaro, G. Malacrida, B. Simionati, N. Cannata, C. Romualdi, D. H. Bartlett, and G. Valle.** 2005. Life at depth: *Photobacterium profundum* genome sequence and expression analysis. *Science* **307:**1459–1461.

49. **Xu, Y., Y. Nogi, C. Kato, Z. Liang, H.-J. Rüger, D. D. Kegel, and N. Glansdorff.** 2003. *Moritella profunda* sp. nov. and *Moritella abyssi* sp. nov., two psychropiezophilic organisms isolated from deep Atlantic sediments. *Int. J. Syst. Evol. Microbiol.* **53:**533–538.

50. **Xu, Y., Y. Nogi, C. Kato, Z. Liang, H.-J. Rüger, D. D. Kegel, and N. Glansdorff.** 2003. *Psychromonas profunda* sp. nov., a psychropiezophilic bacterium from deep Atlantic sediments. *Int. J. Syst. Evol. Microbiol.* **53:**527–532.

51. **Yanagibayashi, M., Y. Nogi, L. Li, and C. Kato.** 1999. Changes in the microbial community in Japan Trench sediment from a depth of 6,292 m during cultivation without decompression. *FEMS Microbiol. Lett.* **170:**271–279.

52. **Yayanos, A. A.** 1995. Microbiology to 10,500 meters in the deep sea. *Annu. Rev. Microbiol.* **49:**777–805.

53. **Yayanos, A. A., A. S. Dietz, and R. Van Boxtel.** 1979. Isolation of a deep-sea barophilic bacterium and some of its growth characteristics. *Science* **205:**808–810.

54. **Zobell, C. E., and F. H. Johnson.** 1949. The influence of hydrostatic pressure on the growth and viability of terrestrial and marine bacteria. *J. Bacteriol.* **57:**179–189.

High-Pressure Microbiology
Edited by C. Michiels, D. H. Bartlett, and A. Aertsen
© 2008 ASM Press, Washington, DC

Chapter 13

Culture-Independent Characterization of Microbial Diversity in Selected Deep-Sea Sediments

Chiaki Kato, Shizuka Arakawa, Takako Sato, and Xiang Xiao

Most of the microbes from the natural environment live in a viable but not cultivable state which makes them difficult to culture in the microbial laboratory. Therefore, only a small percentage of the microbes in these environments can be cultured, isolated, and taxonomically classified. To examine microbial ecosystems in their natural environments, molecular ecological analyses (culture-independent procedures) have been developed. Based upon these approaches, previously unrecognized microbial ecosystems have been identified. In this chapter, we focus on studies of the microbial diversity in deep-sea methane-impacted sediments using culture-independent procedures. We also discuss sulfur and carbon cycling ecosystems in chemosynthetic pathways which are independent of solar-power-dependent energy-generating systems.

MICROBIAL DIVERSITY OF THE DEEP-SEA COLD-SEEP ENVIRONMENTS AROUND JAPAN AND THEIR SULFUR CIRCULATION ECOSYSTEMS

Microbial diversity studies in deep-sea sediments have been performed at depths ranging from ~1,000 m, in Sagami Bay, to ~11,000 m, in the Mariana Trench Challenger Deep (27, 39, 40). Deep-sea "cold-seep" environments have been found worldwide in subduction zones along continental margins (33, 47, 55). The nutrient and energy fluxes from these subseafloor environments promote the presence of abundant and diverse microbial communities. Some of us have studied microbial community structures in cold-seep sediments offshore of Japan, such as the Japan Trench (2, 23, 41), Nankai Trough (4, 37), Sagami Bay (12), and the northeastern Japan Sea (1, 3). These studies have indicated that microbial communities are in most cases composed of anoxic methane-oxidizing archaea (ANME) and sulfate-reducing bacteria (SRB) within the *Euryarchaeota* and *Deltaproteobacteria*, respectively, which are known to be responsible for anaerobic oxidation of

Chiaki Kato, Shizuka Arakawa, and Takako Sato • Extremobiosphere Research Center, Japan Agency for Marine-Earth Science and Technology, Yokosuka 237-0061, Japan. *Xiang Xiao* • Key Laboratory of Marine Biogenetic Resources, Third Institute of Oceanography, State Oceanic Administration, Xiamen 361005, People's Republic of China.

methane (AOM) (for examples, see references 7, 24–26, 38, 51, and 52). However, the distribution of AOM communities associated with fault activity and their effects on diffused methane in the West Pacific Margin remain largely unknown. The features of microbial diversity in four of these deep-sea cold-seep environments (Japan Trench, Nankai Trough, Sagami Bay, and northeastern Japan Sea [Color Plate 2]) are described below.

The Japan Trench Land Slope and the Deepest Cold-Seep Area, at Depths of 5,800 to 7,500 m

The Japan Trench land slope, at depths up to 7,500 m (13, 14, 46), is the deepest cold-seep environment yet identified. It contains several chemosynthesis-based animal communities. *Calyptogena phaseoliformis* (13) and *Maorithyas hadalis* (14, 15) are typical chemosynthesis-based bivalves living in this deep cold-seep environment. Many studies have revealed that chemosynthetic bacteria inhabit the gill surfaces of *Calyptogena* and related bivalves as symbiotic, sulfide-oxidizing bacteria (SOx) which support the growth of their hosts (15, 28, 29, 54). Studies of the microbial communities associated with chemosynthesis-based animal colonies could provide information concerning their diversity and importance in these unique ecosystems, including the specialized functional composition of the bacterial groups present, their utilization of energy sources, and their interactions. In this regard it would be of interest to examine the relationship between the dynamics of these microbial communities and the associated subductional activity. Examples of such studies are presented below, which provide comparisons of the microbial diversity at different depths of cold-seep sediments, from 5,800 to 7,500 m, in the Japan Trench land slope.

Bacteria belonging to the *Deltaproteobacteria* and *Epsilonproteobacteria* (SRB and SOx, respectively) as well as methanogen and ANME members of the archaea were previously concluded to be particularly abundant in the cold-seep communities in the Pacific Ocean (38). An abundance of these microbial communities in the Japan Trench cold-seep sediment have been identified (2, 41). The microbial abundances in the seep sediments from three different water depths between 5,800 and 7,500 m were compared based on clone analysis and phylogenetic considerations. As shown in Fig. 1, *Deltaproteobacteria* and *Epsilonproteobacteria* constituted an increased percentage of the total bacterial population in the sediments obtained from greater water column depths (Fig. 1A), and the group including archaeal methanogens plus ANME was also more abundant at the deeper locations (Fig. 1B). These observations suggest that the microbial activity corresponding to cold-seep activity could be more active in the deeper cold-seep sediments. At a depth of 7,500 m in the Japan Trench off Sanriku, which is located at the bottom of the trench, plate subduction centers might exist. Thus, it is easy to imagine that subductional activities could increase with depth. Cold-seep activity could also be more active in the deeper seep sediments. Therefore, our observations on the abundance of specific cold-seep microbial communities in the deeper sediments might correspond to the prevailing geochemical and geophysical parameters.

In conclusion, the study examining the microbial diversity of Japan Trench cold-seep sediments as a function of water column depth revealed an increased abundance of *Deltaproteobacteria* and *Epsilonproteobacteria* as well as methanogen-plus-ANME archaeal communities at the deeper depths. This was the first report suggesting a correspondence between microbial abundance and cold-seep activities (2).

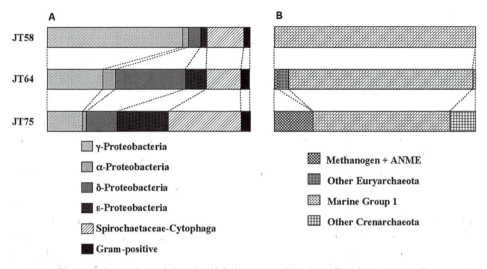

Figure 1. Comparison of the microbial abundance of the Japan Trench cold-seep sediments at different depths (JT58, 5,791 m; JT64, 6,367 m; and JT75, 7,434 m) calculated from the numbers of 16S rRNA gene clones. (A) Bacterial diversity; (B) archaeal diversity.

The Nankai Trough Accretionary Prism at Depths of 600 to 3,300 m

The Nankai Trough is present along the subduction margin between the Shikoku Basin (Philippine Sea plate) and the South-West Honshu arc (Eurasian plate). The current accretionary prism is building from the trench axis and increasing in thickness landward. Several seismic profiles provide excellent images of this prism formation. The Cretaceous to Tertiary Shimanto Belt is exposed from the Ryukyu arc to the middle of the Honshu arc. Considerable geological, geophysical, and geochemical data have been accumulated for this region (31, 34, 53). The Nankai Trough is also one of the areas where cold seeps have been thoroughly investigated. Since 1984, the French-Japanese KAIKO Project has found several cold-seep sites in the accretionary prism by means of submersibles (36). During dive surveys, chemosynthesis-based biological communities served as useful markers for mapping seep sites because anomalies of temperature and geochemistry in cold seeps are more rarely detected than those in hydrothermal areas (36). Gamo et al. (16) reported that there might be a drastic change in pore water chemistry between the interior and exterior of *Calyptogena* communities, because the sedimentary pore water recovered only 0.3 m from the margin of the community showed little indication of in situ sulfate reduction. They also confirmed that surface sediment temperature is higher inside *Calyptogena* communities than outside them. *Calyptogena* is a bivalve associated with endosymbiotic SOx, and researchers have used its communities as markers to investigate the cold-seep environments in the Nankai Trough (5, 47). The description of *Calyptogena* species and their biogeographical properties were reviewed recently by Kojima et al. (32). To study the different cold-seep microbial ecosystems present between the Japan Trench land slope and the Nankai Trough, the microbial diversity of Nankai Trough cold-seep sites at different depths was investigated and correlations were sought between the microbial communities and their geological settings.

Complete cold-seep microbial communities (AOM systems involving SRB-ANME consortia) were found only in the shallowest cold-seep sediments, from the NT06 site at a depth of 615 m, as shown in Fig. 2. Likewise, widespread *Calyptogena* communities were present at the NT06 site, but only a few *Calyptogena* colonies existed at the two deeper sites, NT20 and NT33, at depths of 2,048 and 3,310 m, respectively (4). These observations indicated that the cold-seep activity of the NT06 site could be higher than at the NT20 and NT33 sites. Basically, cold-seep activity might correspond to the existence of active faults and related geological settings (34, 50). Thus, the NT06 site could consist of a more active geological setting than the deeper sites. These results contrast with the case of the Japan Trench described above, where more abundant cold-seep microbial communities are found at the deepest depths of the trench (2), which could be points for fast subduction (about 12 cm/year) by the Pacific Ocean plate into the North American plate. The Nankai Trough is a slower subduction zone (4 cm/year) for the Philippine Sea plate intersecting with the Eurasian plate (34), and the resulting accretionary prism is built from the trench axis and increases in thickness landward. There is a difference in the geological setting of the Japan Trench compared to the Nankai Trough, since no accretionary prism structure was identified in the Japan Trench. Several active faults have been identified in the prism structures by seismic imaging profiles (31, 54). Thus, it is possible that strong cold-seep activity might occur even in the shallower water depths on the prism structures. The NT06 site could be one such area because of the numerous *Calyptogena* colonies observed there (4, 35). It is interesting that complete cold-seep microbial structures (AOM systems) are identified at the shallower depths on the accretionary prism structure in the Nankai Trough, while these were identified in deeper sediments in the Japan Trench lacking prism structures.

In conclusion, the current study of the microbial diversity of the Nankai Trough cold-seep sediments at different depths suggests a relationship between seep microbial diversity and accretionary prism structures.

Sagami Bay, Located near the Center of Japan, at a Depth of 1,200 m

Sagami Bay is located in the northern convergence front along the Sagami Trough among the Japanese Islands and two ocean plates, the Philippine Sea and Pacific Ocean plates. Biological communities of the vesicomyid clam *Calyptogena soyoae* were discovered in the seafloor of the bay in 1984 (19, 48). Colonies of the giant clam extend for about 5 to 7 km along the 1,000-m isobath in the western part of the Bay (18, 45). These communities are apparently supported by methane and hydrogen sulfide seeping out from fault lines in Sagami Bay (18, 21, 45, 61). The Sagami Bay site is perhaps globally the most comprehensively studied seep site, including both in situ and remote-operating-vehicle-based geological, geophysical, geochemical, and biological investigations, and is a natural laboratory for studying cold-seep biogeochemistry and microbiology for several reasons. The Sagami Bay region has a relatively high thermal gradient (1.5°C/20 cm [45]) that is very different from those of other cold-seep sites (e.g., the Gulf of Mexico). The sub-sea bottom temperature inside the *Calyptogena* communities is, on average, 0.7 to 1°C higher than outside the communities (45). Fluxes and chemical compositions of shimmering fluids have been determined (61). Active subsurface venting and mixing among seawater, pore water, and groundwater bring nutrient-, methane-, and hydrogen sulfide-rich water to

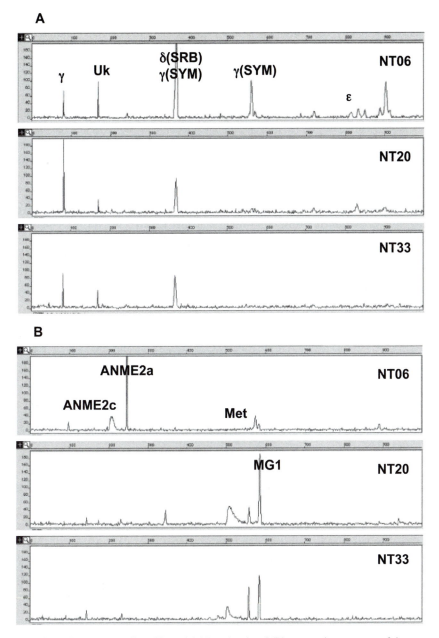

Figure 2. t-RFLP profiles of bacterial (A) and archaeal (B) community structures of the NT06, NT20, and NT33 sites. γ, δ, and ε indicate the corresponding proteobacterial groups. SYM, symbiotic bacteria related to SOx; Met, methanogenic archaea; Uk, unknown. The lengths of fragments (*x* axis) and relative fluorescence intensities of peaks (*y* axis) are also displayed.

the surface and near-surface sea bottom (21, 45, 61). Frequently occurring tectonic events affect the temperature regimen as well as the distribution of biological communities in the region (30). Geochemical measurements further suggest that sulfate reduction and anaerobic methane oxidation are possible microbial processes that support the dense biological communities at the site (43).

We have characterized microbial communities in *Calyptogena* sediment and microbial mats of the Sagami Bay using 16S rRNA gene sequencing and lipid biomarker analysis (12). Characterization of 16S rRNA genes isolated from these samples, the results of clone analyses and terminal restriction fragment length polymorphism (t-RFLP) profiles shown in Fig. 3, suggested a predominance of bacterial phylotypes related to the *Gammaproteobacteria* (57 to 64%) and *Deltaproteobacteria* (SRB; 27 to 29%). The *Epsilonproteobacteria* commonly found in cold seeps and hydrothermal vents were only detected in the microbial mat sample (SBM). There are significantly distinct archaeal phylotypes in the *Calyptogena* sediment (SBC) and microbial mat (SBM); the former contains only clones of *Crenarchaeota* belonging to marine group 1 (MG1 [10]), and the latter contains mostly clones of *Euryarchaeota* (56% of the total archaeal clones of SBM), including the ANME-2a, ANME-2c, and methanogenic archaeal groups. Many of these lineages are as yet uncultured and represent undescribed groups of bacteria and archaea. Phospholipid

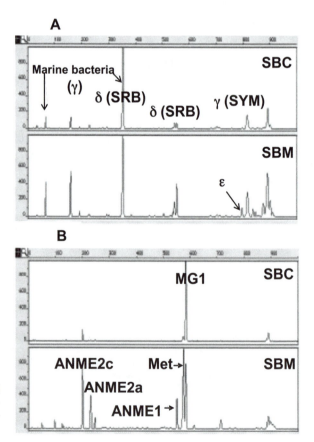

Figure 3. t-RFLP profiles of bacterial (A) and archaeal (B) community structures of the SBC and SBM. Abbreviations are the same as in Fig. 2.

analyses provide complementary insights into microbial biomass and community structure at the cold-seep site (12). The results of these analyses suggest that the microbial community in the Sagami Bay seep site is distinct from those previously characterized in other cold-seep environments, like the Gulf of Mexico, for example (9).

The Northeastern Japan Sea, a New Subduction Zone, at a Depth of 3,000 m

The northeastern margin of the Japan Sea is the location of the convergent plate boundary between the Amurian and Okhotsk plates (66). Large earthquakes (magnitude > 7) have frequently occurred during the past few centuries along the continental slope of this margin, and a few seismic gaps have been reported (49). Many active faults, outcrop collapses, and folds have been observed around this area. The location of the plate boundary and its tectonic mechanism are, hence, of great interest for geologists. Takeuchi and coworkers found widespread microbial mats at the eastern escarpment of the Shiribeshi Trough in the Japan Sea at a depth of 3,145 m during their survey of large earthquakes and bottom disturbances (59). Additionally, several other microbial mat sites were found around this area during a multidisciplinary scientific cruise using the deep-diving submersible *DSV Shinkai 6500* (50). The presence of many microbial mats along the fault indicates the activity of plate movements; in fact, in 1993 a large earthquake occurred near this area, offshore from southwestern Hokkaido (magnitude, 7.8; 42°47′N, 139°12E′).

Preliminary study of the microbial diversity in cold-seep sediments from the Motta Cape site, defined as the M2 site, unexpectedly showed that the sediments harbored diverse microbial communities which did not include methane-consuming archaea (1). To confirm that the absence of AOM community structures was a common feature of this region, in 2003 some of us revisited the northeastern Japan Sea and obtained microbial mat sediment samples from two geologically distinct stations in the Shiribeshi Trough: an active fault scarp at the southern base of the Shiribeshi seamount, defined as the M1 site, at a depth of 2,961 m on the active fault and off the Motta Cape field, the M2 site, at a depth of 3,064 m off the active fault. The comparative results of the microbial community structures at the two different habitats were described. Combining the chemical results of sulfate and methane concentrations and the carbon isotopic compositions of diffusing methane, a correlation between microbial community structures and the geochemical and geological characteristics was established (3).

Microbial communities inhabiting deep-sea cold-seep sediments in the northeastern Japan Sea were characterized by molecular phylogenetic and chemical analyses. White patchy microbial mats were observed along the fault offshore from Hokkaido Island and sediment samples were collected from two stations, the M1 and M2 sites. The phylogenetic and t-RFLP analyses of PCR-amplified 16S rRNA genes revealed that microbial community structures were different between the two sampling stations, as shown in Fig. 4. Members of the ANME-2 archaea and diverse bacterial components, including sulfate reducers within the *Deltaproteobacteria* (SRB), were particularly prevalent at the M1 site (on the active fault), indicating the occurrence of biologically mediated AOM, while microbial communities at the M2 site (off of the active fault) were predominantly composed of members of marine *Crenarchaeota* group 1 (MG1 [10]), SRB of the *Deltaproteobacteria*, and SOx of the *Epsilonproteobacteria*. Chemical analyses of seawater above the microbial mats suggested that the concentrations of sulfate and methane at the M1 site were

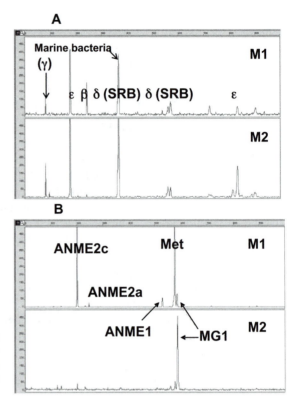

Figure 4. Comparison of t-RFLP profiles of bacterial (A) and archaeal (B) communities between the microbial mats at the M1 and M2 sites. Abbreviations are the same as in Fig. 2.

largely decreased relative to those at the M2 site and that the carbon isotopic composition of methane at the M1 site was increased (^{13}C enriched), which was consistent with the molecular analyses (3). These results suggest that the microbial mat communities in deep-sea cold-seep sediments within the northeastern Japan Sea are responsible for significant sulfur and carbon circulation and that the geological activity caused by the active faults provides unique microbial habitats.

Sulfur and Carbon Cycling Ecosystems in Cold-Seep Environments

A model accounting for the chemistry of cold-seep *Calyptogena* communities has been constructed by Masuzawa et al. (43). This scheme indicates that the generation of hydrogen sulfide occurs through sulfate reduction in a process involving AOM. However, the sites where such sulfate reduction occurs and where methane is generated were not made clear in this study. Our studies on molecular phylogenetic analysis of microorganisms in sediments from the area of the cold-seep environments have expanded that model to include microbial interactions. An ecosystem involving inorganic compound (e.g., sulfur and carbon) circulation in the seep microbial communities is proposed in Fig. 5, updating the previously published model described by Li et al. (41). The abundance of *Gammaproteobacteria* (symbiotic bacteria like SOx), *Deltaproteobacteria* (SRB), and *Epsilonproteobacteria* (SOx) recovered from the sediment of the seep community suggests that

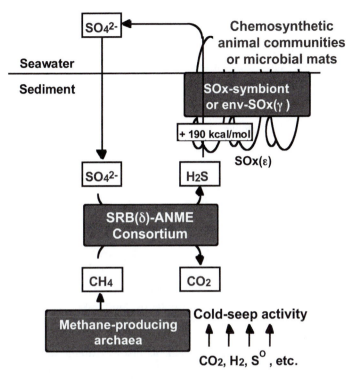

Figure 5. Sulfur and carbon cycling ecosystems within the microbial community in the cold-seep environment.

microbial sulfate reduction and oxidation take place in this community. The recovery of *Euryarchaeota*-related methanogen and ANME groups in the community indicates that high concentrations of methane in the sediment might be produced and the biogenic methane might be oxidized subsequently under anoxic conditions by the AOM systems. Barry et al. (6) indicated that higher concentrations of hydrogen sulfide in the seep sediments favor microbial mats over chemosynthetic animal communities, and our results from the Sagami Bay study support this suggestion. These microbial ecosystems are common in many cold-seep environments, and they could play a role in the circulation of inorganic compounds following sulfur or sulfate reduction as well as sulfide oxidation. At the same time, methane production and anoxic oxidation of methane could provide energy to chemoautotrophic ecosystems independently from solar-power-based photosynthesis-dependent ecosystems.

MICROBIAL DIVERSITY IN DEEP-SEA SEDIMENTS
FROM THE TROPICAL WEST PACIFIC WARM POOL

The tropical West Pacific Warm Pool (WP), with surface water temperatures of >28°C, is a geologically important area for the climate of the earth because it drives the world's most intense atmospheric convection (63). The tropical Pacific serves as a heat engine for the climate of the earth and a vapor source for its hydrological cycle (57). The ecological systems operating in the tropical Pacific are also important components of global ecological processes. Despite its importance, microbial communities, which are recognized as one of the most important constituents in marine ecosystems, have not been sufficiently explored for the WP. Deep-sea sediments that are primarily formed through the continual deposition of particles from the productive ocean surface cover approximately 70% of the earth's surface. The organic matter settled on the seafloor is remineralized by benthic microbial communities which colonize the sediments. The activities of ocean sediment microorganisms thus play important roles in the global cycling of carbon and nutrients (11, 62). The microbes in deep-sea sediments of the WP, particularly those microbes involved in C_1 compound metabolism, have been examined using the tools of molecular microbial ecology (64, 65).

Diversity of Methylotrophs in Deep-Sea Sediments

Methylotrophs are a group of bacteria which can utilize methane (methanotrophs) and/or a variety of other one-carbon (C_1) compounds more reduced than formic acid, such as methanol and methylated amines, as sole carbon and energy sources (17). They play essential roles in carbon cycling on the earth by participating in methane oxidation and C_1 compound metabolism. From deep-sea waters a type I methanotroph, *Methylomonas pelagica,* was previously isolated and partially characterized (56). In the deep sea, methylotrophs are probably one of the key components of this extreme biosystem because of their essential role in carbon cycling.

Conserved functional gene probes are particularly useful in aerobic methylotroph diversity investigations, as methylotrophs span a wide number of taxa. The most commonly utilized functional gene markers for methylotroph and methanotroph diversity studies are the *mxaF* and *pmoA* genes (8, 44). Almost all known methanotrophs contain a functional *pmoA* gene, encoding the alpha subunit of the particulate methane monooxygenase. The *pmoA* gene has been used for methanotroph diversity investigations in various environments (20, 22). However, *pmoA* cannot be used to detect the presence of a group of methylotrophs which use C_1 compounds other than methane as the sole energy and carbon source. This limitation can be overcome by using a second functional gene marker, *mxaF*, which is present in all methylotrophs. All known gram-negative methylotrophs possess the key enzyme methanol dehydrogenase, which is responsible for the oxidation of methanol to formaldehyde, an intermediate of both assimilative and dissimilative metabolism in methylotrophs (17, 44). The large subunit of methanol dehydrogenase is encoded by the 1.8-kb *mxaF* gene. The conserved region of the *mxaF* gene has been utilized as a functional gene marker for methylotroph diversity studies (20, 22, 44).

In WP sediments obtained from water column depths ranging from ~1,900 to 5,000 m, *mxaF* and *pmoA* gene probes were used for detecting aerobic methylotrophs (64, 65). By analyses of the deduced amino acid sequence of the *mxaF* gene product, methylotrophs affiliated with *Hyphomicrobium* and *Methylobacterium*, which use methanol and some other C_1 compounds as the sole carbon sources, were found to be the prevailing methylotrophs in the West Pacific WP, while type II methanotrophs closely related to *Methylocystis* and *Methylosinus* constituted only a small percentage (~5.6%) of the methylotroph community (Fig. 6). However, no type I methanotrophs were detected. These results support the hypothesis that type I methanotrophs are predominant in environments that allow rapid growth of methanotrophic bacteria, while type II methanotrophs are more abundant in environments where growth rates are restricted (22). On the other hand, use of a *pmoA* gene marker could not detect the presence of any methanotrophs in this environment, implying that the *mxaF* gene probe is a more suitable marker in this deep-sea sediment for detecting methylotrophs (including methanotrophs [64]).

The levels of methylotrophs in the sediments of the tropical West Pacific WP were semiquantified by quantitative competitive PCR (64). It was found that the WP contained around 3×10^4 to 3×10^5 molecules of *mxaF* gene copy per gram of sediment. Using this method, the distribution and abundance of methylotrophs in deep-sea sediments from the West Pacific WP were compared with those in east and middle Pacific deep-sea sediments, seashore sediments, and flower garden and rice field soils. From all the samples tested, specific *mxaF* gene fragments could be amplified, indicating the wide distribution of methylotrophs in diverse environments, including deep-sea sediments. It was also found that rice plant soil contained the largest quantity of methylotrophs, with 5×10^5 to 5×10^6 *mxaF* copies per gram of sediment, and the WP site had 10 to 100 times more methylotrophs than the east and middle Pacific sediments. The higher levels of methylotrophs present in the West Pacific WP relative to the middle and east Pacific suggest a higher rate of metabolic activity and carbon cycling in the WP area.

Diversity of Archaea

The population of archaea in this environment was assayed by 16S rRNA gene sequence analyses using universal archaeal primers. The main archaeal community in the deep-sea sediments of the West Pacific WP area was *Crenarchaeota* MG1 (Fig. 7). The MG1 detected in the deep-sea sediments of the West Pacific WP area are nearly identical, and all of them belong to MG1-alpha group (42, 58).

To further detect archaea related to C_1 compound metabolism, PCR amplifications were performed using primers targeting AOM 16S rRNA gene fragments (60). Specific DNA bands were amplified from all five WP sites (WP0, WP1, WP2, WP3, and WP4) and from all selected sediment layers of WP0 by using AOM-specific 16S rRNA gene primers, indicating that these archaea are present in all the WP sites and from the surface to the bottom of the sediment core. PCR products from each sediment layer (1-, 3-, 6-, 10-, and 12-cm layer) of WP0 were cloned and libraries were constructed. From each library, 50 clones were randomly selected and screened by RFLP analysis. No relation could be found between the clone RFLP types and the sediment layers. On the basis of the RFLP types, 15 clones were selected for sequencing. The 16S rRNA gene sequences of all the clones retrieved had relatively low identities with those of known archaea (the highest identities with those of known environmental clones were around 74 to 84%, and there was 71 to 78% identity with those of culturable species). These newly detected archaea were named WPA. The phylogenetic analysis classified all of the sequences into two distinct lines of descent within the *Euryarchaeota* kingdom (Fig. 8). One cluster, WPA-I, formed a *Thermoplasma*-associated branch; however, members of this cluster of archaea were only distantly related to the *Thermoplasma*.

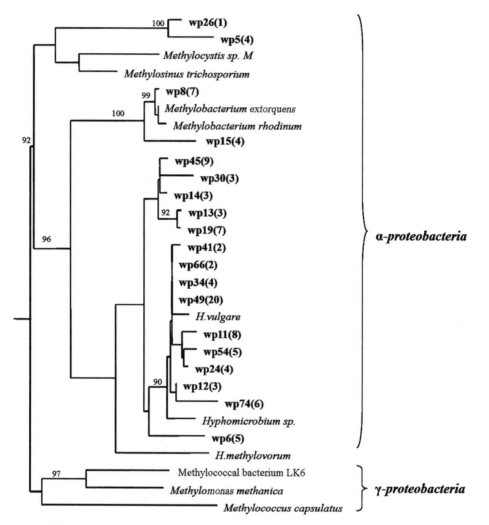

Figure 6. Phylogenetic tree constructed based on deduced partial MxaF amino acid sequences. The MxaF sequences retrieved from the sediment of tropical West Pacific WP and
from the cultured representative methylotrophs, including type I and type II methanotrophs, are involved in the tree construction. Only bootstrap values above 90 from 1,000
replicates are shown. The scale bar represents 0.05 substitution per amino acid site. The
environmental *mxaF* clones from the Pacific WP sediment are designated by "wp"; the
numbers in parentheses are the numbers of clones with identical sequences in the 90 sequenced clones.

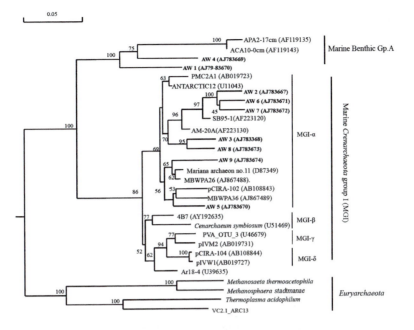

Figure 7. Phylogenetic tree constructed based on 16S rRNA gene sequences of archaea. Archaeal 16S rRNA gene clones retrieved from the Pacific WP sediments were named AW. Nine AW clones together with relative clones in the data bank were used for phylogenetic tree construction. The phylogenetic tree was constructed from a matrix by least-squares distance matrix analysis and the neighbor-joining method using the DNAMAN program. One thousand trials of bootstrap analysis were used to provide confident estimates for phylogenetic tree topologies. Only bootstrap values above 50 are shown. The scale bar represents 0.05 substitution per nucleic acid site.

The intralineage levels of rRNA gene similarity among clones of WPA-I were between 90 and 95%. The other cluster, WPA-II, contained 12 clones distantly related to methanogens, followed by ANME. This cluster could be further divided into six subclades, as shown in Fig. 8. The function of the newly discovered archaea WPA has not yet been determined, but from its phylogenetic position WPA is suggested to function in C_1 compound metabolism in organically depleted deep-ocean sediments. WPA was detected in all the five west Pacific deep-sea sediments, and since these sediments are all geologically distinct, WPA may be a widespread group of archaea present in many deep-sea environments.

The abundances of WPA, archaea, and bacteria along a 12-cm sediment West Pacific WP core were determined by quantitative competitive PCR. The quantification data showed that bacteria dominated over archaea at all depth intervals. The proportion of archaea versus bacteria had a depth-related increasing tendency; it was lowest in the first layer (0.01%) and reached the highest level in the 12-cm layer (10%). WPA constituted only a small proportion of the archaeal community (0.05 to 5%) of the West Pacific WP sediments.

232   Kato et al.

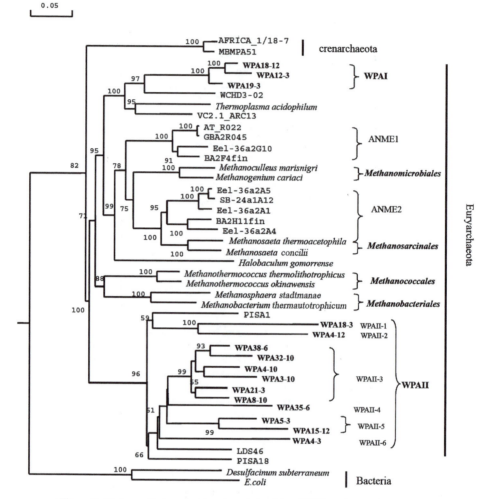

Figure 8. Phylogenetic tree showing the relationship of WPA and related clones and strains. Archaeal 16S rRNA gene clones retrieved by PCR using primers ANMEF and 907R from the Pacific WP sediment were named WPA. Fifteen sequences representing 15 different RFLP types of WPA and 16S rRNA gene sequences of reference clones or strains in the data bank were used for dendrogram construction. The phylogenetic tree was constructed from a matrix by least-squares distance matrix analysis and the neighbor-joining method using the DNAMAN program, and 1,000 trials of bootstrap analysis were used to provide confident estimates for phylogenetic tree topologies. Only bootstrap values above 50 are shown. The scale bar represents 0.05 substitution per nucleic acid site.

CONCLUSIONS

The relationship between the cold-seep ecosystems and their accompanying microbial diversity has been investigated in the Japan Trench, the Nankai Trough, Sagami Bay, and the northeastern portion of the Japan Sea. The results suggest cold-seep activity utilizing sulfur and carbon cycling microbial ecosystems. These communities contained methanogenic archaea, a sulfate-reducing consortium (ANME-SRB), and SOx (microbial-mat-forming and/or chemosynthetic symbionts) which were commonly present in many seep environments. The existence of these ecosystems could be independent from solar-energy-dependent systems (photosynthesis-derived ecosystems).

Generally, open-ocean deep-sea sediments are organically poor, low-methane, low-oxygen, dark, and cold (1 to 2°C) environments. Although microbial activities in the organically depleted deep-sea sediments are very low, considering the large area of the open-sea sediments, microbial communities and their roles in metabolic cycling on earth may be significant. Microbes involved in C_1 compound metabolism in the deep-sea sediments from the West Pacific WP area consist of *Alphaproteobacteria,* including type II methanotrophs and other methylotrophs. Novel archaea, possibly involved in anaerobic production or oxidation of methane, were found to be ubiquitous in these deep-sea environments.

Acknowledgments. Those of us from JAMSTEC express our gratitude to Koki Horikoshi and Xun Xu for their continued support of our deep-sea extremophile studies. We also thank the crews of the research ships and members of the submersible operation division at JAMSTEC, Japan, and SOA, People's Republic of China, for their efforts in collecting samples from the deep-sea environment, as well as Howard K. Kuramitsu for assistance in editing the manuscript.

REFERENCES

1. **Arakawa, S., C. Kato, T. Sato, Y. Nogi, Y. Yoshida, R. Usami, and K. Horikoshi.** 2004. Microbial diversity of bacterial mat samples collected at a depth of 3,100 m in the Japan Sea. *Mar. Biotechnol.* **6:**s185–s189.
2. **Arakawa, S., M. Mori, L. Li, Y. Nogi, T. Sato, Y. Yoshida, R. Usami, and C. Kato.** 2005. Cold-seep microbial communities are more abundant at deeper depths in the Japan Trench land slope. *J. Jpn. Soc. Extremophiles* **4:**50–55.
3. **Arakawa, S., T. Sato, R. Sato, J. Zhang, T. Gamo, U. Tsunogai, A. Hirota, Y. Yoshida, R. Usami, F. Inagaki, and C. Kato.** 2006. Molecular phylogenetic and chemical analyses of the microbial mats in deep-sea cold seep sediments at the northeastern Japan Sea. *Extremophiles* **10:**311–319.
4. **Arakawa, S., T. Sato, Y. Yoshida, R. Usami, and C. Kato.** 2006. Comparison of the microbial diversity in cold-seep sediments from the different water depths in the Nankai Trough. *J. Gen. Appl. Microbiol.* **52:**47–54.
5. **Ashi, J.** 1997. Distribution of cold seepage at the fault scarp of the eastern Nankai accretionary prism. *JAMSTEC J. Deep Sea Res.* **13:**495–501.
6. **Barry, J. P., R. E. Kochevar, and C. H. Baxter.** 1997. The influence of pore-water chemistry and physiology on the distribution of vesicomyid clams at cold seeps in Monterey Bay: implications for patterns of chemosynthetic community organization. *Limnol. Oceanogr.* **42:**318–328.
7. **Boetius, A., K. Ravenschlag, C. J. Schubert, D. Ricket, F. Widdel, A. Gieseke, R. Amann, B. B. Jorgensen, U. Witte, and O. Pfannkuche.** 2000. A marine microbial consortium apparently mediating anaerobic oxidation of methane. *Nature* **407:**623–626.
8. **Bourne, D. B., I. R. Mcdonald, and J. C. Murrell.** 2001. Comparison of *pmoA* PCR primer sets as tools for investigating methanotroph diversity in three Danish soils. *Appl. Environ. Microbiol.* **67:**3802–3809.
9. **Cavanaugh, C. M.** 1994. Microbial symbiosis: patterns of diversity in the marine environment. *Am. Zool.* **34:**79–89.
10. **DeLong, E. F.** 1992. Archaea in coastal marine environments. *Proc. Natl. Acad. Sci. USA* **89:**5685–5689.
11. **D'Hondt, S., S. Rutherford, and A. J. Spivack.** 2002. Metabolic activity of subsurface life in deep-sea sediments. *Science* **295:**2067–2070.
12. **Fang, J., A. Shizuka, C. Kato, and S. Schouten.** 2006. Microbial diversity of cold-seep sediments in Sagami Bay, Japan, as determined by 16S rRNA gene and lipid analyses. *FEMS Microbiol. Ecol.* **57:**429–441.
13. **Fujikura, K., Y. Fujiwara, S. Kojima, and T. Okutani.** 2002. Micro-scale distribution of mollusks occurring in deep-sea chemosynthesis-based communities in the Japan Trench. *Venus* **60:**225–236.
14. **Fujikura, K., S. Kojima, K. Tamaki, Y. Maki, J. Hunt, and T. Okutani.** 1999. The deepest chemosynthesis-based community yet discovered from the hadal zone, 7326m deep, in the Japan Trench. *Mar. Ecol. Prog. Ser.* **190:**17–26.

15. **Fujiwara, Y., C. Kato, N. Masui, K. Fujikura, and S. Kojima.** 2001. Dual symbiosis in the cold-seep thyasirid clam *Maorithyas hadalis* from the hadal zone in the Japan Trench, western Pacific. *Mar. Ecol. Prog. Ser.* **214:**151–159.

16. **Gamo, T., J. Ishibashi, K. Shitashima, T. Nakatsuka, U. Tsunogai, T. Masuzawa, H. Sakai, and K. Mitsuzawa.** 1994. Chemical characteristics of cold seepage at the eastern Nankai Trough accretionary prism (KAIKO-TOKAI Program): a preliminary report of the dive 113 of "Shinkai 6500." *JAMSTEC J. Deep Sea Res.* **10:**343–352.

17. **Hanson, R. S., and T. E. Hanson.** 1996. Methanotrophic bacteria. *Microbiol. Rev.* **60:**439–471.

18. **Hashimoto, J., S. Ohta, K. Fujikura, Y. Fujiwara, and S. Sukizaki.** 1995. Life habit of vesicomyid clam, *Calyptogena soyoae*, and hydrogen sulfide concentration in interstitial waters in Sagami Bay, Japan. *J. Oceanogr.* **51:**341–350.

19. **Hashimoto, J., S. Ohta, T. Tanaka, H. Hotta, S. Matsuzawa, and H. Sakai.** 1989. Deep-sea communities dominated by the giant clam, *Calyptogena soyoae*, along the slope foot of Hatsushima island, Sagami Bay, central Japan. *Paleogeogr. Paleoclimatol. Palaeoecol.* **71:**179–192.

20. **Henckel, T., M. Friedrich, and R. Conrad.** 1999. Molecular analyses of the methane-oxidizing microbial community in rice field soil by targeting the genes of the 16S rRNA, particulate methane monooxygenase, and methanol dehydrogenase. *Appl. Environ. Microbiol.* **65:**1980–1990.

21. **Henry, P., S. Lallemant, K. Nakamura, U. Tsunogai, S. Mazzotti, and K. Kobayashi.** 2002. Surface expression of fluid venting at the toe of the Nanakai wedge and implication for flow paths. *Mar. Geol.* **187:**119–143.

22. **Horz, H. P., M. T. Yimga, and W. Liesack.** 2001. Detection of methanotroph diversity on roots of submerged rice plants by molecular retrieval of *pmoA, mmoX, mxaF,* and 16S rRNA and ribosomal DNA, including *pmoA*-based terminal restriction fragment length polymorphism profiling. *Appl. Environ. Microbiol.* **67:**4177–4185.

23. **Inagaki, F., Y. Sakihama, A. Inoue, C. Kato, and K. Horikoshi.** 2002. Molecular phylogenetic analyses of reverse transcripted bacterial rRNA obtained from deep-sea cold seep sediments. *Environ. Microbiol.* **4:**277–286.

24. **Inagaki, F., U. Tsunogai, M. Suzuki, A. Kosaka, H. Machiyama, K. Takai, T. Nunoura, K. H. Nealson, and K. Horikoshi.** 2004. Characterization of C1-metabolizing prokaryotic communities in methane seep habitats at the Kuroshima Knoll, the Southern Ryukyu arc, by analyzing *pmoA, mmoX, mxaF, mcrA,* and 16S rRNA genes. *Appl. Environ. Microbiol.* **70:**7445–7455.

25. **Kato, C., and S. Arakawa.** 2004. Microbial diversity and uncultivable microorganisms in deep-sea environment, p. 132–147. *In* T. Kudo and M. Ohkuma (ed.), *Novel Technology for Uncultivable Microorganisms—Approach to Unused Bio-Resources.* CMC Press, Tokyo, Japan. (In Japanese.)

26. **Kato, C., S. Arakawa, R. Usami, and T. Sato.** 2005. Microbial diversity in the cold seep sediments and sulfur circulation. *Gekkan Chikyu* **27:**939–948. (In Japanese.)

27. **Kato, C., L. Li, J. Tamaoka, and K. Horikoshi.** 1997. Molecular analyses of the sediment of the 11000 m deep Mariana Trench. *Extremophiles* **1:**117–123.

28. **Kennicutt, M. C., II, J. M. Brooks, R. R. Bridigare, R. R. Fay, T. L. Wade, and S. J. McDonald.** 1985. Vent-type taxa in a hydrocarbon seep region on the Louisiana slope. *Nature* **317:**351–353.

29. **Kennicutt, M. C., II, J. M. Brooks, R. R. Bridigare, S. J. McDonald, and D. L. Adkison.** 1989. An upper slope "cold" seep community: Northern California. *Limnol. Oceanogr.* **34:**635–640.

30. **Kobayashi, K.** 2002. Tectonic significance of the cold seepage zones in the eastern Nankai accretionary wedge—an outcome of the 15 years' KAIKO projects. *Mar. Geol.* **187:**3–30.

31. **Kodaira, S., T. Iidaka, A. Kato, J. O. Park, T. Iwasaki, and Y. Kaneda.** 2004. High pore fluid pressure may cause silent slip in the Nankai Trough. *Science* **304:**1295–1298.

32. **Kojima, S., K. Fujikura, and T. Okutani.** 2004. Multiple trans-Pacific migrations of deep-sea vent/seep-endemic bivalves in the family Vesicomyidae. *Mol. Phylogenet. Evol.* **32:**396–406.

33. **Kulm, L. D., E. Suess, J. C. Moore, B. Carson, B. T. Lewis, S. D. Ritger, D. C. Kadko, T. M. Thornburg, R. W. Embly, W. D. Rugh, G. J. Massoth, M. G. Langseth, G. R. Cochrane, and R. L. Scamman.** 1986. Oregon subduction zone: venting, fauna, and carbonates. *Science* **231:**561–566.

34. **Kuramoto, S., J. Ashi, J. Greinert, S. Gulic, T. Ishimura, S. Morita, K. Nakamura, M. Okada, T. Okamoto, D. Rickert, S. Saiyo, E. Suess, U. Tsunogai, and T. Tomosugi.** 2001. Surface observations of subduction related mud volcanoes and large thrust sheets in the Nankai subduction margin; report on YK00-10 and YK01-04 cruises. *JAMSTEC J. Deep Sea Res.* **19:**131–139.

35. **Kuramoto, S., and M. Joshima.** 1998. Precise gravity measurements for gas hydrate layer. *JAMSTEC J. Deep Sea Res.* **14:**371–377.

36. **Le Pichon, X., T. Iiyama, J. Boulegue, J. Charvet, M. Faure, K. Kano, S. Lallemant, H. Okada, C. Rangin, A. Taira, T. Urabe, and S. Uyeda.** 1987. Nankai Trough and Zenisu Ridge: a deep-sea submersible survey. *Earth Planet Sci. Lett.* **83:**285–299.

37. **Li, L., J. Guenzennec, P. Nichols, P. Henry, M. Yanagibayashi, and C. Kato.** 1999. Microbial diversity in Nankai Trough sediments at a depth of 3843 m. *J. Oceanogr.* **55:**635–642.

38. **Li, L., and C. Kato.** 1999. Microbial diversity in the sediments collected from cold-seep areas and from different depths of the deep-sea, p. 55–58. *In* K. Horikoshi and K. Tsujii (ed.), *Extremophiles in Deep-Sea Environments.* Springer-Verlag, Tokyo, Japan.

39. **Li, L., C. Kato, and K. Horikoshi.** 1999. Bacterial diversity in deep-sea sediments from different depths. *Biodivers. Conserv.* **8:**659–677.

40. **Li, L., C. Kato, and K. Horikoshi.** 1999. Microbial communities in the world's deepest ocean bottom, the Mariana Trench, p. 17–20. *In* H. Ludwig (ed.), *Advances in High Pressure Bioscience and Biotechnology.* Springer-Verlag, Heidelberg, Germany.

41. **Li, L., C. Kato, and K. Horikoshi.** 1999. Microbial diversity in sediments collected from the deepest cold-seep area, the Japan Trench. *Mar. Biotechnol.* **1:**391–400.

42. **Massana, R., E. F. DeLong, and C. Pedrós-Alió.** 2000. A few cosmopolitan phylotypes dominate planktonic archaeal assemblages in widely different oceanic provinces. *Appl. Environ. Microbiol.* **66:**1777–1787.

43. **Masuzawa, T., N. Handa, H. Kitagawa, and M. Kusakabe.** 1992. Sulfate reduction using methane in sediments beneath a bathyal "cold seep" giant clam community off Hatsushima Island, Sagami Bay, Japan. *Earth Planet Sci. Lett.* **110:**39–50.

44. **Mcdonald, I. R., and J. C. Murrell.** 1997. The methanol dehydrogenase structural gene *mxaF* and its use as a functional gene probe for methanotrophs and methylotrophs. *Appl. Environ. Microbiol.* **63:**3218–3224.

45. **Momma, H., R. Iwase, K. Mitsuzawa, Y. Kaiho, and Y. Fujiwara.** 1998. Preliminary results of a three-year continuous observation by a deep seafloor in Sagami Bay, central Japan. *Phys. Earth Planet. Interiors* **108:**263–274.

46. **Ogawa, Y., K. Fujioka, K. Fujikura, and Y. Iwabuchi.** 1996. En echelon patterns of *Calyptogena* colonies in the Japan Trench. *Geology* **24:**807–810.

47. **Ohta, S., and L. Laubier.** 1987. Deep biological communities in the subduction zone of Japan from bottom photographs taken during "Nautile" dives in the Kaiko project. *Earth Planet Sci. Lett.* **83:**329–342.

48. **Ohta, S., H. Sakai, A. Taira, K. Ohwada, T. Ishii, M. Maeda, K. Fujioka, Y. Saino, K. Kogure, T. Gamo, Y. Shirayama, T. Furuta, T. Ishizuka, K. Endow, T. Sumi, H. Hotta, J. Hashimoto, N. Handa, T. Masuzawa, and M. Horikoshi.** 1987. Multidisciplinary investigation of the *Calyptogena* communities at the Hatsushima site, Sagami Bay, central Japan. *JAMSTEC J. Deep-Sea Res.* **3:**51–60.

49. **Ohtake, M.** 1995. A seismic gap in the eastern margin of the Sea of Japan as inferred from the time-space distribution of past seismicity. *Island Arc* **4:**156–165.

50. **Okamura, Y., K. Satake, A. Takeuchi, T. Gamo, C. Kato, Y. Sasayama, F. Nakayama, K. Ikehara, and T. Kodera.** 2002. Techtonic, geochemical and biological studies in the eastern margin of the Japan Sea—preliminary results of Yokosuka/Shinkai 6500 YK01-06 Cruise. *JAMSTEC J. Deep Sea Res.* **20:**77–114.

51. **Orphan, V. J., K. U. Hinrichs, W. Ussler III, C. K. Paull, L. T. Taylor, S. P. Sylva, J. M. Hayes, and E. F. DeLong.** 2001. Comparative analysis of methane-oxidizing archaea and sulfate-reducing bacteria in anoxic marine sediments. *Appl. Environ. Microbiol.* **67:**1922–1934.

52. **Orphan, V. J., C. H. House, K. U. Hinrichs, K. D. McKeegan, and E. F. DeLong.** 2001. Methane-consuming archaea revealed by directly coupled isotopic and phylogenetic analysis. *Science* **293:**484–487.

53. **Park, J. O., T. Tsuru, S. Kodaira, P. R. Cummins, and Y. Kaneda.** 2002. Splay fault branching along the Nankai subduction zone. *Science* **297:**1157–1160.

54. **Paull, C. K., B. Hecker, R. Commeau, R. P. Freeman-Lynde, C. Neumann, W. P. Corso, S. Golubic, J. E. Hook, E. Sikes, and J. Curray.** 1984. Biological communities at the Florida Escarpment resemble hydrothermal vent taxa. *Science* **226:**965–967.

55. **Sibuet, M., S. K. Juniper, and G. Pautot.** 1988. Cold-seep benthic communities in the Japan subduction zones: geological control of community development. *J. Mar. Res.* **46:**333–348.

56. **Sieburth, J. M., P. W. Johnson, M. A. Eberhardt, M. E. Sieracki, M. Lidstrom, and D. Laux.** 1987. The first methane-oxidizing bacterium from the upper mixing layer of the deep ocean: *Methylomonas pelagica* sp. nov. *Curr. Microbiol.* **14:**285–293.

57. **Stott, L., C. Poulsen, S. Lund, and R. Thunell.** 2002. Super ENSO and global climate oscillations at millennial time scales. *Science* **297:**222–226.

58. **Takai, K., H. Oida, Y. Suzuki, H. Hirayama, S. Nakagawa, T. Nunoura, F. Inagaki, K. H. Nealson, and K. Horikoshi.** 2004. Spatial distribution of marine *Crenarchaeota* group I in the vicinity of deep-sea hydrothermal systems. *Appl. Environ. Microbiol.* **70:**2404–2413.

59. **Takeuchi, A., Y. Okamura, Y. Kato, K. Ikehara, J. Zhang, K. Satake, T. Nagao, M. Hirano, and M. Watanabe.** 2000. Large earthquakes and bottom disturbances in the Okushiri ridge along the eastern margin of Japan Sea. *JAMSTEC J. Deep Sea Res.* **16:**29–46.

60. **Thomsen, T. R., K. Finster, and E. B. Ramsing.** 2001. Biochemical and molecular signatures of anaerobic methane oxidation in a marine sediment. *Appl. Environ. Microbiol.* **67:**1646–1656.

61. **Tsunogai, U., J. Ishibashi, H. Wakita, T. Gamo, T. Masuzawa, T. Natatsuka, Y. Nojiri, and T. Nakamura.** 1996. Freshwater seepage and pore water recycling on the seafloor: Sagami Trough subduction zone. *Earth Planet Sci. Lett.* **138:**157–168.

62. **Vetriani, C., H. W. Jannasch, B. J. MacGregor, D. A. Stahl, and A. L. Reysenbach.** 1999. Population structure and phylogenetic characterization of marine benthic Archaea in deep-sea sediments. *Appl. Environ. Microbiol.* **65:**4375–4384.

63. **Visser, K., R. Thunell, and L. Stott.** 2003. Magnitude and timing of temperature change in the indo-Pacific Warm Pool during deglaciation. *Nature* **421:**152–155.

64. **Wang, P., F. Wang, M. Xu, and X. Xiao.** 2004. Molecular phylogeny of methylotrophs in a deep-sea sediment from a tropical west Pacific Warm Pool. *FEMS Microbiol. Ecol.* **47:**77–84.

65. **Wang, P., X. Xiao, and F. P. Wang.** 2005. Phylogenetic analysis of Archaea in the deep-sea sediments of west Pacific Warm Pool. *Extremophiles* **9:**209–217.

66. **Wei, D., and T. Seno.** 1998. Determination of the Amurian Plate motion, p. 337–346. *In* M. J. Flower et al. (ed.), *Mantle Dynamics and Plate Interactions in East Asia.* Geodynamics series, vol. 27. American Geophysical Union, Washington, DC.

High-Pressure Microbiology
Edited by C. Michiels, D. H. Bartlett, and A. Aertsen
© 2008 ASM Press, Washington, DC

Chapter 14

Deep-Sea Geomicrobiology

Jiasong Fang and Dennis A. Bazylinski

The exploration for and discoveries of microorganisms in extreme environments have greatly expanded the range of known habitats where microbial life flourishes on this planet as well as demonstrated the vastness of the diversity of microbial life on Earth. Geomicrobiology, a subdiscipline arising from interdisciplinary studies of microbiology, geology, chemistry, and other disciplines, is the study of the interactions between microorganisms and their environments in the geosphere. The rapid development and growth of geomicrobiology can partially be attributed to discoveries in the last several decades of unique extremophiles present in many different harsh environments that play key roles in the biogeochemistry that occurs in these environments. It has been clearly demonstrated that life's boundaries far exceed the conditions required for human existence (93, 98). Microorganisms inhabit a large, diverse variety of extreme environments defined by a wide range of environmental factors, including temperature, pressure, pH, water availability, salinity, radioactivity, and nutrient source and concentrations. For example, these habitats can exhibit extremes (48) in temperature (-20 to $121°C$), pH (0 to 11), salinity ($>25\%$ NaCl), high pressure (up to 110 MPa, or megapascal), etc. Microorganisms thriving in these environments are generally referred to as extremophiles (67), which encompass thermo-, psychro-, alkali-, acido-, halo-, and piezophiles (20, 106). Indeed, many, if not most, extremophiles do not just tolerate environmental extremes but actually require extreme conditions for growth (67, 69). The new and rapidly developing studies of extremophiles and geomicrobiology have revolutionized our views of the origin of life and how life has evolved since its origin, as well as the diversity and activity of microbial life on ancient and present-day Earth. The development of new technologies and experimental approaches in geomicrobiology and in the studies of extremophiles has spawned a revolution that will surely have profound social and economic impact now and in the future (1, 3, 13, 57, 67).

Jiasong Fang • Department of Geological and Atmospheric Sciences, Iowa State University, Ames, IA 50011-0001. *Dennis A. Bazylinski* • School of Life Sciences, University of Nevada, Las Vegas, Las Vegas, NV 89154-4004.

PIEZOPHILY, PIEZOPHYSIOLOGY, AND DIVERSITY OF PIEZOPHILIC BACTERIA

This chapter focuses on the piezophilic members of the extremophiles, with an emphasis on geomicrobiological considerations. Phylogenetic analysis of a number of piezophilic bacteria has been performed using 5S rRNA or 16S rRNA sequence analyses (22, 29, 30, 43, 57–59, 62, 63, 66, 82–87). Additional taxonomic information comes from analyses of cellular fatty acids, in particular the spectrum of polyunsaturated fatty acids (PUFA) and the presence or absence of eicosapentaenoic acid (EPA; 20:5ω3 *cis*-5,8,11,14,17) and docosahexaenoic acid (DHA; 22:6ω3 *cis*-4,7,10,13,16,19) (Fig. 1). Detailed phylogenetic descriptions of piezophilic bacteria are provided by Kato et al. in chapter 12. Table 1 lists the characteristics of a variety of piezophiles.

Piezophily and Piezophysiology of Piezophilic Bacteria

Piezophilic bacteria in the deep sea are adapted not to a particular temperature (T) or pressure (P) but to a condition defined by both T and P (126, 129), and thus, these two environmental parameters are interrelated (for examples, see references 59, 126, and 127). The T range of growth is a function of the P at which it is determined and vice versa (128). Therefore, piezophilic bacteria grow over a P, T domain. A bacterial isolate is piezophilic if it has a greater generation time at some high pressure than it does at atmospheric pressure when tested at its habitat temperature (1, 127). Generally, deep-sea bacteria show the strongest piezophilic response to pressure at their upper temperature for growth (typically 15°C). Piezophilic strains become more piezophilic at higher temperatures (57, 125). For example, *Shewanella benthica* strain DB172F exhibits piezotolerant growth at 4°C (identical growth rates at 50 and 0.1 MPa) but shows piezophilic growth (70 MPa) at its optimal growth temperature of 10°C. Therefore, true piezophiles are rare (131), and the effects of temperature and pressure on piezophily are addressed concurrently in laboratory culture experiments (4, 23, 24). These studies indicate that all piezophilic isolates are obligately piezophilic above the temperature at which growth occurs at atmospheric pressure. This means that the upper temperature limit for growth can be extended by high pressure (125). Likewise, piezophilic bacteria reproduce more rapidly at a lower temperature (such as 2°C) when the pressure is less than that at their capture depth. The doubling time of piezophiles at pressures near those at the depth from where they were collected increases

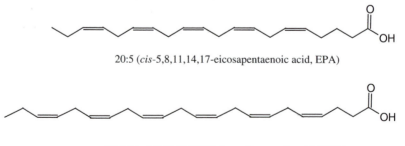

20:5 (*cis*-5,8,11,14,17-eicosapentaenoic acid, EPA)

22:6 (cis-4,7,10,13,16,19-docosahexaenoic acid, DHA)

Figure 1. Chemical structures of PUFA.

Table 1. Piezophilic microorganisms isolated from various sources[a]

Organism(s)	Source (depth, m)	Optimal growth conditions	TEAP	Reference(s)
Piezotolerant bacteria (0.1–10 MPa)				
Sporosarcina sp. strain DSK25	Japan Trench (6,500)	0.1 MPa, 35°C	FAN	58
Moritella japonica DSK1	Japan Trench (6,353)	10–50 MPa, 15°C	FAN, R&F	84
Piezophilic bacteria (10–50 MPa)				
Psychromonas profunda 2825[T]	Atlantic (2,770)	25 MPa, 10°C	FAN, R&F	122
Moritella profunda	Atlantic (2,815)	25 MPa, 10°C	FAN, R&F	121
Moritella abyssi	Atlantic (2,815)	30 MPa, 10°C	FAN, R&F	121
Shewanella strain SC2A	E. Pacific Ocean	14 MPa, 20°C	FAN	133
Shewanella benthica WHB46	Weddell Sea (4,995)	40 MPa, 5°C	FAN?	22
Shewanella benthica F1A	N. Atlantic	41 MPa, 8°C	FAN	22
Photobacterium profundum DSJ4	Ryukyu Trench (5,110)	10 MPa, 10°C	FAN, R&F	81
Photobacterium profundum SS9	Sulu Sea (2,551)			24
Shewanella violacea DSS12	Ryukyu Trench (5,110)	30 MPa, 8°C	FAN, R&F	87
Moritella strain PE36	E. Pacific Ocean (3,600)	41 MPa, 15°C	FAN, R&F	22
Shewanella benthica DB5501	Suruga Bay (2,485)	50 MPa, 10°C	FAN, R&F	87
Shewanella benthica DB6101	Ryukyu Trench (5,110)	50 MPa, 10°C	FAN, R&F	87
Shewanella benthica DB6705	Japan Trench (6,356)	50 MPa, 10°C	FAN, R&F	87
Shewanella benthica DB6906	Japan Trench (6,269)	50 MPa, 10°C	FAN, R&F	87
Psychromonas strain CNPT3	N. Pacific (5,700)	52 MPa, 8°C	FAN	22
Shewanella benthica PT48	Philippine Trench		FAN	22
UM40	Puerto Rico Trench (5,920)		FAN, R&F	29
UM145	South Atlantic Ocean (4,575)		FAN, R&F	29
Hyperpiezophilic bacteria (>50 MPa)				
Psychromonas kaikoae JT7301 and JT7304	Japan Trench (7,304)	50 MPa, 10°C	FAN, R&F	83
Shewanella benthica DB172F	Izu-Bonin Trench (6,499)	70 MPa, 10°C	FAN, R&F	29
Shewanella benthica DB172R	Izu-Bonin Trench (6,499)	70 MPa, 10°C	FAN, R&F	29
Shewanella benthica DB21MT-2	Mariana Trench (10,898)	70 MPa, 10°C	FAN, R&F	60, 85
Moritella yayanosii DB21MT-5	Mariana Trench (10,898)	70 MPa, 10°C	FAN, R&F	85
Colwellia piezophila Y223G[T] and Y251E	Japan Trench (6,278)	60 MPa, 10°C	FAN, R&F	86
Colwellia hadaliensis BNL1	Puerto Rico Trench (7,410)	85 MPa, 10°C	FAN	30
Shewanella benthica PT99	Philippine Trench (8,600)	62 MPa, 8°C	FAN	22
Strain PT64	N. Pacific	90 MPa, 9°C		132
Strain MT199	N. Pacific	90 MPa, 13°C		132
Colwellia strain MT41	Mariana Trench	103 MPa, 8°C		22

[a]Based on references 1, 9, and 37. TEAP, terminal electron-accepting process; FAN, facultatively anaerobic; R&F, respiratory and fermentative metabolism.

with increasing temperature within the range of 6 to 10°C (125). It also appears to be true as a general rule that the pressure at which the rate of reproduction at 2°C is maximal may reflect the true habitat depth of an isolate (125, 134). The degree of piezophily increases with increasing collection depth or pressure (125). Therefore, each piezophilic isolate would have a single maximum growth rate, k_{max} (k = ln 2/doubling time [in hours]) at ($P_{k_{max}}, T_{k_{max}}$) on a *PTk* diagram (Fig. 2). Notice that $T_{k_{max}}$ and $P_{k_{max}}$ are temperatures and pressures where the growth rate has its highest value and are not P_{max} and T_{max}, which are cardinal temperatures and pressures.

The *PTk* diagram illustrates the known envelope of life of piezophiles (Fig. 2): the habitat conditions, the $T_{k_{max}}$ and $P_{k_{max}}$, and the response of k to T and P for a particular piezophilic isolate under a specific set of nutrient conditions (125, 126, 128, 130). The envelope is defined by a set of elliptical curves; each curve shows the k of a combined T and P. The k increases as T and P approach the habitat T and P (Fig. 2). It is apparent that the pressure range of growth, $P_{max} - P_{min}$, changes with temperature and vice versa. Typically, psychrophilic piezophiles have a $P_{k_{max}}$ that is $\approx P_{habitat}$ and a $T_{k_{max}}$ that is $\approx T_{habitat}$ plus 6 to 10°C (126).

Clearly, deep-sea piezophilic bacteria are stenothermal (126). The temperature range ($T_{max} - T_{min}$) is roughly 10 to 20°C (125, 135), whereas the pressure range is about 40 MPa for bacteria captured at a depth of less than 3,600 m (134) and 80 MPa for isolates of depths greater than 5,000 m (for example, see reference 130). Thus, piezophilic bacteria are bacteria with a $P_{k_{max}}$ of >0.1 MPa and can be defined as follows based on $P_{k_{max}}$ (Table 1): piezo-tolerant bacteria are those with a $P_{k_{max}}$ of 0.1 to10 MPa, piezophilic bacteria are those with a $P_{k_{max}}$ of 10 to 50 MPa, and hyperpiezophilic bacteria are those with a $P_{k_{max}}$ of >50 MPa.

Is There a Functionally Dominant Piezophilic Bacterial Community in the Deep Sea?

Few studies have been done on the ecological distribution of piezophilic bacterial populations in the deep sea. However, studies conducted hitherto suggest the existence and dominance of metabolically active, functionally dominant piezophilic bacterial popula-

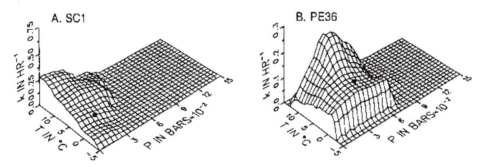

Figure 2. Three-dimensional views showing the temperature (T)-pressure (P) dependence of the exponential growth rate constant (k) of strain SC1 (A) and *Moritella* strain PE36 (B) from the North Pacific Ocean. The *PTk* diagram allows a reasonable, unambiguous determination of k_{max}, $T_{k_{max}}$, and $P_{k_{max}}$ of piezophilic bacterial growth (adapted from reference 125).

tions in the deep sea. These studies were done based on phospholipid fatty acid analysis, microbial utilization of substrates supplemented to deep-sea sediments incubated at in situ and atmospheric pressure, and cultivation-independent rRNA gene surveys.

Laboratory and Field Culturing and Growth Experiments

Isolation and growth of pure cultures at high pressures and inocula from a variety of deep-sea samples (water, sediment, and intestinal tracts and decaying parts of invertebrates) have demonstrated that piezophilic bacteria are ubiquitous (11, 23, 24, 26–30, 56, 57, 59, 79, 89, 109, 110, 129, 133). Based on hundreds of growth experiments with more than 20 isolates from the Pacific Ocean, Yayanos (125) concluded that piezophily is a common feature and a signal characteristic of true deep-sea bacteria. The threshold depth where piezophily appears is about 2,000 m. Given that the mean ocean depth is 3,800 m, the majority of marine bacteria likely show piezophily (125–127).

Baird et al. (8) determined microbial biomass and community structure of surface sediments (0 to 1 cm) of the Venezuela Basin and the Puerto Rico Trench (depth from 3,937 to 8,375 m) by analyzing ester-linked phospholipid fatty acids. Numerous fatty acids, including EPA and DHA and a mid-methyl-branched fatty acid (10Me-16:0), were detected in sediments. The concentrations of DHA ranged from 3.6 to 5.6 mol% of the total fatty acids. Given the fact that PUFA are labile and those produced by plankton in surface waters are preferentially degraded in the water column (21, 116), the high abundances of EPA and DHA in sediments of the Venezuela Basin and Puerto Rico Trench can be attributed at least partly to a piezophilic bacterial source. Indeed, Baird et al. (8) suggested that the detection of 10Me-16:0 in the sediments indicates the presence of sulfate-reducing piezophilic bacteria.

Wakeham and Canuel (115) studied fatty acid profiles of suspended particulate matter in the eastern tropical North Pacific Ocean. Samples were collected in the euphotic zone (60 m) and at a depth of 1,500 m using two different types of apparatus: Woods Hole in situ pumps, which have a nominal size of 1 μm, and particle interceptor traps, which have a 1-cm grid at the cone opening. Surprisingly, the relative percentage of EPA and DHA in the Woods Hole in situ pump samples collected at 1,500 m (18% of the total fatty acids) was essentially the same as that (20%) in samples collected in the euphotic zone. In contrast, the proportions of EPA and DHA in samples collected at the same site using particle interceptor traps decreased from 12% at the euphotic zone to about 2% at 1,500 m (115). Because of the high reactivity and labile nature of PUFA, they are preferentially degraded over the water column if we assume that all PUFA are produced in surface waters by phytoplankton. The unexpected high abundances of PUFA at depth led the authors to conclude that the PUFA were from piezophilic bacterial production at low-temperature, high hydrostatic pressure conditions (115, 117).

Piezophilic bacterial activity in abyssal sediments was tested and confirmed based on substrate utilization in a number of laboratory and field studies (25, 28, 33, 89, 91). Deming and Colwell (28) examined microbial activity in box cores and sediment trap samples collected in the Demerara abyssal plain in the South Atlantic Ocean (4,470 and 4,850 m). Samples were supplemented with low levels (<10% above natural abundance) of [^{14}C]glutamate and incubated at 3°C and in situ and atmospheric pressure. In both sediment and sinking-particulate samples examined, microbial utilization of [^{14}C]glutamate was enhanced by incubation at in situ pressure (Fig. 3 and 4), suggesting that

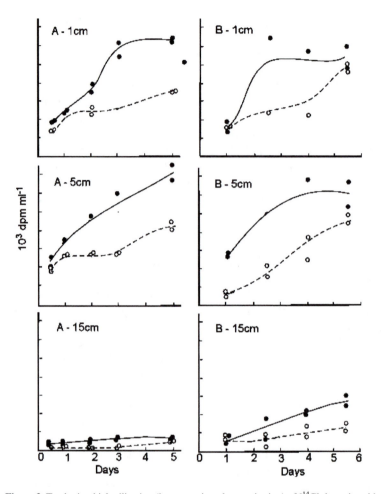

Figure 3. Total microbial utilization (incorporation plus respiration) of [^{14}C]glutamic acid at 3°C and in situ (44 MPa) pressure (filled circles) or atmospheric pressure (open circles) in sediment suspensions prepared from depths of 1, 5, and 15 cm in boxcores from stations A (depth, 4,470 m) and B (depth, 4,850 m). Respiration accounted for 89 to 94% of total substrate utilization at both pressures (adapted from reference 28).

indigenous piezophilic bacteria, not the piezosensitive microbes that originated in shallow surface waters, were metabolically active and functionally predominant in the cycle of naturally low levels of organic matter in the abyssal sediments (28). Similar results were obtained in a study conducted on sediment from the Bay of Biscay (4,300 to 4,800 m) (93) and the Porcupine abyssal plain of the Atlantic Ocean (33). Thus, it is highly likely that piezophilic bacterial activity may be widespread in the deep sea. However, the proportion of the metabolically and functionally dominant piezophilic bacterial population over the total microbial population in the deep sea is unknown (for example, see reference 123).

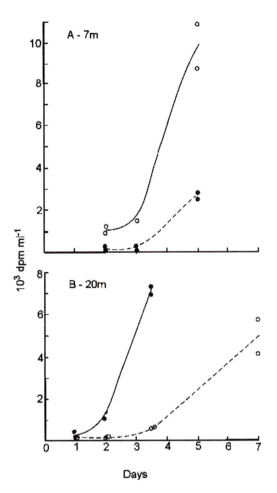

Figure 4. Total microbial utilization (incorporation plus respiration) of [^{14}C]glutamic acid at 3°C and in situ (44 MPa) pressure (filled circles) or atmospheric pressure (open circles) in seawater suspensions of particulates from temperature-compromised sediment trap sample A-7 (depth, 4,463 m) and cold trap sample B-20 (depth, 4,830 m). Respiration accounted for 84 to 89% of total substrate utilization at both pressures (adapted from reference 28).

Abundance and Distribution of Marine Bacteria in the Deep Sea

The number of bacteria in deep-sea sediments has been determined using a number of different techniques: a lipid-based approach, epifluorescence microscopy, and whole-cell fluorescent in situ hybridization. Bacterial cells in surficial sediments (0 to 5 cm) of the deep sea are estimated at a range of 10^6 to 10^9 cells/g. Schwartz et al. (102) reported 3.6×10^6 cells/g (wet weight) of sediment collected in the Atlantic Ocean at 4,940 m. Tabor et al. (107) reported bacterial counts of 2.1×10^8 to 4.7×10^8 cells/g of sediment again collected in the Atlantic Ocean (3,800 to 5,200 m). Deming and Colwell (28) determined bacterial abundance in deep-sea cores from the Demerara abyssal plain. Cell numbers at 0 to 5 cm of sediment ranged from 4.58×10^8 to 8.29×10^8 bacteria/g (dry weight) of sediment at a 4,470-m water depth and from 2.39×10^8 to 3.10×10^8 bacteria/g at 4,850 m. Bacterial abundance at 15 cm from the sediment-water interface decreased to 1.7×10^7 cells/g (31). Rowe and Deming (94) studied microbial abundance and the role of bacteria in the cycling of organic matter in sediments of the Bay of Biscay. Bacterial counts ranged from 2.02×10^8 to 2.82×10^8 cells/g of sediment at 4,100 m. Bacterial abundance generally

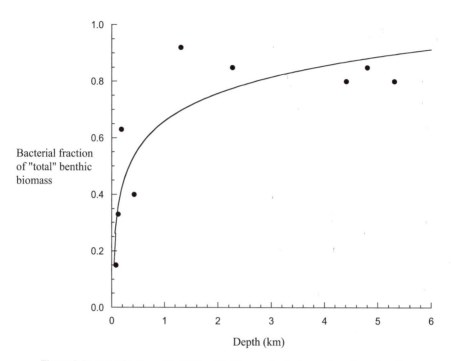

Figure 5. Bacterial fraction of "total" benthic biomass (excluding protozoa) as a function of depth in the ocean (adapted from reference 95).

decreases with depth by an order of magnitude. These numbers are in accord with cell numbers estimated from lipid analysis (for examples, see references 8 and 45).

Bacterial populations in the water column of the deep sea are orders of magnitude lower than those in sediments (10^4 to 10^7 cells/ml) (55, 78, 120). For example, Deming (25) reported 1.44×10^5 bacteria/ml at 1,850 m in the North Atlantic. Because of the apparently smaller sizes of deep-sea bacteria, Tabor et al. (108) used 0.22-μm-pore-size filters and reported 1.5×10^5 to 1.3×10^8 bacteria/ml in waters at depths of 1,700 to 8,160 m from a number of localities. Microbial abundance generally increased near the sediment-water interface, which has been observed in the subarctic and equatorial Pacific (78) and in the North Pacific gyre (55). Indeed, Rowe et al. (95) reported the increasing abundance of the bacterial fraction of the total benthic biomass with ocean depth, suggesting that bacteria dominate the benthic biomass of the deep sea (Fig. 5).

METABOLISM OF PIEZOPHILIC BACTERIA IN THE DEEP SEA

Microbial populations and their metabolic rates depend on substrate diversity and availability (20). The deep sea in general is an oligotrophic environment, except in areas of cold seeps and hydrothermal vents. Thus, it has been asserted by many that deep-sea microorganisms exist mostly in a dormant state (26). In research based on more than 60 bacterial isolates from depths of 1,957 to 10,500 m of the Pacific Ocean, Yayanos et al.

showed that the mean generation times of piezophilic bacteria typically range from 7 to 35 h at 2°C and at deep-sea pressures in laboratory cultures of nutrient-rich media (125, 134). Similar generation times were obtained in in situ experiments under nutrient-rich conditions (6 to 10 h), and generation times were much longer (>200 h) under nutrient-limited conditions (26).

However, other studies suggest that bacteria operate extremely efficient catabolic systems. The classical view among microbiologists and biogeochemists is that microorganisms only inhabit environments that are thermodynamically favorable to them. The minimum quantum of free energy in the environment that can be biochemically converted is -20 kJ mol^{-1}. However, recent studies suggest that as little as -4.5 kJ of free energy mol^{-1} can support bacterial growth and that bacterial metabolism can proceed near thermodynamic equilibrium in syntrophic associations (50). The reduction of the thermodynamic constraints on microbial life is significant for the deep-sea environment, where, generally, energy is expected to be limiting.

The deep-sea water column is mostly oxic, so aerobic respiration would dominate. In the upper sediments of the seafloor, oxygen is rapidly depleted and other electron acceptors (TAs), including nitrate and sulfate, that diffuse downward from the water column are utilized by facultative and obligate anaerobic bacteria for metabolism. These TAs are used in a predictable sequential series, according to the free-energy yield of the redox reactions: NO_3^- reduction and denitrification, dissimilatory Mn(IV) and Fe(III) reduction, sulfate reduction, and methanogenesis (autotrophic, fermentation, and acetoclastic). However, the predicted succession of the redox reactions is based on standard-state conditions (25°C and 1 bar). To address the effects of pressure and temperature on redox reactions in the deep sea, we computed the redox potential (E_h^o) and electron activity ($p\varepsilon^o$) of half-reactions at 2°C and various pressures based on the following equations (6):

$$E_h^o = \frac{-\Delta G_r^o}{nF} \quad \text{and} \quad p\varepsilon^o = \frac{FE_h^o}{2.303RT}$$

where ΔG_r^o is the Gibbs free energy of reaction at the temperature and pressure of interest, n is the number of moles of electrons in the reaction, and F is the Faraday constant. $p\varepsilon^o$ is further converted to $p\varepsilon^{o\prime}$, electron activity at the biological standard state (6):

$$p\varepsilon^{o\prime} = p\varepsilon^o + \frac{n_H}{2} \log K_w$$

where n_H is the number of moles of protons per moles of electrons in the half-reaction and K_w is the equilibrium constant at the temperature and pressure of interest for the water dissociation reaction ($H_2O \rightarrow H^+ + OH^-$).

It is generally believed that the effect of pressure on metabolic processes is secondary to that of temperature (for example, see reference 5). However, based on thermodynamic calculations using SUPCRT92 (5, 51), pressure does have a significant effect on microbially mediated redox reactions. It can be seen that pressure reduces the energy yield of redox reactions under high-pressure conditions (Fig. 6a, b, and d). For

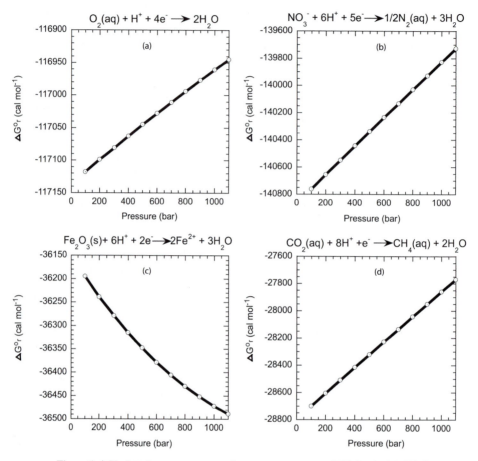

Figure 6. $\Delta G°$ plotted versus pressure and constant temperature (2°C) for O_2 (a), NO_3^- (b), Fe_2O_3 (s) (c), and CO_2 (d) reduction reactions.

example, the ΔG_r^o decreases from $-117,118$ cal mol^{-1} at 2°C and 10 MPa to $-116,945$ cal mol^{-1} at 2°C and 110 MPa for O_2 reduction. Similar patterns have been observed for other redox reactions (NO_3^- reduction, sulfate reduction, etc.). The only exception is the reduction of Fe_2O_3, where ΔG_r^o increases with pressure, from $-36,195$ cal mol^{-1} to $-36,489$ cal mol^{-1} (Fig. 6c). The electron towers were constructed based on electron activity at the biological standard state ($p\varepsilon^{o\prime}$) at 25°C and 0.1 MPa (Fig. 7a) and 2°C and 40 MPa (Fig. 7b). It is clear that the additive effect of temperature and pressure has raised the $p\varepsilon^{o\prime}$ values of the first four reactions commonly taking place at

Figure 7. Sequence of microbially mediated reduction reactions based on values of electron activity at biological standard state ($p\varepsilon^{o\prime}$) at 25°C and 1 bar (10^5 Pa) (a) and 2°C and 400 bar (4×10^7 Pa) (b). The $p\varepsilon^{o\prime}$ values are calculated per Johnson et al. (51) and Amend and Teske (6). TEAP, terminal electron-accepting process.

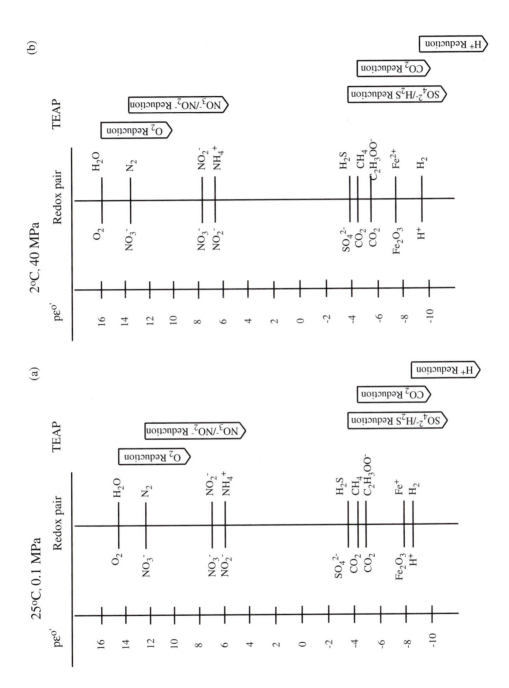

more positive redox potentials and lowered the $p\varepsilon^{o\prime}$ values for six other reactions that take place at more reducing conditions (Fig. 7b). Thus, redox reactions prevailing under more oxidative conditions (O_2, NO_3^-/NO_2^- reduction, etc.) yield slightly more energy in the deep-sea low-temperature and high-pressure conditions, whereas those that dominate in more reducing conditions yield relative less energy than in surface environments (25°C, 0.1 MPa).

BACTERIAL ADAPTATION TO DEEP-SEA LOW-TEMPERATURE AND HIGH-PRESSURE ENVIRONMENTS

Effects of Hydrostatic Pressure on Biological Membranes

On a molecular scale, hydrostatic pressure exerts significant effects upon the molecular order of biological membrane bilayers. Fundamentally, the effects of hydrostatic pressure are the change in system volume that accompanies a physiological or biochemical process (2, 9, 37, 77, 105). The response of biological systems to pressure is governed by the principle of Le Chatelier. The equations describing the effects of hydrostatic pressure on equilibrium and rate constants are

$$\left(\frac{\partial \ln K}{\partial P}\right)_T = -\frac{\Delta V}{RT}$$

$$\left(\frac{\partial \ln K}{\partial P}\right)_T = -\frac{\Delta V^{\ddagger}}{RT}$$

where K and k are the equilibrium and rate constants, respectively, P is the pressure, T is the absolute temperature (in Kelvin), and R is the gas constant. ΔV is the volume change in the system that accompanies the reaction. $\Delta V^{\ddagger}$ is the activation volume, the change in system volume representing the difference in volume between the reactants and the transition state. Pressure affects volume change-dependent reactions in cells. Pressure favors processes that are accompanied by negative volume changes. When a process occurs with an increase in system volume, pressure inhibits the process. The pressure-ordering effect imposed on a lipid bilayer may be evaluated in the context of a decrease in temperature. For example, the effect of a hydrostatic pressure of 1,000 atm (100 MPa) is equivalent to that of a 20°C decrease in transition temperature (15, 64). Thus, the effective temperature at the bottom of the Mariana Trench, for example, would be approximately −18°C. An isothermal phase transition may be brought about if sufficient pressure is applied to a lipid bilayer. If we assume that the molecular order of bacterial membranes in the deep sea is similar to those of bacteria in the shallow sea or surface environments, it is then reasonable to predict that the biochemical compositions of the membranes of piezophilic bacteria are modified to offset or compensate for the combined effects of low temperature and high pressure. High hydrostatic pressure affects the molecular organization of membrane lipid bilayers and results in the tight packing of acyl chains and decreased membrane fluidity (10, 65). Piezophilic bacteria have been found to change the fatty acid composition of their membranes to maintain optimal membrane fluidity and function at high pressure (4, 23, 24, 39).

Homeoviscous and Homeophasic Adaptation

The bilayers of phospholipid molecules are one of the major constituents in bacterial membranes. They constitute about 40 to 70% of the total bacterial membrane on a dry-weight basis (92). The dynamic states of lipids (the fluidity and the order) correlate strongly with the functions of biological membranes (18, 19). While pressure affects all biomolecules, lipids are more sensitive to pressure than, for example, proteins (9, 119). Increasing pressure, like a reduction in temperature, tends to solidify or "freeze-out" phospholipids (or the transition from a liquid-crystalline state to the gel state) (65), which leads to a disruption of biological functions of lipid-based cell membranes (76). In this way, the combination of high hydrostatic pressure and low temperature can create especially severe problems for deep-sea organisms.

Biological maintenance of membrane fluidity at low temperature and high pressure may be achieved by two mechanisms: homeoviscous adaptation and homeophasic adaptation. Homeoviscous adaptation maintains lipid physical state and membrane order (104), while homeophasic adaptation prevents the formation of nonbilayer phases and preserves the liquid-crystalline phase for membrane functionality (47, 71, 114). Homeoviscous adaptation involves the maintenance of the liquid-crystalline phase through alterations in the degree of acyl chain saturation, branching, and acyl chain length of the membrane lipids. In homeophasic adaptation, on the other hand, membranes are prevented from undergoing a phase transition that would compromise their structural integrity (for example, see reference 32).

DeLong and Yayanos (23) were the first to test the responses of a bacterium to pressures of 30 to 50 MPa (at 2°C) using the gram-negative, facultative anaerobic bacterial strain CNPT3. Fatty acids detected in the lipids of strain CNPT3 lipids include 14:0, 14:1, 16:0, 16:1, 18:0, and 18:1, with 16:0, 16:1, and 18:1 being the most abundant. The concentration of saturated fatty acids decreased from 34 to 25% with pressure, whereas the concentration of unsaturated fatty acids increased from 45 to 75%. The authors concluded that the pressure-induced changes in fatty acid composition are comparable to those induced by temperature changes and that homeoviscous adaptation of membrane lipids occurs in piezophilic bacteria in response to pressure.

DeLong and Yayanos (24) were the first to report PUFA in piezophilic bacteria. They determined the PUFA of 11 piezophilic bacteria (collected from depths of 1,200 to 10,476 m). All organisms contained either EPA (20:5ω3) or DHA (22:6ω3). These methylene-interrupted PUFA had previously only been found in microeukaryotes. The detection of these PUFA in piezophiles suggests that these apparently unique fatty acids (to prokaryotes) were at least partially involved in facilitating growth of piezophilic bacteria at high pressures (24).

Allen et al. (4) investigated the changes in fatty acid composition of the piezotolerant bacterium *Photobacterium profundum* strain SS9 in response to hydrostatic pressure (0.1, 28, and 50 MPa). *P. profundum* SS9 contained saturated fatty acids, monounsaturated fatty acids, and terminally branched fatty acids (TBFA) (*i*16:0) as well as EPA, similar to fatty acids observed in other piezophilic bacteria (24, 124). Fatty acid compositions were similar for the outer membrane and inner membrane except that the outer membrane contained higher concentrations of the hydroxy fatty acid 3OH-12:0. Concentrations of EPA increased with decreasing temperature and increasing pressure; pressure increases resulted in more significant increases in EPA and the monounsaturated fatty acid 18:1. Addition of the antibiotic cerulenin (2,3-epoxy-4-oxo-7,10-dodecadienamide) inhibited the biosynthesis of monounsaturated fatty acids with concurrent increases in EPA, suggesting a role for EPA in modulating membrane fluidity. When both cerulenin and 18:1 were added to growth media, there was a marked growth enhancement at high pressure (28 MPa), indicating that cells could take up exogenous 18:1 from the growth medium. The authors suggested that EPA was not required for the growth of strain SS9 (4, 9), and indeed, not all piezophilic bacteria contain PUFA.

These investigations demonstrate an important point: the vital roles of unsaturated fatty acids in regulating membrane fluidity under high pressure are due to their low melting temperature and unique molecular geometry (e.g., the formation of kinks in the fatty acid chain) in biomembranes. In other words, to prevent the solidification of membranes at high hydrostatic pressure, piezophilic bacteria respond by synthesizing high percentages of unsaturated, branched, shorter-chain fatty acids. However, an intriguing question is why piezophilic bacteria synthesize 20:5 and 22:6 if these fatty acids were only for the modulation of membrane fluidity, since tri- and tetraenoic fatty acids are much the same as 20:5 or 22:6 in enhancing membrane fluidity. In fact, Coolbear et al. (17) showed that introducing additional double bonds into diunsaturated 18:2 fatty acids in phosphatidylcholine causes a slight increase in melting point. PUFA may be needed for the maintenance of the correct phase of membrane lipids (homeophasic adaptation), because many bacterial phospholipids favor the formation of nonbilayer phases which would disrupt membrane packing (97). Therefore, these fatty acids play a dual role in bacterial adaptation to deep-sea low-temperature, high-pressure environments by lowering the phase transition temperature (thereby keeping the membrane fluid) and by providing a higher degree of packing order (thereby preventing the formation of nonbilayer phases) (9, 47, 97). It is clear that piezophilic bacteria also contain higher concentrations of *iso-* and *anteiso-*branched fatty acids (39, 124). The biosynthesis of abundant TBFA may be related to bacterial adaptation to the deep-sea low-temperature, high-pressure environment. The phase transition temperatures of phospholipids with TBFA are significantly lower than that of phospholipids with normal fatty acids of the same carbon number. For example, di-a15:0-phosphatidylcholine has a phase transition temperature of $-16.5°C$, compared with $34.2°C$ for di-15:0-phosphatidylcholine (103). In sum, lipids play key roles in piezophily.

Regulation of membrane fluidity depends on fatty acid composition as well as the polar head group of the phospholipids (90, 96). Fang et al. (36) determined the phospholipid profiles of the hyperpiezophilic bacteria *Shewanella benthica* strain DB21MT-2 and *Moritella yayanosii* strain DB21MT-5 originally from the Mariana Trench. Most phospholipids in both piezophilic strains were in the lipid classes of phosphatidylglycerol (PG), phosphatidylethanolamine, and its derivatives phosphatidylmethylethanolamine and phosphatidyldimethylethanolamine. The majority of the fatty acids were unsaturated, with one, five, or six double bonds. EPA and DHA were distributed on almost every PG molecule, mostly at the *sn*-2 position. It was suggested that the synthesis of high concentrations of PUFA and PG is a part of the mechanism employed by these bacteria to adapt to low temperature and high hydrostatic pressure in the deep sea. Given the same fatty acid composition, the phase transition temperature of PG is 20 to 30°C lower than that of phosphatidylethanolamine (70). Presumably, the larger head group of PG would be expected to cause greater disruption in acyl chain packing within the membrane bilayer and thereby lower the transition temperature in response to the additive effects of low temperature and high pressure (36).

Biosynthesis versus Dietary Uptake of PUFA

The regulation of membrane fluidity via changing the fatty acid composition of phospholipids entails the de novo biosynthesis and dietary uptake of fatty acids. To understand the role of and the possible requirement for PUFA in bacterial adaptation to high pressure, Fang et al. (39) examined biosynthesis and cellular uptake of PUFA in the moderately piezophilic bacterium *Shewanella violacea* strain DSS12 and the hyperpiezophilic bacteria *S. benthica* strain DB21MT-2 and *Moritella yayanosii* strain DB21MT-5. These strains were grown at pressures of 50 and 100 MPa in media containing marine broth 2216 supplemented with arachidonic acid (sodium salt) and/or the antibiotic cerulenin. In the absence of cerulenin, cells of strains DB21MT-2 and DB21MT-5 took up and incorporated exogenous arachidonic acid (14.7 and 1.4% of total fatty acids, respectively). No uptake by strain DSS12 was observed. When cells were treated with cerulenin, all three strains incorporated arachidonic acid into cell membranes (13 to 19%). The biosynthesis of monounsaturated fatty acids was significantly inhibited (10 to 37%) by the addition of cerulenin, whereas the concentrations of PUFA increased two to four times. These results suggest that piezophilic bacteria biosynthesize and/or incorporate dietary PUFA that may be important for their growth and piezoadaptation.

USE OF LIPIDS AND STABLE CARBON ISOTOPES IN DEEP-SEA GEOMICROBIOLOGY

Lipids and stable carbon isotopes preserved in lipids have proven to be excellent biosignatures applied to deep-sea geomicrobiology. For biologically mediated reactions, the isotopic signature can reflect the characteristics of specific enzymes in biochemical pathways (46). This is especially true in deep-sea microbial systems, where environmental conditions (temperature and pressure) may exert a greater influence on carbon isotope fractionation in the biosynthesis of fatty acids and therefore the carbon isotopic ratios of individual compounds.

Lipid Profiles of Piezophilic Bacteria

The fatty acid profiles of a number of piezophilic bacteria have been determined (4, 23, 24, 36, 39–41, 56, 58, 121, 124). Fang et al. (40) reported detailed fatty acid compositions of cells of *Moritella japonica* strain DSK1, *Shewanella violacea* strain DSS12, *S. benthica* strains DB6705 and DB21MT-2, and *M. yayanosii* strain DB21MT-5 grown on marine broth 2216 (Table 2). Characteristics of phospholipid fatty acids of piezophilic bacteria can be summarized as follows.

(i) Piezophilic bacteria biosynthesize typical bacterial fatty acids: C_{14-19} saturated, monounsaturated, terminal methyl-branched, hydroxyl, and cyclopropane fatty acids (Table 2; Fig. 8).

(ii) Piezophilic bacteria contain abundant monounsaturated fatty acids with multiple positions of unsaturation and geometric configuration (*cis* and *trans*). The proportions of monounsaturated fatty acids can be up to 65% of the total fatty acids.

Table 2. Fatty acid compositions of piezophilic bacteria[a]

Fatty acid	Concn, μg/g (dry wt) (% of total fatty acids)			
	DSK1	DSS12	DB6705	DB21MT-2
i13:0	—	803 (3.5)	—	—
3-OH-12:0	—	134 (0.6)	10 (0.5)	—
14:1Δ^7c	222 (0.8)	229 (1.0)	—	ND
14:1Δ^7t	1,821 (6.9)	207 (0.9)	ND	—
14:1Δ^9	458 (1.7)	—	ND	ND
14:0	3,550 (13.2)	2,148 (9.5)	90 (4.8)	83 (2.8)
3-OH-13:0	ND	400 (1.8)	25 (1.3)	—
i15:0	—	2,029 (9.0)	140 (7.5)	111 (3.8)
a15:0	463 (1.8)	671 (3.0)	—	ND
15:1Δ^7	217 (0.8)	—	ND	ND
15:0	301 (1.1)	288 (1.3)	40 (2.1)	40 (1.4)
16:1Δ^9c	1,929 (7.3)	448 (2.0)	12 (0.6)	1,101 (37.5)
16:1Δ^9t	9,109 (34.6)	5,238 (23.2)	698 (37.5)	—
16:1Δ^{11}	147 (0.6)	—	—	ND
16:0	2,764 (10.5)	2,454 (10.8)	370 (19.9)	494 (16.8)
17:1Δ^9	150 (0.6)	—	20 (1.1)	63 (2.1)
17:1Δ^{11}	—	—	—	26 (0.9)
17:0	—	—	8.9 (0.5)	21 (0.7)
18:3	ND	ND	104 (5.6)	ND
18:2	365 (1.4)	216 (1.0)	—	ND
18:1Δ^9	2,706 (10.3)	793 (3.5)	38 (2.0)	93 (3.2)
18:1Δ^{11}	431 (1.6)	893 (3.9)	191 (10.3)	579 (19.7)
18:0	—	—	14 (0.8)	50 (1.7)
22:6 (DHA)	1,679 (6.4)	—	—	—
20:5 (EPA)	ND	6,209 (27.4)	91 (4.9)	237 (8.1)
20:2	ND	ND	11 (0.6)	41 (1.4)
SFA	25.1	21.6	28.1	23.4
MUSA	65.3	34.5	51.5	63.4
PUFA	7.8	28.4	11.1	9.5
BFA	1.8	15.5	7.5	3.8
Total fatty acids (μg/g [dry wt] cell)	26,983	23,896	1,901	3,014
TUFA/SFA	2.9	2.9	2.2	3.1

[a]Based on reference 40. ND, not detected; —, <0.5%. SFA, saturated fatty acids; MUFA, monounsaturated fatty acids; BFA, branched fatty acids; TUFA, total unsaturated fatty acids. Fatty acids are designated by the total number of carbon atoms:number of double bonds (i.e., a 16-carbon alkanoic acid is 16:0).

(iii) Piezophilic bacterial species of the genera of *Shewanella*, *Moritella*, and *Photobacterium* all contain β-hydroxyl fatty acids. All piezophilic bacterial isolates examined thus far are gram negative, and the presence of hydroxyl fatty acids (Fig. 8) in piezophilic bacteria seems to be consistent with their gram-negative nature.

(iv) Piezophiles biosynthesize large amounts of TBFA (*iso* and *anteiso*) (Table 2). The concentrations of TBFA can be as high as 15% of the total fatty acids. Generally, the *iso*-branched fatty acids are in greater concentrations than *anteiso*-branched fatty acids. TBFA are typically found in gram-positive bacteria (e.g., *Bacillus*) (54). The presence of these branched fatty acids suggests that they have a functional role in piezoadaptation.

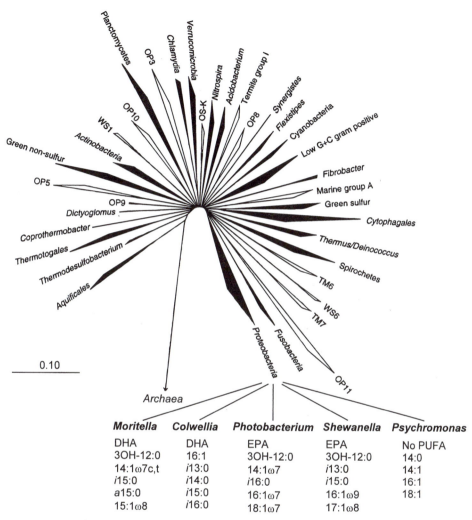

Moritella	Colwellia	Photobacterium	Shewanella	Psychromonas
DHA	DHA	EPA	EPA	No PUFA
3OH-12:0	16:1	3OH-12:0	3OH-12:0	14:0
14:1ω7c,t	i13:0	14:1ω7	i13:0	14:1
i15:0	i14:0	i16:0	i15:0	16:1
a15:0	i15:0	16:1ω7	16:1ω9	18:1
15:1ω8	i16:0	18:1ω7	17:1ω8	

Figure 8. Representative fatty acids of the five piezophilic genera *Moritella, Colwellia, Photobacterium, Shewanella,* and *Psychromonas.* Evolutionary distance tree of the domain *Bacteria* was taken from reference 49.

(v) Piezophilic bacteria contain abundant long-chain PUFA, EPA (20:5ω3) and DHA (22:6ω3) (Fig. 8).

Production of PUFA is a characteristic of piezophilic bacteria (4, 24, 35, 36, 39, 56, 58). Some psychrophilic bacteria also synthesize the same type of PUFA (for example, see reference 81). Marine PUFA (EPA or DHA)-producing bacteria are distributed in two distinct phylogenetic lineages (Fig. 8): the marine genera of the *Gammaproteobacteria* (*Shewanella, Colwellia, Moritella, Psychromonas,* and *Photobacterium*) and the two genera (*Flexibacter* and *Psychroserpens*) of the *Cytophaga-Flavobacterium-Bacteroides* group (80, 81). PUFA producers are piezophilic, psychrophilic, or halophilic. Bacteria in the genera *Shewanella, Colwellia, Moritella, Psychromonas,* and *Photobacterium* are true psychrophiles and/or piezophiles (59, 97) and may be the major PUFA producers in the oceans (36). The deep-sea members of these genera (e.g., *Shewanella*) are different from their surface water counterparts of mesophilic and piezosensitive (growth inhibited by increasing pressure) species in that they produce larger amounts of unsaturated fatty acids, particularly PUFA (58). Thus, the production of PUFA appears to be a unique trait of piezophilic and psychrophilic bacteria. *Flexibacter* and *Psychroserpens* species are psychrophilic and halophilic but not piezophilic, and their distributions are limited to the permanent cold areas of the Arctic (53) and Antarctica (12). Some species of these genera also produce PUFA (12). Thus, EPA and DHA can be used as an informative (but not exclusive) signature for detecting piezophilic bacteria in deep-sea sediment/water columns.

Psychropiezophilic bacteria probably possess unique pathways for the biosynthesis of fatty acids (Fig. 9) (72, 74, 75, 97, 118). Specifically, two independent fatty acid biosynthetic systems are shown to operate in piezophilic bacteria: the fatty acid synthase (FAS)-

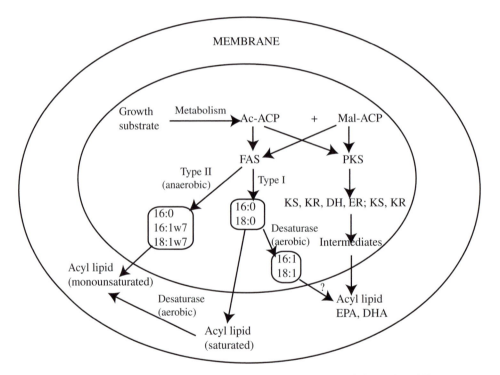

Figure 9. Summary of possible biosynthetic pathways of fatty acyl chains in piezophilic bacterial membrane lipids (modified from references 97 and 118 with permission of the publishers). The saturated and monounsaturated fatty acids are synthesized by the FAS pathway common to members of the domain *Bacteria,* which include the aerobic (type I) and anaerobic (type II) branches. The PUFA found in piezophilic bacteria are probably synthesized via the PKS pathway, which appears to be unique to marine bacteria. Biosynthesis of PUFA by an aerobic mechanism through sequential elongation and desaturation reactions appears less likely to occur in piezophilic bacteria. Ac-ACP, acetyl-acyl carrier protein; Mal-ACP, malonyl-acyl carrier protein; DH, dehydrase; ER, enoyl reductase; KR, 3-ketoacyl reductase; KS, 3-ketoacylsynthase.

based and polyketide synthase (PKS)-based pathways. The former is the biosynthetic pathway common to the bacteria which synthesizes typical bacterial fatty acids. The latter is a fundamentally different pathway which involves PKSs (72) which catalyze the biosynthesis of long-chain PUFA (Fig. 9). The PKS pathway apparently acts independently of FAS, elongase, and desaturase activities to synthesize EPA and DHA without any reliance on fatty acyl intermediates such as 16:0-acyl carrier protein (136). The PKS pathway appears to be widely distributed in marine bacteria (118), as genes with high homology to the *Shewanella* EPA gene cluster (*Shewanella* sp. strain SCRC-2738) (136) have been found in *Photobacterium profundum* strain SS9, which synthesizes EPA (4), and in *Moritella marina* strain MP-1, which contains DHA (111). For many years, it was believed that prokaryotes were unable to produce PUFA (34, 113), and PUFA in the environment have been ascribed exclusively to the de novo biosynthesis of microeukaryotes (44). In the determination of the sources of organic matter in marine sediments, PUFA (e.g., 20:5ω3 and 22:6ω3) have frequently been used as biomarkers of production of surface water plankton (for example, see reference 16). Clearly, piezophilic bacteria (4, 24, 36, 124) provide another de novo source of PUFA to marine sediments (36, 41). Therefore, we must review the role and potential importance of PUFA-producing bacteria in marine food chains (for example, see reference 79) and in deep-sea biogeochemistry (36, 37, 41). Thus, the reconstructions of paleoceanographic environments and biological activity using fatty acid biomarkers must be approached with caution because of the apparent bacterial origin of PUFA (for example, see reference 115).

Stable Carbon Isotope Signature of Lipids

Our current understanding of microbial carbon isotope fractionation is based on studies on lipid biosynthesis by mesophilic, nonpiezophilic bacteria. The models and parameters of lipid and carbon isotope biogeochemistry derived from these organisms may be significantly different from those of the deep-sea piezophilic bacteria and may not be applicable to solving marine biogeochemical problems. Kinetic isotope fractionation factors are sensitive to environmental conditions and to specific organisms and enzymes.

Fang et al. (35) determined carbon isotopic compositions of fatty acids isolated from the hyperpiezophilic bacteria *Shewanella benthica* strain DB21MT-2 and *Moritella yayanosii* strain DB21MT-5 grown on marine broth 2216. The variations of the $\delta^{13}C$ values between fatty acids were nearly 8 and 14‰ for each strain, respectively. Despite the fact that the two strains were grown on the same medium and under the same temperature and pressure, DB21MT-2 showed a systematic enrichment of ^{13}C in fatty acids compared to DB21MT-5 on a molecule-to-molecule basis. PUFA (EPA and DHA) exhibited the most depleted $\delta^{13}C$ values in both strains. All fatty acids except the odd-carbon-numbered ones from DB21MT-2 were depleted in ^{13}C relative to bacterial growth substrate (marine broth 2216). Fang et al. (35) concluded that the same type of microorganisms could have different $\delta^{13}C$ values under the same growth conditions, and that sedimentary fatty acids with distinct $\delta^{13}C$ values do not necessarily have to originate from different organisms.

Recently, Fang et al. (41) examined carbon isotope fractionation during fatty acid biosynthesis in cells of *Moritella japonica* strain DSK1 grown on a defined, noncomplex substrate (glucose) at 0.1, 10, 20, and 50 MPa. The $\delta^{13}C$ values of bacterial cell biomass were -13.1, -14.8, and $-17.6‰$ at 0.1, 10, and 50 MPa (Table 3), suggesting that carbon

Table 3. Stable carbon isotopic composition of fatty acids in *M. japonica* DSK1
grown on glucose at various pressures[a]

Compound	Stable carbon isotopic composition (‰) at indicated culture pressure (MPa)			
	0.1	10	20	50
14:1	−10.9		−27.7	−28.3
14:0	−12.9	−26.1	−27.4	−28.3
15:0	−15.1		−28.4	−29.8
16:1ω7c	−12.6	−25.0	−25.5	−27.5
16:0	−6.5	−13.7	−17.0	−26.6
cy17:0	−16.0	−24.6	−26.8	−30.3
17:0	−19.5			
18:1ω9	−9.9	−24.8	−27.0	−27.3
18:1ω7	−15.6	−28.5	−28.3	
18:0	−11.0		−11.1	
22:6ω3	−21.7	−36.7	−32.4	
Biomass	−13.1	−14.8		−17.6

[a]Based on reference 41.

isotope fractionation relative to the carbon source ($\delta^{13}C_{glucose} = -9.9‰$) is pressure dependent. The resultant carbon fractionations were −3.2, −4.9, and −7.7‰ at 0.1, 10, and 50 MPa, respectively. The differences in $\delta^{13}C$ among fatty acids were as much as −15.2, −23.0, −21.3, and −3.7‰ at 0.1, 10, 20, and 50 MPa, respectively. PUFA had much more negative $\delta^{13}C$ values than other short-chain saturated and monounsaturated fatty acids (Table 3).

A strong and consistent dependence of carbon fractionations on pressure was observed for individual fatty acids. Carbon isotopic fractionation between short-chain fatty acids (excluding DHA) and glucose ($\Delta\delta_{FA\text{-}glucose}$, average = −3.6‰) at 0.1 MPa was comparable to or slightly higher than fractionations observed on nonpiezophilic bacteria, e.g., *Escherichia coli* (73) and *Shewanella putrefaciens* (112). However, the fractionations at high pressures were much higher than that for nonpiezophilic bacteria, on average, −13.9, −14.5, and −18.3‰ at 10, 20, and 50 MPa, respectively. A strong linear correlation was observed between carbon isotopic fractionation and hydrostatic pressure (Fig. 10). Fang et al. (41) hypothesized that the observed isotope fractionation may be the result of the effects of high hydrostatic pressure on the kinetics of enzymatic reactions. Fatty acids are biosynthesized from the basic C_2 unit acetyl coenzyme A. The magnitude of fractionation is determined by a kinetic isotopic effect (ε_{PDH}). The magnitude of bacterial substrate utilization can be calculated (99) by

$$\varepsilon_{FA\text{-}substrate} = (1 - f)\, \varepsilon_{PDH}$$

where f is the fraction of pyruvate flowing to acetyl coenzyme A (73). The substantially higher carbon isotopic fractionation high pressures, −13.9, −15.6, and −18.0‰ at 10, 20, and 50 MPa, invalidate the equation ($f < 0$). This suggests that the ε_{PDH} in biosynthesis of fatty acids of piezophilic bacteria is greater than 23‰, a value commonly observed at atmospheric pressure of nonpiezophilic bacteria. Given the low lipid content at high pressures (5- to 10-fold lower than at atmospheric pressure [Table 2]), the f value may be as

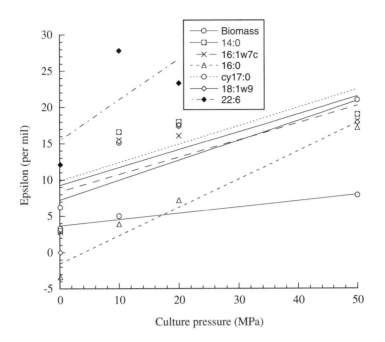

Figure 10. The calculated carbon isotopic fractionation, ε ($\varepsilon = \alpha_{s/p} - 1$) $\times$ 1,000, of whole cell biomass and selected fatty acids biosynthesized by *Moritella japonica* DSK1 at 0.1, 10, 20, and 50 MPa, where α is defined as $\alpha_{s/p} = (1{,}000 + \delta_s)/(1{,}000 + \delta_p)$, δ_s is the carbon isotopic ratio of substrate (glucose), and δ_p is that of product (cell biomass and fatty acids).

low as 0.1. Thus, the corresponding ϵ_{PDH} can be calculated at 31, 35, and 40‰ at 10, 20, and 50 MPa for even-carbon-numbered fatty acids. Therefore, carbon isotopic fractionation in the biosynthesis of fatty acids is pressure dependent. PUFA were much more depleted in ^{13}C. This was attributed to the operation of two different fatty acid biosynthetic systems in piezophilic bacteria: the FAS- and PKS-based pathways (38).

Isotope data are nearly always interpreted with the assumption that the isotopic composition of fatty acids is indicative of the isotopic composition of whole cells of the bacterium. Presumably, this can then reveal the carbon source utilized by the bacterium and thus its position within the paleo-ecosystem or present ecosystem. Furthermore, carbon isotopic data of fatty acids isolated from marine environments have been interpreted based on theories derived from nonpiezophilic bacteria. Piezophilic bacteria fractionate carbon isotopes significantly (14 to 18‰) more than surface heterotrophic bacteria. Thus, the recycling and resynthesis of fatty acids by piezophilic bacteria utilizing organic matter originating from primary production will greatly alter the carbon isotope signature of both short-chain bacterial and long-chain planktonic fatty acids in oceanic environments and marine sediments. For example, if the carbon isotopic composition of phytoplankton-derived organic matter is $-22‰$ (112), fatty acids synthesized by piezophilic bacteria would have δ^{13}C values of -36 to $-40‰$. These depleted δ^{13}C values of fatty acids could be falsely interpreted as having a terrestrial origin or as being from bacteria utilizing an isotopically light carbon source (e.g., methane). Moreover, DHA biosynthesized by piezophilic bacteria would have much more negative $\delta^{13}C$ values than DHA produced by surface plankton (for example, see reference 101).

Fang et al. (41) estimated the relative strength of carbon isotope signature of EPA and DHA from deep-sea piezophilic bacteria and surface phytoplankton. The total mass of marine primary producers is 2×10^{15} g (42). Assuming that phytoplankton contains 5 to 15% fatty acids on a dry-weight basis, the total amount of fatty acids produced by marine primary producers would be 0.1×10^{14} to 0.3×10^{14} g (based on 90% water content of phytoplankton). The percentage of EPA and DHA in total fatty acids of phytoplankton is about 15% (102). Thus, the total amount of EPA and DHA produced from marine primary production is 0.15×10^{13} to 0.45×10^{13} g. These compounds are preferentially degraded in the water column during transport (21); only a small proportion (0.01 to 0.02%) of the surface water-produced PUFA reaches the bottom water of the oceans (21, 117). Therefore, the amount of EPA and DHA that is produced in the surface ocean and that may potentially reach the sediment/water interface of the deep sea is 0.15×10^{9} to 0.45×10^{9} g.

The volumes of the deep sea and the top sediment layer (0 to 20 cm) are 1.028×10^{24} and 7.356×10^{19} cm^3, respectively (41). Assuming that bacterial abundance is 4.6×10^{8} cells/ml in deep-sea surface sediment and 0.5×10^{5} cells/ml in the water column (120), the total abundances of bacteria in the deep-sea surface sediment and water column are 3.39×10^{28} and 5.14×10^{28} cells, respectively. The total number of cells in the deep sea is 8.53×10^{28}, which is equivalent to 2.39×10^{16} g (dry weight) (2.8×19^{-13} g/bacterial cell [68]). Assuming that 20% of the deep-sea bacteria are metabolically active piezophilic bacteria that produce EPA and DHA at a rate of 3.0 µg/g (dry weight) (Table 2), the total amount of EPA and DHA from piezophilic bacterial production is 0.143×10^{11} g. Thus, the amount of EPA and DHA from deep-sea piezophilic bacteria is nearly 2 orders of magnitude higher than that from marine primary producers. Therefore, the carbon isotope signature of fatty acids preserved in marine sediments may be derived mostly from piezophilic bacteria whose contributions may easily override that of surface phytoplankton. The role of piezophilic bacteria in recycling and resynthesis of marine organic matter and in contributing biosignatures of these processes to the deep-sea sediments may be more important than hitherto recognized.

CONCLUSIONS AND DIRECTIONS OF FUTURE RESEARCH

The deep-sea piezobiosphere is an extreme environment where life has adapted in many unique ways. The evidence obtained thus far suggests that microbial life in the deep sea is diverse and that microbial biomass there can be significant but that microbial activity is diffuse. Despite significant progress in the past several decades, many questions in deep-sea geomicrobiology remain unanswered. Some of the most fundamental questions are as follows. What is the proportion of piezophilic bacteria in a given deep-sea microbial population? Are there functionally dominant piezophilic bacterial communities or species in the deep sea? What portions of the piezophilic and nonpiezophilic communities are actively reproducing and metabolizing in the deep sea? What are the extent and activity of the deep-sea piezobiosphere? What roles do piezophiles play in biogeochemical cycling in the oceans' interior and surface sediments? What is the extent of carbon isotope fractionations during biosynthesis of lipids which can be useful in deep-sea geomicrobiology? All of the piezophilic isolates reported thus far are facultatively anaerobic, some demonstrating anaerobic respiration as well as fermentation (Table 1). It has been shown that the degradation of organic matter by fermentation and anaerobic respiration is the principal energy generation process in both surface and subsurface marine sediments (31, 52, 88). Thus, marine bacteria must play an important role in the overall ocean carbon cycle (7, 14, 61).

Our knowledge and understanding of psychropiezophilic archaea are nearly nonexistent. Recent studies utilizing culture-independent, whole-cell fluorescent in situ hybridization suggested that the abundance of archaeal cells increases with depth in the ocean and exceeds that of bacteria below 1,000 m, and that most pelagic deep-sea microorganisms are metabolically active (55, 100).

The field of deep-sea geomicrobiology now begs for an increased understanding of the influence of piezophilic communities on the global ocean environment and on biogeochemical cycling occurring in the deep sea. These microorganisms surely influence the surface of Earth by changing the chemistry of the ocean and by affecting the rate of organic carbon burial, with consequences for both the marine carbon cycle and global climate.

Acknowledgment. We are grateful to Harold C. Helgeson of the University of California, Berkeley, for providing the computer program SUPCRT92 for thermodynamic calculations of microbial reactions.

REFERENCES

1. **Abe, F., and K. Horikoshi.** 2001. The biotechnological potential of piezophiles. *Trends Biotechnol.* **19:**102–108.
2. **Abe, F., C. Kato, and K. Horikoshi.** 1999. Pressure-regulated metabolism in microorganisms. *Trends Microbiol.* **7:**447–453.
3. **Aguilar, A.** 1996. Extremophile research in the European Union: from fundamental aspects to industrial expectations. *FEMS Microbiol. Rev.* **18:**89–92.
4. **Allen, E. E., D. Facciotti, and D. H. Bartlett.** 1999. Monounsaturated but not polyunsaturated fatty acids are required for growth of the deep-sea bacterium *Photobacterium profundum* SS9 at high pressure and low temperature. *Appl. Environ. Microbiol.* **65:**1710–1720.
5. **Amend, J. P., and E. L. Shock.** 2001. Energetics of overall metabolic reactions of thermophilic and hyperthermophilic archaea and bacteria. *FEMS Microbiol. Rev.* **25:**175–243.
6. **Amend, J. P., and A. Teske.** 2005. Expanding frontiers in deep subsurface microbiology. *Paleogeogr. Peloclimatol. Paleoecol.* **219:**131–155.
7. **Azam, F., and R. A. Long.** 2001. Sea snow microcosms. *Nature* **414:**495–498.
8. **Baird, B. H., D. E. Nivens, J. H. Parker, and D. C. White.** 1985. The biomass, community structure and spatial distribution of the sedimentary microbiota from a high-energy area of the deep sea. *Deep-Sea Res.* **32:**1089–1099.
9. **Bartlett, D. H.** 2002. Pressure effects on in vivo microbial processes. *Biochim. Biophys. Acta* **1595:**367–381.
10. **Bartlett, D. H., and K. A. Bidle.** 1999. Membrane-based adaptions of deep-sea piezophiles, p. 503–512. *In* J. Sechbach (ed.), *Enigmatic Microorganisms and Life in Extreme Environments.* Kluwer Academic Publishers, Boston, MA.
11. **Bartlett, D. H., E. Chi, and T. J. Welch.** 1996. High pressure sensing and adaptation in the deep-sea bacterium *Photobacterium* species strain SS9, p. 29–36. *In* R. Hayashi and C. Balny (ed.), *High Pressure Bioscience and Biotechnology.* Elsevier Science B. V., Amsterdam, The Netherlands.
12. **Bowman, J. P., S. A. McCammon, T. Lewis, J. H. Skerratt, J. L. Brown, D. S. Nichols, and T. A. McMeekin.** 1998. *Psychroflexus torquis* gen. nov., sp. nov., a psychrophilic species from Antarctic sea ice, and reclassification of *Flavobacterium gondwanense* (Dobson et al. 1993) as *Psychroflexus gondwanense* gen. nov., comb. nov. *Microbiology* **144:**1601–1609.
13. **Burton, S. K., and H. M. Lappin-Scott.** 2005. Geomicrobiology, the hidden depths of the biosphere. *Trends Microbiol.* **13:**401.
14. **Cho, B. C., and F. Azam.** 1988. Major role of bacteria in biogeochemical fluxes in the ocean's interior. *Nature* **332:**441–443.

15. **Chong, P. L.-G., and A. R. Cossins.** 1983. A differential polarized fluorimetric study of the effects of high hydrostatic pressure upon the fluidity of cellular membranes. *Biochemistry* **22:**409–415.

16. **Colombo, J. C., N. Silverberg, and J. N. Gearing.** 1996. Lipid biogeochemistry in the Laurentian Trough. I. Fatty acids, sterols and aliphatic hydrocarbons in rapidly settling particles. *Org. Geochem.* **25:**211–226.

17. **Coolbear, K. P., C. B. Berde, and K. M. W. Keough.** 1983. Gel to liquid crystalline phase transitions of aqueous dispersion of polyunsaturated mixed-acid phosphatidylcholines. *Biochemistry* **22:**1466–1473.

18. **Cossins, A. R., and A. G. MacDonald.** 1984. Homeoviscous theory under pressure. II. The molecular order of membranes from deep-sea fish. *Biochim. Biophys. Acta* **776:**144–150.

19. **Cossins, A. R., and M. Sinensky.** 1984. Adaptation of membranes to temperature, pressure and exogenous lipids, p. 1–6. *In* M. Shinitzky (ed.), *Physiology of Membrane Fluidity*, vol. II. CRC Press, Boca Raton, FL.

20. **Cowan, D. A.** 1998. Hot bugs, cold bugs and sushi. *Trends Biotechnol.* **16:**241–242.

21. **De Barr, H. J. W., J. W. Farrington, and S. G. Wakeham.** 1983. Vertical flux of fatty acids in the North Atlantic Ocean. *J. Mar. Res.* **41:**19–41.

22. **DeLong, E. F., D. G. Franks, and A. A. Yayanos.** 1997. Evolutionary relationship of cultivated psychrophilic and barophilic deep-sea bacteria. *Appl. Environ. Microbiol.* **63:**2105–2108.

23. **DeLong, E. F., and A. A. Yayanos.** 1985. Adaptation of membrane lipids of a deep-sea bacterium to changes in hydrostatic pressure. *Science* **228:**1101–1103.

24. **DeLong, E. F., and A. A. Yayanos.** 1986. Biochemical function and ecological significance of novel bacterial lipids in deep-sea prokaryotes. *Appl. Environ. Microbiol.* **51:**730–737.

25. **Deming, J. W.** 1985. Bacterial growth in deep-sea sediment trap and boxcore samples. *Mar. Ecol. Prog. Ser.* **25:**305–312.

26. **Deming, J. W., and J. A. Baross.** 2000. Survival, dormancy, and nonculturable cells in extreme deep-sea environments, p. 147–198. *In* R. R. Colwell and D. J. Grimes (ed.), *Nonculturable Microorganisms in the Environment.* ASM Press, Washington, DC.

27. **Deming, J. W., and R. R. Colwell.** 1982. Barophilic bacteria associated with the digestive tracts of abyssal holothurians. *Appl. Environ. Microbiol.* **44:**1222–1230.

28. **Deming, J. W., and R. R. Colwell.** 1985. Observation of barophilic microbial activity in samples of sediment and intercepted particulates from the Demerara abyssal plain. *Appl. Environ. Microbiol.* **50:**1002–1006.

29. **Deming, J. W., H. Hada, R. R. Colwell, K. R. Luehrsen, and G. E. Fox.** 1984. The ribonucleotide sequence of 5S rRNA from two strains of deep-sea barophilic bacteria. *J. Gen. Microbiol.* **130:**1911–1920.

30. **Deming, J. W., L. K. Somers, W. L. Strauble, and M. T. McDonell.** 1988. Isolation of an obligately barophilic bacterium and description of a new genus, *Nogi, Y., Colwellia* gen-nov. *Syst. Appl. Microbiol.* **10:**152–155.

31. **D'Hondt, S., B. B. Jørgensen, D. J. Miller, A. Batzke, R. Blake, B. A. Cragg, H. Cypionka, G. R. Dickens, T. Ferdelman, K.-U. Hinrichs, N. G. Holm, R. Mitterer, A. Spivack, G. Wang, B. Bekins, B. Engelen, K. Ford, G. Gettemy, S. D. Rutherford, H. Sass, C. G. Skilbeck, I. W. Aiello, G. Guèrin, C. H. House, F. Inagaki, P. Meister, T. Naehr, S. Niitsuma, R. J. Parkes, A. Schippers, D. C. Smith, A. Teske, J. Wiegel, C. N. Padilla, and J. L. S. Acosta.** 2004. Distributions of microbial activities in deep subseafloor sediments. *Science* **306:**2216–2221.

32. **Drobnis, E. Z., L. M. Crowe, and T. Berger.** 1993. Cold shock damage is due to lipid phase transitions in cell membranes: a demonstration using sperm as a model. *J. Exp. Zool.* **265:**432–437.

33. **Eardly, D. F., M. W. Carton, J. M. Gallagher, and J. W. Patching.** 2001. Bacterial abundance and activity in deep-sea sediments from the eastern North Atlantic. *Prog. Oceanogr.* **50:**245–259.

34. **Erwin, E., and K. Bloch.** 1964. Biosynthesis of unsaturated fatty acids in microorganisms. *Science* **143:**1006–1012.

35. **Fang, J., M. J. Barcelona, T. A. Abrajano, Jr., C. Kato, and Y. Nogi.** 2002. Isotopic composition of fatty acids isolated from the extremely piezophilic bacteria from the Mariana Trench at 11,000 meters. *Mar. Chem.* **80:**1–9.

36. **Fang, J., M. J. Barcelona, C. Kato, and Y. Nogi.** 2000. Biochemical function and geochemical significance of novel phospholipids isolated from extremely barophilic bacteria from the Mariana Trench at a depth of 11,000 meters. *Deep-Sea Res.* **47:**1173–1182.

37. **Fang, J., and C. Kato.** 2002. Piezophilic bacteria: taxonomy, diversity, adaptation, and potential biotechnological applications, p. 47–80. *In* M. Fingerman (ed.), *Recent Advances in Marine Biotechnology,* vol. 8. Science Publishers, Inc., Enfield, NH.

38. **Fang, J., and C. Kato.** FAS or PKS, lipid biosynthesis and stable carbon isotope fractionation in deep-sea piezophilic bacteria, p. 190–200. *In* A. Méndez-Vilas (ed.), *Communicating Current Research and Educational Topics and Trends in Applied Microbiology,* vol. 1. The Formatex Microbiology Book Series, Formatex Center, Badajoz, Spain.

39. **Fang, J., C. Kato, T. Sato, O. Chan, and D. S. McKay.** 2004. Polyunsaturated fatty acids in piezophilic bacteria: biosynthesis or dietary uptake? *Comp. Biochem. Physiol. B* **137:**455–461.

40. **Fang, J., C. Kato, T. Sato, O. Chan, T. Peeples, and K. Niggemeyer.** 2003. Phospholipid fatty acid profiles of piezophilic bacteria from the deep sea. *Lipids* **38:**885–887.

41. **Fang, J., M. Uhle, K. Billmark, D. H. Bartlett, and C. Kato.** 2005. Fractionation of carbon isotopes in biosynthesis of fatty acids by a piezophilic bacterium *Moritella japonica* DSK1. *Geochim. Cosmochim. Acta* **70:**1753–1760.

42. **Garrison, T.** 2005. *Oceanography: an Invitation to Marine Science,* 5th ed. Brooks/Cole, Belmont, CA.

43. **Gauthier, G., M. Gauthier, and R. Christen.** 1995. Phylogenetic analysis of the genera *Alteromonas, Shewanella,* and *Moritella* using genes coding for small-subunit rRNA sequences and division of the genus *Alteromonas* into two genera, *Alteromonas* (emended) and *Pseudoalteromonas* gen. nov., and proposal of twelve new species combinations. *Int. J. Syst. Bacteriol.* **45:**755–761.

44. **Gonzalezbaro, M. D., and R. J. Pollero.** 1998. Fatty acid metabolism of *Macrobrachium borellii*—dietary origin of arachidonic and eicosapentaenoic acids. *Comp. Biochem. Physiol.* **119:**747–752.

45. **Harvey, H. R., M. D. Richardson, and J. S. Patton.** 1984. Lipid composition and vertical distribution of bacteria in aerobic sediments of the Venezuela Basin. *Deep-Sea Res.* **31:**403–413.

46. **Hayes, J. M.** 1993. Factors controlling ^{13}C contents of sedimentary compounds: principles and evidence. *Mar. Geol.* **13:**111–125.

47. **Hazel, J. R.** 1995. Thermal adaptation in biological membranes: is homeoviscous adaptation the explanation? *Annu. Rev. Physiol.* **57:**19–42.

48. **Horneck, G.** 2000. The microbial world and the case for Mars. *Planet. Space Sci.* **48:**1053–1063.

49. **Hugenholtz, P., B. M. Goebel, and N. R. Pace.** 1998. Impact of culture-independent studies on the emerging phylogenetic view of bacterial diversity. *J. Bacteriol.* **180:**4765–4774.

50. **Jackson, B. E., and M. J. McInerney.** 2002. Anaerobic microbial metabolism can proceed close to thermodynamic limits. *Nature* **415:**454–456.

51. **Johnson, J. W., E. H. Oelkers, and H. C. Helgeson.** 1992. SUPCRT92: a software package for calculating the standard molal thermodynamic properties of minerals, gases, aqueous species, and reactions from 1 to 500 bar and 0 to 1000°C. *Comput. Geosci.* **18:**899–947.

52. **Jørgensen, B. B.** 1982. Mineralization of organic matter in the sea bed—the role of sulfate reduction. *Nature* **296:**643–645.

53. **Jøstensen, J.-P., and B. Landfald.** 1997. High prevalence of polyunsaturated-fatty-acid producing bacteria in arctic invertebrates. *FEMS Microbiol. Lett.* **151:**95–101.

54. **Kaneda, T.** 1991. *Iso*-fatty and *anteiso*-fatty acids in bacteria: biosynthesis, function, and taxonomic significance. *Microbiol. Rev.* **55:**288–302.

55. **Karner, M., E. F. DeLong, and D. M. Karl.** 2001. Archaeal dominance in the mesopelagic zone of the Pacific Ocean. *Nature* **409:**507–510.

56. **Kato, C., L. Li, Y. Nogi, Y. Nakamura, J. Tamaoka, and K. Horikoshi.** 1998. Extremely barophilic bacteria isolated from the Mariana Trench, Challenger Deep, at a depth of 11,000 meters. *Appl. Environ. Microbiol.* **64:**1510–1513.

57. **Kato, C., N. Masui, and K. Horikoshi.** 1996. Properties of obligately barophilic bacteria isolated from a sample of deep-sea sediment from the Izu-Bonin Trench. *J. Mar. Biotechnol.* **4:**96–99.

58. **Kato, C., and Y. Nogi.** 2001. Correlation between phylogenetic structure and function: examples from deep-sea *Shewanella. FEMS Microbiol. Ecol.* **35:**223–230.

59. **Kato, C., T. Sato, and K. Horikoshi.** 1995. Isolation and properties of barophilic and barotolerant bacteria from deep-sea mud samples. *Biodivers. Conserv.* **4:**1–9.

60. **Kato, C., H. A. Tamegai, R. Ikegami, R. Usami, and K. Horikoshi.** 1996. Open reading frame 3 of the barotolerant bacterium strain DSS12 is complementary with *cydD* in *Escherichia coli: cydD* functions are required for cell stability at high pressure. *J. Biochem.* **120:**301–305.

61. **Kiorboe, T., and G. A. Jackson.** 2001. Marine snow, organic solute plumes, and optimal chemosensory behavior of bacteria. *Limnol. Oceanogr.* **46:**1309–1318.

62. **Li, L., C. Kato, Y. Nogi, and K. Horikoshi.** 1999. Microbial diversity in sediments collected from the deepest cold-seep area, the Japan Trench. *Mar. Biotechnol.* **1:**391–400.

63. **Liesack, W., H. Weyland, and E. Stackebrandt.** 1991. Potential risks of gene amplification by PCR as determined by 16S rDNA analysis of a mixed culture of strict barophilic bacteria. *Microb. Ecol.* **21:**191–198.

64. **Macdonald, A. G.** 1975. *Physiological Aspects of Deep Sea Biology.* Cambridge University Press, Cambridge, United Kingdom.

65. **Macdonald, A. G.** 1984. The effects of pressure on the molecular structure and physiological function of cell membranes. *Philos. Trans. R. Soc. Lond. B.* **304:**47–68.

66. **MacDonell, M. T., and R. R. Colwell.** 1985. Phylogeny of the Vibrionaceae, and recommendation for two new genera, *Listonella* and *Shewanella. Syst. Appl. Microbiol.* **6:**171–182.

67. **Madigan, M. T., and B. L. Marrs.** 1997. Extremophiles. *Sci. Am.* **276:**82–87.

68. **Madigan, M. T., J. M. Martinko, and J. Parker.** 2003. *Brock Biology of Microorganisms,* 10th ed. Prentice Hall, Upper Saddle River, NJ.

69. **Madigan, M. T., and A. Oren.** 1999. Thermophilic and halophilic extremophiles. *Curr. Opin. Microbiol.* **2:**265–269.

70. **Marsh, D.** 1990. *Handbook of Lipid Bilayers.* CRC Press, Boca Raton, FL.

71. **McElhaney, R. N.** 1984. The structure and function of the Acholeplasma laidlawii plasma membrane. *Biochim. Biophys. Acta* **779:**1–42.

72. **Metz, J. G., P. Roessler, D. Facciotti, C. Leverine, F. Dittrich, M. Lassner, R. Valentine, K. Lardizabal, F. Domergue, A. Yamada, K. Yazawa, V. Knauf, and J. Browse.** 2001. Production of polyunsaturated fatty acids by polyketide synthases in both prokaryotes and eukaryotes. *Science* **293:**290–293.

73. **Monson, K. D., and J. M. Hayes.** 1980. Biosynthetic control of the natural abundance of carbon 13 at specific positions within fatty acids in *Escherichia coli. J. Biol. Chem.* **255:**11435–11441.

74. **Morita, N., M. Tanaka, and H. Okuyama.** 2000. Biosynthesis of fatty acids in the docosahexaenoic acid-producing bacterium *Moritella marina* MP-1. *Biochem. Soc. Trans.* **28:**943–945.

75. **Morita, N., A. Ueno, M. Tanaka., S. Ohgiya, K. Kawasaki, I. Yumoto, K. Ishizaka, and H. Okuyama.** 1999. Cloning and sequencing of clustered genes involved in fatty acid biosynthesis from the docosahexaenoic acid-producing bacterium, *Vibrio marinus* strain MP-1. *Biotechnol. Lett.* **21:**641–644.

76. **Morita, R. Y.** 1986. Pressure as an extreme environment, p. 171–185. *In* R. A. Herbert and G. A. Codd (ed.), *Microbes in Extreme Environments.* Academic Press, London, United Kingdom.

77. **Mozhaev, V. V., K. Heremans, J. Frank, P. Masson, and C. Balny.** 1994. Exploring the effects of high hydrostatic pressure in biotechnological applications. *Trends Biotechnol.* **12:**493–501.

78. **Nagata, T., H. Fukuda, R. Fukuda, and I. Koike.** 2000. Bacterioplankton distribution and production in deep Pacific waters: large-scale geographic variations and possible coupling with sinking particle fluxes. *Limnol. Oceanogr.* **45:**426–435.

79. **Nakayama, A., Y. Yano, and K. Yoshida.** 1994. New method for isolating barophiles from intestinal contents of deep-sea fishes retrieved from the abyssal zone. *Appl. Environ. Microbiol.* **60:**4210–4212.

80. **Nichols, D. S.** 2003. Prokaryotes and the input of polyunsaturated fatty acids into the marine food web. *FEMS Microbiol. Lett.* **219:**1–7.

81. **Nichols, D. S., and T. A. McMeekin.** 2002. Biomarker techniques to screen for bacteria that produce polyunsaturated fatty acids. *J. Microbiol. Methods* **48:**161–170.

82. **Nogi, Y., S. Hosoya, C. Kato, and K. Horikoshi.** 2004. Colwellia piezophila sp. nov., a novel piezophilic *Colwellia* species from deep-sea sediments of the Japan Trench. *Int. J. Syst. Evol. Microbiol.* **54:**1627–1631.

83. **Nogi, Y., and C. Kato.** 1999. Taxonomic studies of extremely barophilic bacteria isolated from the Mariana Trench, and *Moritella yayanosii* sp. nov., a new barophilic bacterial species. *Extremophiles* **3:**71–77.

84. **Nogi, Y., C. Kato, and K. Horikoshi.** 1998. Taxonomic studies of deep-sea barophilic *Shewanella* species, and *Shewanella violacea* sp. nov., a new barophilic bacterial species. *Arch. Microbiol.* **170:**331–338.

85. **Nogi, Y., C. Kato, and K. Horikoshi.** 1998. *Moritella japonica* sp. nov., a novel barophilic bacterium isolated from a Japan Trench sediment. *J. Gen. Appl. Microbiol.* **44:**289–295.

86. **Nogi, Y., C. Kato, and K. Horikoshi.** 2002. *Psychromonas kaikoi* sp. nov., isolation of novel piezophilic bacteria from the deepest cold-seep sediments in the Japan Trench. *Int. J. Syst. Evol. Microbiol.* **52:**1527–1532.

87. **Nogi, Y., N. Masui, and C. Kato.** 1998. *Photobacterium profundum* sp. nov., a new, moderately barophilic bacterial species isolated from a deep-sea sediment. *Extremophiles* **2:**1–7.

88. **Parkes, R. J., G. Webster, B. A. Cragg, A. J. Weightman, C. J. Newberry, T. G. Ferdelman, J. Kallmeyer, B. B. Jørgensen, I. W. Aiello, and J. C. Fry.** 2005, Deep sub-seafloor prokaryotes stimulated at interfaces over geologic time. *Nature* **436:**390–394.

89. **Patching, J. W., and D. Eardly.** 1997. Bacterial biomass and activity in the deep waters of the eastern Atlantic—evidence of a barophilic community. *Deep-Sea Res.* **44:**1655–1670.

90. **Pluschke, G., and P. Overath.** 1981. Function of phospholipids in *Escherichia coli. J. Biol. Chem.* **256:**3207–3212.

91. **Poremba, K., D. Eardly, and J. W. Patching.** 1994. Dynamics of microbial abundance and activity in deep-sea sediment of the Northeast Atlantic. *Microbiol. Eur.* **2:**22–25.

92. **Quinn, P. J.** 1976. *Molecular Biology of Cell Membranes.* McMillan, London, United Kingdom.

93. **Rothschild, L. J., and R. L. Mancinelli.** 2001. Life in extreme environments. *Nature* **409:**1092–1101.

94. **Rowe, G. T., and J. W. Deming.** 1985. The role of bacteria in the turnover of organic carbon in deep-sea sediments. *J. Mar. Res.* **43:**925–950.

95. **Rowe, G. T., M. Sibuet, J. W. Deming, A. Khripounoff, J. Tietjen, S. Macko, and R. Theroux.** 1991. "Total" sediment biomass and preliminary estimates of organic carbon residence time in deep-sea benthos. *Mar. Ecol. Prog. Ser.* **79:**99–114.

96. **Russell, N. J.** 1988. Functions of lipids: structural roles and membrane functions, p. 279–365. *In* C. Ratledge and S. G. Wilkinson (ed.), *Microbial Lipids,* vol. II. Academic Press, London, United Kingdom.

97. **Russell, N. J., and D. S. Nichols.** 1999. Polyunsaturated fatty acids in marine bacteria—a dogma rewritten. *Microbiology* **145:**767–779.

98. **Rusterholtz, K., and M. Pohlschroder.** 1999. Where are the limits of life? *Cell* **96:**469–470.

99. **Sakata, S., J. M. Hayes, A. R. McTaggart, R. A. Evans, K. J. Leckrone, and R. K. Togasaki.** 1997. Carbon isotopic fractionation associated with lipid biosynthesis by a cyanobacterium: relevance for interpretation of biomarker records. *Geochim. Cosmochim. Acta* **61:**5379–5389.

100. **Schippers, A., L. N. Neretin, J. Kallmeyer, T. G. Ferdelman, B. A. Cragg, R. J. Parkes, and B. B. Jørgensen.** 2005. Prokaryotic cells of the deep sub-seafloor biosphere identified as living bacteria. *Nature* **433:**861–864.

101. **Schouten, S., W. C. K. Breteler, P. Blokker, N. Schogt, W. I. C. Rijpstra, K. Grice, M. Baas, and J. S. Sinninghe Damsté.** 1998. Biosynthetic effects on the stable carbon isotopic compositions of algal lipids: implications for deciphering the carbon isotopic biomarker record. *Geochim. Cosmochim. Acta* **62:**1397–1406.

102. **Schwartz, J. R., J. D. Walker, and R. R. Colwell.** 1974. Growth of deep-sea bacteria on hydrocarbons at ambient and in situ pressure. *Dev. Ind. Microbiol.* **15:**239–249.

103. **Silvius, J. R.** 1982. Thermotropic lipid phase transitions of pure lipids in model membranes and their modifications by membrane proteins, p. 239–281. *In* P. C. Cost and O. H. Griffith (ed.), *Lipid-Protein Interactions,* vol. 2. John Wiley and Sons, Inc., New York, NY.

104. **Sinensky, M.** 1974. Homeoviscous adaptation—a homeostatic process that regulates viscosity of membrane lipids in *Escherichia coli. Proc. Natl. Acad. Sci. USA* **71:**522–525.

105. **Somero, G. N.** 1992. Adaptations to high hydrostatic pressure. *Annu. Rev. Physiol.* **54:**557–577.

106. **Stetter, K. O.** 1999. Extremophiles and their adaptation to hot environments. *FEBS Lett.* **452:**22–25.

107. **Tabor, P. S., J. W. Deming, K. Ohwada, and R. R. Colwell.** 1982. Activity and growth of microbial populations in pressurized deep-sea sediments and animal gut samples. *Appl. Environ. Microbiol.* **44:**413–422.

108. **Tabor, P. S., K. Ohwada, and R. R. Colwell.** 1981. Filterable marine bacteria found in the deep sea: distribution, taxonomy, and response to starvation. *Microb. Ecol.* **7:**67–83.

109. **Takami, H., A. Inoue, F. Fuji, and K. Horikoshi.** 1997. Microbial flora in the deepest sea mud of the Mariana Trench. *FEMS Microbiol. Lett.* **152:**279–285.

110. **Tamegai, H., C. Kato, and K. Hirokoshi.** 1998. Pressure-regulated respiratory system in barotolerant bacterium, *Shewanella* sp. strain DSSI2. *J. Biochem. Mol. Biol. Biophys.* **1:**213–220.

111. **Tanaka, M. A. Ueno, K. Kawasaki, I. Yumoto, S. Ohgiya, T. Hoshino, K. Ishizaki, H. Okuyama, and N. Morita.** 1999. Isolation of clustered genes that are notably homologous to the eicosapentaenoic acid biosynthesis gene cluster from the docosahexaenoic acid-producing bacterium *Vibrio marinus* strain MP-1. *Biotechnol. Lett.* **21:**939–945.

112. **Teece, M. A., M. L. Fogel, M. E. Dollhopf, and K. H. Nealson.** 1999. Isotopic fractionation associated with biosynthesis of fatty acids by a marine bacterium under oxic and anoxic conditions. *Org. Geochem.* **30:**1571–1579.

113. **Tornabene, T. G.** 1985. Lipid analysis and the relationship to chemotaxonomy. *Methods Microbiol.* **18:**209–234.

114. **Veld, G. I., A. J. Driessen, and W. N. Konings.** 1993. Bacterial solute transport proteins in their lipid environment. *FEMS Microbiol. Rev.* **12:**293–314.

115. **Wakeham, S. G., and E. A. Canuel.** 1988. Organic geochemistry of particulate matter in the eastern tropical North Pacific Ocean: implications for particle dynamics. *J. Mar. Res.* **46:**183–213.

116. **Wakeham, S. G., J. H. Hedges, C. Lee, M. L. Peterson, and P. I. Hernes.** 1997. Compositions and transport of lipid biomarkers through the water column and surficial sediments of the equatorial Pacific Ocean. *Deep-Sea Res. Part II* **44:**2131–2162.

117. **Wakeham, S. G., and C. Lee.** 1993. Production, transport, and alteration of particulate organic matter in the marine water column, p. 145–169. *In* M. H. Engel and S. A. Macko (ed.), *Organic Geochemistry, Principles and Applications.* Plenum Press, New York, NY.

118. **Wallis, J. G., J. L. Watts, and J. Browse.** 2002. Polyunsaturated fatty acid synthesis: what will they think of next? *Trends Biol. Sci.* **27:**467–473.

119. **Weber, G., and H. G. Drickamer.** 1983. The effect of high pressure upon proteins and other biomolecules. *Q. Rev. Biophys.* **16:**89–112.

120. **Whitman, W. B., D. C. Coleman, and J. W. Wiebe.** 1998. Prokaryotes: the unseen majority. *Proc. Natl. Acad. Sci. USA* **95:**6578–6583.

121. **Wirsen, C. O., H. W. Jannasch, S. G. Wakeham, and E. A. Canuel.** 1987. Membrane lipids of a psychrophilic and barophilic deep-sea bacterium. *Curr. Microbiol.* **14:**319–322.

122. **Xu, Y., Y. Nogi, C. Kato, Z. Liang, H.-J. Rüger, D. D. Kegel, and N. Glansdorff.** 2003. *Moritella profunda* sp. nov. and *Moritella abyssi* sp. nov., two psychropiezophilic organisms isolated from deep Atlantic sediments. *Int. J. Syst. Evol. Microbiol.* **53:**533–538.

123. **Yanagibayashi, M., Y. Nogi, L. Li, and C. Kato.** 1999. Changes in the microbial community in Japan Trench sediment from a depth of 6292 m during cultivation without decompression. *FEMS Microbiol. Lett.* **170:**271–279.

124. **Yano, Y., A. Nakayama, and K. Yoshida.** 1997. Distribution of polyunsaturated fatty acids in bacteria presented in intestines of deep-sea fish and shallow-sea poikilothermic animals. *Appl. Environ. Microbiol.* **63:**2572–2577.

125. **Yayanos, A. A.** 1986. Evolutional and ecological implications of the properties of deep-sea barophilic bacteria. *Proc. Natl. Acad. Sci. USA* **83:**9542–9546.

126. **Yayanos, A. A.** 1995. Microbiology to 10500 meters in the deep sea. *Annu. Rev. Microbiol.* **49:**777–805.

127. **Yayanos, A. A.** 1998. Empirical and theoretical aspects of life at high pressure in the deep sea, p. 47–92. *In* K. Horikoshi and W. D. Grant (ed.), *Extremophiles, Microbial Life in Extreme Environments.* John Wiley and Sons, Inc., New York, NY.

128. **Yayanos, A. A.** 1999. The influence of nutrition on the physiology of piezophilic bacteria. *In* C. R. Bell, M. Brylinsky, and P. Johnson-Green (ed.), *Microbial Biosystems: New Frontiers.* Proceedings of the 8th International Symposium on Microbial Ecology. Atlantic Canada Society for Microbial Ecology, Halifax, Canada, 1999.

129. **Yayanos, A. A.** 2001. Deep-sea piezophilic bacteria. *Methods Microbiol.* **30:**615–635.

130. **Yayanos, A. A., and E. F. DeLong.** 1987. Deep-sea bacterial fitness to environmental temperatures and pressures, p. 17–32. *In* H. W. Jannasch, R. E. Marquis, and A. M. Zimmerman (ed.), *Current Perspectives in High Pressure Biology.* Academic Press, London, United Kingdom.

131. **Yayanos, A. A., and A. S. Dietz.** 1982. Death of a hadal deep-sea bacterium after decompression. *Science* **220:**497–498.

132. **Yayanos, A. A., A. S. Dietz, and R. Van Boxtel.** 1979. Isolation of a deep-sea barophilic bacterium and some of its growth characteristics. *Science* **205:**808–810.

133. **Yayanos, A. A., A. S. Dietz, and R. Van Boxtel.** 1981. Obligately barophilic bacterium from the Mariana Trench. *Proc. Natl. Acad. Sci. USA* **78:**5212–5215.

134. **Yayanos, A. A., A. S. Dietz, and R. Van Boxtel.** 1982. Dependence of reproduction rate on pressure as a hallmark of deep-sea bacteria. *Appl. Environ. Microbiol.* **44:**1356–1361.

135. **Yayanos, A. A., R. Van Boxtel, and A. S. Dietz.** 1984. High-pressure-temperature gradient instrument: use for determining the temperature and pressure limits of bacterial growth. *Appl. Environ. Microbiol.* **48:** 771–776.

136. **Yazawa, K.** 1996. Production of eicosapentaenoic acid from marine bacteria. *Lipids* **31:**S297–S300.

High-Pressure Microbiology
Edited by C. Michiels, D. H. Bartlett, and A. Aertsen
© 2008 ASM Press, Washington, DC

Chapter 15

Deep-Sea Fungi

Chandralata Raghukumar and Samir Damare

The deep-sea environment, despite being rich in mineral nutrients and near-saturation levels of oxygen, because of its darkness is not conducive to photosynthesis, the process that sustains life on earth. If a high abundance and diversity of life yet persist in this ecosystem, it is because of the transport of organic matter formed by photosynthesis several hundreds to thousands of meters above, which sinks to it. The exception is the hydrothermal vents. Organic carbon of photosynthetic primary producers in the euphotic zone, as a waste product of such organisms, rains down to the deep sea in what is described as the "biological pump" that transports such organic matter to the depths (30). A substantial portion of the atmospheric CO_2 is sequestrated into the deep sea by this process. The deep sea is the largest sink for organic carbon on the earth (84). The universal process of recycling such organic matter depends on microorganisms, which use part of the carbon and minerals for their biomass and release the rest as carbon dioxide and minerals through chemoheterotrophic processes. Bacteria and fungi are the major groups that are capable of accomplishing this.

Several studies have shown intense bacterial activities in deep-sea sediments (93). Total bacterial numbers and their biomass from several sites in the world oceans have been estimated (20). In contrast, almost no information exists in relation to fungi in the deep sea. This is in contrast to terrestrial ecosystems, where fungi are recognized to be major players in decomposition and remineralization of organic matter (54, 81). Studies from forest and prairie soils have shown that fungi have lower rates of remineralization but higher C assimilation efficiency than bacteria and thus may be more significant in terms of carbon sequestration (5, 8). In light of this, the diversity, abundance, and role of fungi in deep-sea sediments may form an important link in the global C biogeochemistry. This review focuses on issues related to collection and isolation of deep-sea fungi, direct detection in deep-sea sediments, diversity and biomass, growth and physiology, adaptations, and their biotechnological applications.

HISTORICAL BACKGROUND

The presence of fungi in oceanic waters and the deep sea has been sporadically reported in the past. Their presence in shells collected from deep-sea waters at a depth of 4,610 m (28, 29) was the first report on deep-sea fungi. This was followed by isolation of fungi

Chandralata Raghukumar • National Institute of Oceanography, Dona Paula, Goa 403 004, India. *Samir Damare* • Institut für Biotechnologie, Biotechnikum, Walter Rathenau Strasse 49A, 17489 Greifswald, Germany.

from water samples collected from the subtropical Atlantic Ocean, from the surface to a depth of 4,500 m using sterile van Dorn bags or Niskin samplers (79). Deep-sea fungi were obtained by directly submerging wooden panels at depths of 1,615 to 5,315 m (40). These fungi were not cultured and only preserved specimens are available (41). Four of these fungi were found growing on wooden panels and one on the chitin of hydrozoa (Table 1). Mycelial fungi growing inside shells of mollusks at a depth of 4,830 m in the Atlantic were documented (69). A study spanning over 10 years showed a distinct succession pattern of fungi as an endolithic community in molluskan shells (68). Several filamentous

Table 1. Fungi reported or isolated from deep-sea sources

Organism(s)	Source(s)	Depth (m) and location	Reference	Remarks
Abyssomyces hydrozoicus	Hydrozoa, chitin	631–641, Atlantic	41	Only preserved
Bathyascus vermisporus		1,615 and 1,720, Pacific		specimens
Oceanitis scuticella	Wood	3,975, Atlantic		
Allescheriella bathygena		1,720		
Periconia abyssa		3,975 and 5,315, Atlantic		
Aureobasidium pullulans	Water	1,000–4,500, subtropical Atlantic	79	CFU
Cladosporium spp.				
Alternaria spp.				
Aspergillus sydowii				
Nigrospora spp.				
Penicillium solitum				
Aspergillus ustus	Calcareous shells	965, Bay of Bengal	73	CFU
Penicillium citrinum				
Cladosporium sp.				
Scopulariopsis sp.				
Nonsporulating fungus				
Aspergillus fumigatus				
Cladosporium herbarum				
Rhodotorula mucilaginosa	Sediments	10,500, Mariana Trench	89	In culture
Penicillium lagena				
Gymnascella marismortui	Water	Depth not mentioned, Dead Sea	14	In culture
Phoma pomorum				
Penicillium westlingii				
Aspergillus sydowii	Sediments	5,100, Central Indian Ocean	74	In culture
Nonsporulating unidentified sp.				
Aspergillus sp.	Sediments	~5,000, Central Indian Basin	18	CFU
Aspergillus terreus				
A. restrictus				
A. sydowii				
Penicillium sp.				
Cladosporium sp.				
Curvularia sp.				
Fusarium sp.				
Nonsporulating fungi				
Unidentified fungi				
Aureobasidium sp.				
Unidentified yeasts				

fungi were isolated from surface-sterilized calcareous fragments collected from a depth of 300 to 860 m in the Bay of Bengal (73). These fungi were isolated using 1/5-diluted malt extract medium prepared with seawater. It was observed that "tests for the tolerance of high pressures and low temperatures can indicate whether the isolated fungal species are indigenous deep-sea forms or aliens from other habitats" (41). In accordance with this, it was demonstrated that conidia of *Aspergillus restrictus* isolated from the calcareous sediments germinated at a pressure of 30 MPa in Czapek-Dox medium and on shells suspended in seawater (73). Detection of fungal filaments in formalin-preserved calcareous fragments obtained from a depth of 965 m in the Arabian Sea (72) further confirmed that they were actively growing in these shells. These calcareous fragments were treated with EDTA to dissolve the calcium carbonate and subsequently stained with the fluorescent brightener calcofluor to visualize fungal filaments under an epifluorescence microscope (Fig. 1 and 2). Calcofluor was originally used to detect fungal infections of phytoplankton in natural waters (58). The marine yeasts *Debaryomyces hansenii, Rhodotorula rubra,* and *Rhodosporidium sphaerocarpum* were cultured at temperatures from 7 to 34°C and pressures from 0.1 to 80 MPa (50). Germination of fungal spores under simulated deep-sea conditions of low temperature and elevated hydrostatic pressures was also demonstrated (98). Fungi have been retrieved from sediment samples of the Mariana

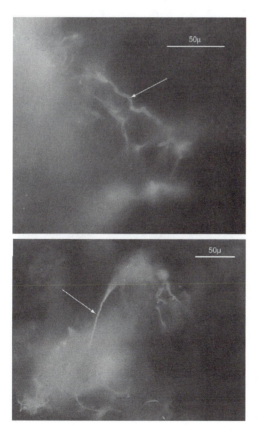

Figures 1 and 2. Calcofluor-stained fungal hyphae (arrows) visible under an epifluorescence microscope after dissolution of a carbonate shell with EDTA.

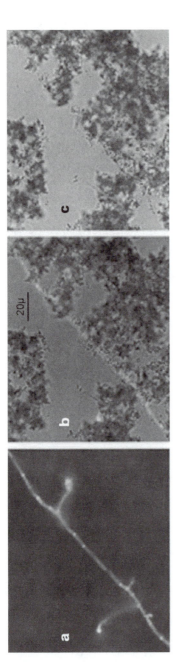

Figure 3. (a) Direct visualization of a fungal hypha after treating a deep-sea sediment with the optical brightener calcofluor; (b) epifluorescence and bright-field microscopy showing a fungal hypha and sediment particles; (c) bright-field microscopy highlights only the sediment and masks the fungal hyphae (73).

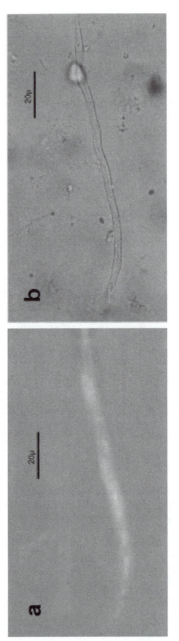

Figure 4. (a) Detection of hypha of *A. terreus* (isolate A 4634) from a deep-sea sediment treated with fluorescein isothiocyanate-tagged polyclonal antibodies raised against the fungus; (b) the same hypha as in panel a, viewed under bright-field microscopy.

Trench in the Pacific Ocean from a depth of 10,500 m (90). Sequences belonging to fungi have been reported from the aphotic zone at a depth of 250 to 3,000 m in the Antarctic polar front (47). These reports were based on direct amplification of small-subunit rRNA genes from water.

Recently, isolation and direct detection of fungi were reported from deep-sea sediments in the Chagos Trench in the Indian Ocean at a depth of 5,900 m (74). The age of the sediments from which fungi were isolated and detected was estimated to range from >0.18 to 0.43 million years, the oldest recorded age for recovery of culturable fungi. Fungal hyphae could not be detected by bright-field microscopy but were visible by epifluorescence microscopy. The relation between sediment particles and fungal hyphae was clear when these were viewed under the combination of bright-field and epifluorescence light (Fig. 3). These studies indicated that presence of fungi in deep-sea sediments might have gone undetected by conventional microscopy.

Several filamentous fungi and yeasts were isolated from sediments collected at depths of ~5,000 m in the Central Indian Basin by baiting dilute nutrient medium with sediments and incubating them under elevated pressures and low temperatures (18). Such a selective isolation technique yielded fungi that showed good growth under low temperatures and elevated hydrostatic pressures. In order to detect one such fungus directly in the deep-sea sediments, a polyclonal immunofluorescence probe was developed by raising antibodies against it in male New Zealand White rabbits (18). The fungus could be detected in the subsection from which it was isolated with a fluorescein isothiocyanate-tagged polyclonal antibody (Fig. 4). This chronology of the events in the research methodologies and strategies clearly indicates the increasing attention being paid to the presence of fungi in the deep sea.

METHODOLOGIES FOR COLLECTION AND ISOLATION OF DEEP-SEA FUNGI

Mycelial fungi pervade solid substrates by networks of apically growing filaments. They draw nutrients from organic matter in the sediment and produce hyphae, which spread out and grow in between the sediment particles and often adhere to them strongly (76). Unlike bacteria, mycelial fungi are not adapted to grow as free-living entities in water. Thus, it is appropriate to look for fungi in sediments and other solid substrata. In the water column, they might be found in association with particles, macroaggregates, transparent exopolymeric substances, and marine snow.

Numerous methodologies have been employed for collecting water and sediment samples from the deep sea. Fungi in deep-sea waters may be collected using routine oceanographic samplers such as Niskin or Nansen bottles or a Rosette or ZoBell sampler (Fig. 5 and 6), without maintaining in situ hydrostatic pressure. The first deep-sea-pressure-retaining sampler was designed to collect water samples for isolating piezophilic bacteria (34; Fig. 7). The Deep Star Group, working on extremophilic microorganisms at the Japan Agency for Marine-Earth Science and Technology, designed and developed a deep-sea workstation for retrieving, isolating, and culturing deep-sea organisms under simulated deep-sea conditions of temperature and pressure (26). Later, the manned submersible Shinkai 6500 was deployed for collecting sediment samples with sterile corers (38). The unmanned submersible Kaiko was used for collecting sediment samples from the Mariana Trench at a

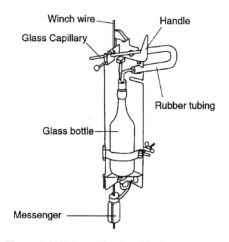

Figure 5. Rosette sampler with several bottles fitted to a conductivity, temperature, and depth measuring device (CTD) frame for collecting oceanic water samples from different depths.

Figure 6. ZoBell sampler for collecting water samples for microbiological studies. Reprinted with kind permission from Lorenz (48).

depth of 11,000 m in the Pacific Ocean with a sterilized sediment sampler without any microbiological contamination derived from overlying waters (37).

For routine microbiological sampling of deep-sea sediments, multiple corers or box corers were used (36; Fig. 8 and 9). We collected subsamples from the center of a box corer with sterilized polyvinyl chloride cylinders 4 to 5 cm in diameter (Fig. 10). Subsections of 0.5 to 2 cm were extruded from these sediment cores down to the desired depth directly into sterile plastic bags to avoid any contamination by airborne fungi (71, 74). The bags were tightly secured with rubber bands or clips, and fungal isolations were carried out immediately (18).

For isolating fungi from the water column, water samples were filtered through sterile cellulose-ester membranes with a 0.45-μm pore size (79). These membranes with retained fungal hyphae were placed on various mycological media prepared with seawater and incorporated with broad-spectrum antibiotics to curtail bacterial growth. Generally, a 50-ml sample is sufficient for isolating fungi from nearshore waters, whereas from offshore stations about 100 ml of water is required.

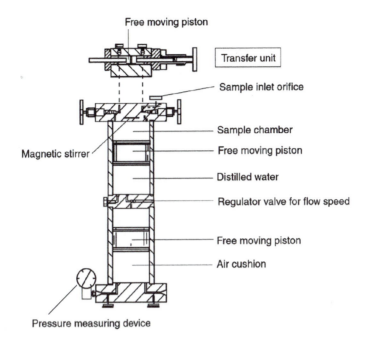

Figure 7. Water sampler for collecting water samples with in situ pressure and a transfer unit. Original figure (34) modified by Lorenz and reproduced with kind permission from Lorenz (48).

Figure 8. Multiple corer for collecting deep-sea sediment cores. Reprinted from the Oktopus Gmbh catalog with kind permission from G. Schriever.

Figure 9. Box corer for collecting a large volume of sediments. Courtesy A. B. Valsangkar, Geological Oceanography Division, National Institute of Oceanography, Dona Paula, Goa, India.

The possible sources of fungi in the deep sea are sediments, calcareous shells, phytodetritus, and benthic fauna associations. Fungal cultures were obtained from spread-plating crushed calcareous shells or by directly plating surface-sterilized fine pieces of calcareous shells or meiofauna on 1/5-diluted malt extract agar medium prepared with seawater and supplemented with streptomycin and penicillin to suppress bacterial growth (18, 73). Dilution plating of decompressed deep-sea sediments is another method for isolating fungi from sediments. The plates are incubated at 5°C. One of the culture techniques that enhanced recovery of fungi in a terrestrial ecosystem was particle plating (13). Here particles that are retained in the 100- to 200-μm range by sieving sediment slurry under sterile conditions are plated on different mycological media and incubated at 5°C aboard the research vessel. The particle filtration procedure brought into culture a wide range of taxa, many of which might have originated from the interior of fine particles (18).

Recently, we have employed a technique where the decompressed sediments were resuspended in dilute malt extract broth or plain seawater and recompressed at the desired hydrostatic pressure (18). The pressurized vessels were incubated at 5°C. All these activities were carried out aboard the research vessel immediately after retrieving the sediments and with minimum exposure of the samples to avoid aerial contamination. Airborne fungi in the vicinity of the collection area and laboratory space were routinely ascertained by exposing medium plates for 10 min or more during each sampling. Our experience has

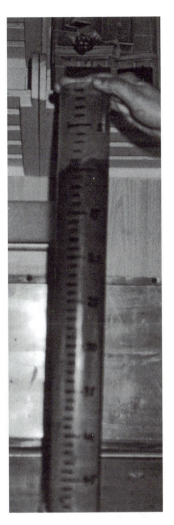

Figure 10. A graduated polyvinyl chloride cylinder for collecting sediment subsamples from a box corer. Courtesy A. B. Valsangkar, Geological Oceanography Division, National Institute of Oceanography, Dona Paula, Goa, India.

shown that on such oceanic cruises far away from the mainland, the aerial population of fungi is almost absent.

Colony counts of fungi that develop on plates should be taken when they are extremely small, before sporulation, so as to avoid error due to autoinoculation. It is further recommended that subculturing of fungi to get pure cultures also be carried out aboard the research vessel during long voyages. The fungi obtained by these various methods may be examined for their barotolerance by germination of conidia, growth of fungal hyphae, and growth yield by incubation under simulated deep-sea conditions.

Fungi isolated by employing different techniques and nutrient media were examined for spore germination and mycelial growth under elevated hydrostatic pressure (18, 72, 74) by the following methods. (i) The plates with test fungi growing on suitable agar media were flooded gently with sterile seawater to collect the spores. The spore suspension was shaken with glass beads on a shaker after adding Tween 80 at a 0.5 to 1.0% concentration to separate the clumped spores. This was diluted appropriately after hemocytometer counts, inoculated in diluted medium or sediment extract in pouches made with sterile gas-permeable polypropylene sheets, and sealed without trapping air bubbles. The pouches were suspended in a deep-sea culture vessel (Tsurumi & Seiki Co., Yokohama, Japan) filled with sterile water and then pressurized to the desired pressure and incubated at the desired temperature (74). Similarly prepared bags were incubated at 0.1 MPa for comparison. After 7 to 10 days of incubation, the deep-sea culture vessels were depressurized gradually (5 MPa and 15 min), and the percentage of germinated conidia was counted microscopically. (ii) For raising the mycelial biomass of the test fungi, cultures were grown in an appropriate liquid medium and vegetative mycelium prior to the onset of sporulation was homogenized with sterile glass beads. A known weight of the finely broken mycelial suspension was inoculated in 20 ml of diluted liquid medium and incubated under pressure as described above. After 20 days, the contents of the bags were filtered over preweighed filter papers and dried to a constant weight, and the difference between the initial and final biomasses was determined as the mycelial dry weight.

Alternatively, the homogenized spore suspension can be spread on a thin layer of seawater agar medium. Disks 5 mm in diameter can be cut from such plates and introduced into culture bags containing sterile seawater and incubated in a pressure vessel. Fine pieces of calcareous shells dipped in spore suspension were put into gas-permeable bags containing sterile seawater and incubated under elevated hydrostatic pressure. The rough crevices on such shells trap fungal spores, and they are easily viewed under a microscope after incubation under pressure (73). These methods are appropriate for quantitative assessment of spore germination under elevated hydrostatic pressure.

Polypropylene hypodermic injection syringes with a 2- to 5-ml capacity, modified by sawing off their nozzles, also served as culture vessels. The cylinders thus obtained are closed at either end by sterile rubber septa and tightly secured with Parafilm. They are suspended in culture vessels filled with sterile water. Eppendorf tubes with their lids snapped off and covered with Parafilm were also used as culture tubes under elevated pressure (72).

Occasionally a problem of providing sufficient oxygen in the culture medium for long-term experiments may arise, and therefore some workers have used a fluorocarbon compound, FC-77, as hydraulic fluid; it also served as an oxygen reservoir (49). FC-77 has a high oxygen solubility (49). Oxygen consumption during cultivation is compensated for by diffusion of oxygen from the hydraulic fluid through the gas-permeable plastic foil into the culture medium. Sufficient oxygen is thus available over a long period. These workers devised high-pressure equipment containing four interconnected, thermostatically controlled high-pressure vessels. An inlet at the bottom and an outlet in the lid of each of the vessels permit exchange of the hydraulic fluid without changes in pressure. The temperature is maintained by a water bath connected to thermostatically controlled jackets of the vessel. Jannasch et al. (33) devised a pressurized chemostat using modified high-pressure chromatographic equipment for continuous culturing of barophilic bacteria at pressures up to 71 MPa.

DIRECT DETECTION OF FUNGI IN SEDIMENTS

Reports of terrestrial species of fungi in the sea have been based on cultural studies. Isolation of a terrestrial species of fungus in a culture from a marine substrate may come from dormant spores and does not guarantee that it was active there. Direct detection of hyphae in marine sediments, as has been reported from fungal growth on wood or leaf litter (56, 59, 82), to a large extent removes these doubts. For this purpose it is best to fix sediments in buffered formalin solution immediately after retrieval (18). Aliquots of these fixed sediments are stained with 0.5% solution of sterile filtered calcofluor (Sigma Chemicals, St. Louis, Mo.). After washing the excess stain, the sediments are examined with an epifluorescence microscope under a UV light filter with an excitation wavelength of 330 to 385 nm (77). The presence of fungal hyphae and germinating conidia in calcareous shells and deep-sea sediments was demonstrated by staining with the fluorescent brightener calcofluor (72), which enhances fluorescence of the cellulose and chitin, the latter being a fingerprint molecule of mycelial fungi (58). A total of 48 out of 255 sediment samples examined by us showed the presence of fungi by this method (18).

Immunofluorescence has been widely used to detect specific fungi in terrestrial and in a few marine substrates (10, 18, 22, 77). We used this for detecting one of the commonly isolated fungi, *Aspergillus terreus* (isolate A 4634), from deep-sea sediments of the Central Indian Basin (18). However, using this method for detection of a specific fungus in sediments was not very easy. Molecular probes for detecting commonly occurring fungi would help to resolve the issue of direct detection of fungi in deep-sea sediments.

DIVERSITY AND BIOMASS

Fungi in the marine environment are either obligate or facultative marine fungi. Obligate marine fungi are defined "as those which grow and sporulate exclusively under marine conditions" (41). In order to accommodate the existence of terrestrial species active in the sea, Kohlmeyer and Kohlmeyer (41) offered a definition for facultative marine fungi "as those originating from freshwater or terrestrial environment that are also capable of growth and sporulation in the sea." Spores and hyphal fragments carried with the terrestrial runoff, sedimentation, and current and those blown with the wind from distant places find their way into the deep-sea sediments. Dry spores of terrestrial fungi can easily be transported by wind during dry periods and can reach fresh habitats. It has been recently reported that spores of *Aspergillus sydowii* are carried from the Saharan deserts across the Atlantic Ocean during dust storms to the Caribbean islands and cause aspergillosis disease in seafans (87). Fungi have been found in cloud water, fog, and precipitation. Aerosolization is thus an important mode of transport of microorganisms to remote environments (12). It is imaginable that fungal spores and mycelial fragments from the ocean surface sink to the bottom via micro- and macroaggregates, acclimatize, and acquire capabilities to grow and multiply. However, some of them may get eliminated as a result of natural selection.

The five obligate marine fungi reported (40) were found to grow on wooden panels suspended in the sea. These were not cultured, and only preserved specimens are available for examination (41). They are *Bathyascus vermisporus* Kohlm, *Oceanitis scuticella* Kohlm, *Allescheriella bathygena* Kohlm, *Periconia abyssa* Kohlm, and *Abyssomyces hydrozoicus* Kohlm (41; Table 1). These were reported based on a single observation. On the other hand, the common terrestrial fungi, also called "geofungi," have been repeatedly isolated using different methods of isolation and media (Tables 1 and 2). Many of these genera (Table 1) were isolated from the deep subtropical Atlantic waters (79) and shallow depths (30 to 70 m) of the Bay of Fundy in Canada (53). Fungi are known to exist under microaerophilic, xerophilic, and oligophilic conditions; therefore, there is no reason that the

Table 2. Comparison of different methods of isolation and media to culture fungi from deep-sea sediments[a]

Organism(s)	Frequency of isolation in indicated medium									
	Cruise AAS 46			Cruise AAS 61						
	Dilution plating	Particle plating	Pressure incubation	Particle plating			Pressure incubation			
	MEA	MEA	MEA	CMA	MEA	SDA	CMA	MEA	SDA	MEB
Aspergillus sp.	1	4		2	5	4				3
Aspergillus terreus	1	1	4	1	1	1	4			4
Aspergillus restrictus				1	1		2		1	15
Aspergillus sydowii			1				1			
Penicillium sp.			1		2		1			2
Cladosporium sp.				1	2	1				
Curvularia sp.				1						4
Fusarium sp.				1						
Nonsporulating fungi		4		2	4	4	1			4
Unidentified sporulating fungi				2	7	3	4	1		8
Unidentified ascomycetes					1					
Aureobasidium sp.		1								
Unidentified yeasts										3

[a]AAS 46 and AAS 61 refer to the cruises on board the Russian research vessel Sidorenko to the Central India Basin in June 2002 and March 2003, respectively. Numbers indicate frequencies of isolation of these fungi by the method indicated from various sediment subsections. MEA, malt extract agar medium; CMA, cornmeal agar medium; SDA, Sabouraud dextrose agar medium; MEB, malt extract broth.

"geofungi" cannot adapt and exist under marine conditions. Deep-sea sediments with an average organic carbon content of 2.5 to 5.0 mg g^{-1} of dry sediment (32) and minerals offer conducive environments for growth of saprophytic fungi. Thus, it is not surprising that most of the fungi reported to have been isolated from deep-sea sediments are terrestrial fungi, i.e., geofungi (Table 1).

A total of 43 fungal species were obtained from oceanic waters of the Atlantic (79), out of which 14 species were obtained from waters collected below a depth of 1,000 m. The most common oceanic species reported by these authors were the black yeast *Aureobasidium pullulans* and the filamentous fungi *Cladosporium, Alternaria, Aspergillus sydowii, Nigrospora,* and *Penicillium solitum.* The fungi isolated from the Bay of Fundy (53) were from shallow depths, 30 to 70 m, and therefore are not included in Table 1. These authors reported 48 species of fungi; a majority of the isolates were geofungi, although a number of them were marine or have marine affinities, such as *Monodictys pelagica, Dendryphiella salina, Gliomastix inflata, Gliocladium roseum,* and *Phialophora fastigiata.* Three species of filamentous fungi were isolated from the hypersaline Dead Sea waters (14). Among them one was a new species, *Gymnascella marismortui;* the other two were the known species *Phoma pomorum* and *Penicillium westlingii* (Table 1). Paradoxically, the terrestrial fungus *Aspergillus terreus* and other species of *Aspergillus* were the most dominant among the fungi isolated from the Central Indian Ocean and were found to be distributed in the sediment cores down to a depth of 35 to 40 cm (Fig. 11). We have isolated such common geofungi from the Bay of Bengal and the Arabian Sea (70). Filamentous fungi and unicellular yeasts have been isolated from a depth of 10,500 m from the Mariana Trench in the Pacific Ocean. These were *Penicillium lagena* (a terrestrial mycelial fungus) and the unicellular yeast *Rhodotorula mucilaginosa* (Table 1; 89). These reports indicate that terrestrial species in the deep sea are quite common.

Quantification of fungal biomass and carbon has been standardized for several systems in the terrestrial ecosystem (52). Fungal biomass is calculated from fungal biovolume. Fungal biovolume is estimated microscopically by measuring the hyphal lengths and diameter of fungi in calcofluor-stained sediments and derived by the equation $\pi r^2 \times$ total hyphal length (39). Fungal biomass carbon is estimated by assuming a carbon content of 0.13 pg of C μm^{-3} (94). Based on the fungal biovolume, the organic carbon contribution in deep-sea sediments of the Central Indian Ocean was in the range of 2.3 to 5.1 μg g^{-1} of dry sediment (16). The bacterial carbon contribution in the same site ranged from 1 to 40 μg g^{-1} of dry sediment (71, 75). The results obtained by us could be conservative estimates, since the fungal biomass estimated by hyphal length may often underestimate the true value, as opposed to estimates based on phospholipid content (9).

Fungal biomass has also been measured in terrestrial ecosystems using biochemical proxies such as hexosamine or ergosterol estimations. Hexosamine is the monomer of the chitin polymer, which is present in cell walls of fungi. The hexosamine method yields total biomass, since chitin is not readily lysed upon fungal death (24). However, the hexosamine method cannot be used in marine sediments, as chitin is a component of several marine shell-bearing arthropods. Small amounts of ergosterol can be found in algae and protozoa, but generally it is safe to use it as a specific biomarker for fungi (see reference 52 and references therein). The ergosterol method was developed for determination of fungal contaminants in cereals (85) and for determination of biomass in soil (88), mycorrhizae (55), sediments (57), and plant material (46, 60). The ergosterol method yields living biomass,

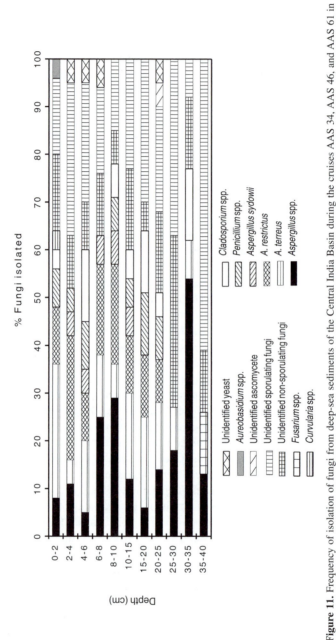

Figure 11. Frequency of isolation of fungi from deep-sea sediments of the Central India Basin during the cruises AAS 34, AAS 46, and AAS 61 in April 2001, June 2002, and March 2003, respectively.

since ergosterol is a membrane component and membranes are easily lysed upon fungal death (61). However, photochemical degradation of ergosterol is significant and can lead to about a 40% decrease in ergosterol content within 24 h, and thus caution should be taken to use ergosterol as a biomarker for living fungi (52). This method for estimating fungal biomass in deep-sea sediments is worth exploring in the future.

An expanded version of the ergosterol method was developed to estimate fungal productivity (63). The method involved incubating samples containing fungi with radiolabeled acetate ([^{14}C] acetate), which is the precursor in ergosterol synthesis. Subsequently, the labeled ergosterol is separated by liquid chromatography and measured. This is used for calculating the rate of production of fungal biomass (62). A modified version of this method for monitoring acetate incorporation into fungal lipids has also been used (6).

As a considerable amount of ergosterol is lost during extraction from soil and purification, alternative methods have been used for estimation of fungal biomass. Phospholipid fatty acid analysis has potential as a sensitive biochemical indicator for estimation of fungal biomass (81). The phospholipid 18:2ω6,9c is dominant in saprotrophic fungi and has been standardized for determination of fungal biomass in an estuarine detrital system (97) and terrestrial soil (80).

GROWTH AND PHYSIOLOGY

Truly piezophilic mycelial fungi have not been reported to date. This, however, does not rule out their existence as nonculturable forms. Culturable fungi have been isolated from a depth of ~5,500 m in the Chagos Trench in the Central Indian Ocean (74) to a depth of 10,897 m in the Mariana Trench in the Pacific Ocean (91). They have been isolated by suspending decompressed sediments in dilute media or ambient seawater and incubating them under elevated hydrostatic pressure and low temperature (18). Presuming that fungi in the deep sea have arrived from land in the form of mycelial fragments or spores attached to organic detritus with land runoffs and/or wind blown, their active growth under simulated deep-sea conditions of elevated hydrostatic pressure and low temperatures can prove their active participation in biogeochemical processes in the deep sea. Twenty-five fungi randomly chosen out of the 181 fungi isolated by us from the Central Indian Basin showed growth under simulated deep-sea conditions (Table 3) when the growth was initiated with mycelial fragments (18). Surprisingly, shallow-water and terrestrial isolates also showed growth in simulated deep-sea conditions (Table 3). These results are substantiated by the fact that the deep-sea bacterial and archaeal piezophiles obtained in cultures are closely related to shallow-water microbes which are not piezophiles, indicating that high-pressure selection has not required the evolution of dramatically different lineages of life (11). Two shallow-water marine fungi, *Dendryphiella salina* and *Asteromyces cruciatus,* showed no growth above 10 MPa (48). The ability of some species to grow over a wide range of pressures will enhance their role in the remineralization process in the deep sea and influence the geochemical cycle in the deep sea.

Diminished activities of surface microbes attached to sinking particles with depth were observed, and it was concluded that only the resident benthic microbes play a major role in deep-sea processes (92). Our results indicate that at least some terrestrial fungi could have acquired a "resident status" in deep-sea sediments and could be involved in dissolution of calcium carbonate and remineralization of organic matter (72).

Table 3. Comparison of growth yields of terrestrial and deep-sea fungi under atmospheric pressure and simulated deep-sea conditions

Culture	Biomass produced by the fungi (mg)[a]	
	0.1 MPa and 30°C	20 MPa and 5°C
Deep-sea isolates		
A. terreus (A 4634)	11.2	9.6
Aspergillus sp. (A 61P4)	4.6	2.0
Unidentified (A 3415)	10.5	5.0
Aspergillus sp. (A 6128)	10.0	1.6
Cladosporium sp. (A 6136)	3.7	1.5
Nonsporulating (A 3428)	1.9	0.6
Unidentified orange yeast (A 61P63)	1.9	1.3
Unidentified yeast (A 344)	6.6	1.7
Terrestrial and shallow-water isolates		
Aspergillus terreus MTCC 279	12.7	6.6
Aspergillus terreus MTCC 479	10.6	10.4
Aspergillus sydowii MTCC 635	12.6	6.3
Shallow-water nonsporulating form 1	16.1	9.3
Shallow-water nonsporulating form 2	17.1	5.7
Shallow-water nonsporulating form 3	13.3	3.4

[a]The cultures were grown in 20 ml of malt extract broth medium for 20 days.

Spores of the deep-sea fungal isolate A 4634 obtained from the Central Indian Basin did not germinate in diluted malt extract broth at 20 MPa and 5°C, but 66% of the spores germinated at 20 MPa and 30°C (18). The spores of terrestrial fungal isolates did not germinate under any of these conditions. Spores of terrestrial as well as deep-sea fungal isolates germinated when incubated in deep-sea sediment extracts at 20 MPa and 30°C, whereas at 5°C they failed to germinate (unpublished data). Thus, it appears that for the fungal spores, temperature, and not hydrostatic pressure, is the limiting factor. Similarly, a higher percentage of spores of *Aspergillus ustus* isolated from calcareous shells at 860 m in the Arabian Sea and a *Graphium* sp. isolated at a depth of 965 m in the Bay of Bengal germinated only at 20 MPa and 30°C and not at 20 MPa and 10°C (72). Delayed germination of fungal spores under simulated deep-sea conditions cannot be ruled out, as was demonstrated for some of the fungi (98). Kriss and Zaichkin (44) demonstrated inhibition of spore germination in several terrestrial fungi at 30 MPa.

Several workers have reported extensive fungal borings of calcareous structures from deep-sea habitats (66, 67, 73). Fungal endoliths that penetrate and dwell inside hard mineral substrates, carbonates, and phosphates participate in many geologically significant processes, including bioerosion of limestone and other calcareous substrates. During this process they produce fine-grain sediment and bring about modification of sediment grains by micritization (25). Active dissolution of calcium carbonate by one such endolithic fungus, *Aspergillus restrictus,* was shown by measuring the amount of calcium dissolved during growth of this isolate on calcareous pieces under a pressure of 10 MPa (73).

Several species of *Aspergillus* showed abnormal morphology immediately after isolation. These showed extremely long conidiophores with vesicles that were covered with long hyphae, instead of phialides or metulae or conidia, as are typical of the genus *Aspergillus* (Fig. 12). Some of the fungi when grown under elevated pressure showed abnormal mor-

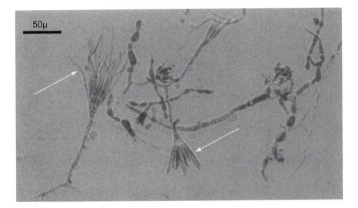

Figure 12. *Aspergillus* species isolated from deep-sea sediments with abnormal morphology, showing hyphae in place of metulae (arrows).

phology in the form of hyphal swellings (Fig. 13). Occasionally, germinating spores showed unusually enlarged germ tubes (Fig. 14). Initially, this phenomenon was considered a response to decompression of sediment, but in natural formalin-fixed samples of calcium carbonate shells, the endolithic fungi displayed abnormal beaded hyphal structures (66, 72). Strangely, these abnormalities were absent when the fungi were grown in sediment extract prepared with deep-sea sediment (16). Nutritional factors appear to counter the adverse impact of elevated hydrostatic pressure, at least in some of the deep-sea fungi.

Several cold-adapted microorganisms are reported to produce cold-active enzymes (51). Out of 221 cultures of fungi obtained from the deep sea by us, 33% showed production of

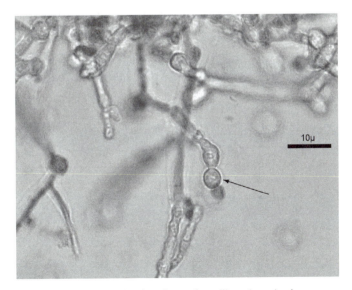

Figure 13. Deep-sea fungus showing abnormal swellings (arrow) when grown under 20 MPa.

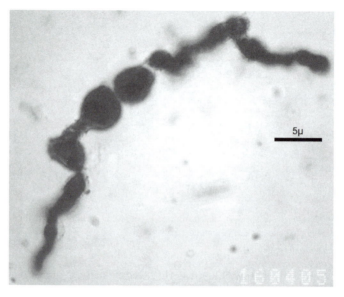

Figure 14. Germinating conidia showing unusual enlargements.

protease active at 5°C and pH 9.0 (17). In contrast, out of 22 fungi isolated from shallow water, only 14% showed low-temperature-active protease production. The deep-sea fungi when grown under elevated pressure synthesized extracellular protease, albeit in very low quantities in comparison with that produced under 0.1 MPa (Fig. 15). As the extracellular enzyme secretion in fungi is governed by mycelial biomass, reduced growth at 5°C and elevated hydrostatic pressures appeared to be responsible for low-level enzyme synthesis. The enzyme activity per se under elevated hydrostatic pressure was reduced by 50% in two of the test fungi (Fig. 16). This might be due to a change in affinity of the enzyme for the substrate which is brought about by changes in the tertiary or quaternary structure of enzymes under pressure (45). This was demonstrated by an increased K_m for the enzyme serine protease, derived from a deep-sea fungus, for its substrate, azocasein, when estimated under increasing hydrostatic pressure (17, Table 4). Reduced growth and metabolic rates under elevated pressure are reported for bacteria (45), and therefore, obtaining similar results with filamentous fungi was not surprising.

Figure 15. Protease synthesis by two deep-sea fungi, *A. ustus* and a *Graphium* sp., when grown at different pressure and temperature combinations.

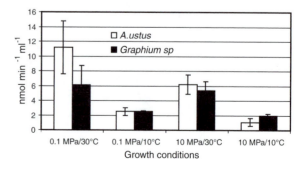

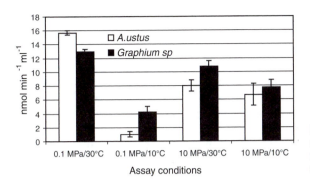

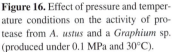

Figure 16. Effect of pressure and temperature conditions on the activity of protease from *A. ustus* and a *Graphium* sp. (produced under 0.1 MPa and 30°C).

ADAPTATIONS

It has been postulated that life originated in the deep sea and, more precisely, under conditions similar to those of hydrothermal vents (15). Therefore, high-pressure-adapted mechanisms of gene expression, protein synthesis, or physiology could represent features associated with early life forms. It is expected that high-pressure genes may be present in terrestrial forms also (4). Bacterial and archaeal piezophiles in culture are closely related to shallow-water microbes which are not piezophilic (11). It can be argued further that as terrestrial life forms evolved, they deadapted from the deep-sea conditions. Therefore, the geofungi that we repeatedly isolated from deep-sea sediments (18) appear to have inherently retained the capacity to endure the deep-sea environment. The 1-atm-adapted organisms that come down with sinking aggregates and marine snow from the surface waters may also form part of the deep-sea microbial population. In such cases special adaptations to exist in the extremophilic environment of the deep sea may be necessary. Pressure adaptations in mesophilic bacteria have received enormous attention in efforts to understand pressure biology. It has been shown that membrane fluidity plays an important role in survival and growth under high-pressure conditions. Appropriate membrane fluidity in prokaryotes is maintained by increasing the ratio of unsaturated fatty acids to saturated ones in membrane phospholipids (19). The transmembrane protein ToxR in *Photobacterium profundum* strain SS9 is reported to play a key role in pressure-induced expression of genes (95). A substantial amount of work has been carried out to understand the effect of pressure on the eukaryotic yeast *Saccharomyces cerevisiae* (3). Abe and colleagues have postulated that the effect of hydrostatic pressure on yeast cells is due to volume change in

Table 4. K_m of the purified protease of the deep-sea fungus NIOCC20 measured under different conditions[a]

Conditions for assay of enzyme activity	K_m (mg ml^{-1} of azocasein)
0.1 MPa and 45°C	2.0
0.1 MPa and 5°C	2.9
5 MPa and 45°C	2.2
10 MPa and 45°C	2.4
20 MPa and 45°C	3.3
30 MPa and 45°C	5.0

[a]The optimum activity of protease was at 45°C.

the vacuole which is a result of increased acidification. Pressure-induced vacuole acidification is caused by production of carbon dioxide. The ionization of H_2CO_3 is facilitated by elevated pressure. As a result, a large number of protons are produced in the cytoplasm, which tends to become acidic under elevated pressures.

To our knowledge no true piezophilic fungi have been reported so far, and therefore only adaptations of piezotolerant fungi are discussed here. One such adaptation may be morphological changes. A distinct morphological change that was observed in fungi grown under elevated hydrostatic pressure was swollen hyphae and conidia (Fig. 13 and 14). A similar phenomenon was observed in an *Aspergillus* sp. grown under 20 MPa (48). This might help the fungi to increase surface area for quicker nutrient absorption. The type of nutrients used for culturing fungi also influences their morphology. For example, when the deep-sea fungi were grown in dilute malt extract broth, the fungal hyphae showed several interhyphal swellings, whereas they were perfectly normal when grown in sediment extract. Two of the deep-sea fungi showed microcyclic conidiation when grown at 10 MPa (Fig. 17). In this, immediate conidiation occurs following conidial germination without an intervening phase of prolonged vegetative mycelium. Microcyclic conidiation is reported to occur under nutrient-limiting conditions and might help the fungi to complete their life cycles in a shorter time (83). This phenomenon appears to depend upon an arrest of apical growth followed by a lateral differentiation of conidium-producing cells (78). Species of *Aspergillus* grown under 20 MPa showed production of microcolonies in high-pressure incubation bags (48). The significance of nutrition was further evident when spores of several terrestrial fungi showed germination at 20 MPa and 30°C in deep-sea sediment extract medium but not in glucose solution or dilute malt extract broth (unpublished data). In *Saccharomyces cerevisiae* under high-pressure conditions, the uptake of tryptophan via the high-affinity tryptophan permease Tat2 is impaired and the expression of Tat2 is down-regulated, leading to growth inhibition (1). Addition of excess tryptophan or overexpression of Tat2 protein enables *S. cerevisiae* cells to grow at 25 MPa and also at the low temperature of 10 or

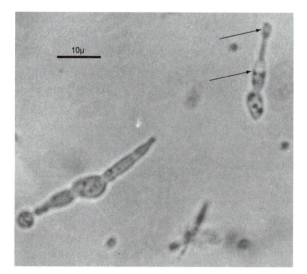

Figure 17. Microconidiation in a *Graphium* sp., where a conidium gives rise to conidia (arrows) directly without an intervening phase of hyphal growth (72).

15°C. These results suggest that the uptake of tryptophan is one of the most pressure-sensitive processes in living yeast cells.

Interestingly, spores of deep-sea-derived fungi or terrestrial fungi did not germinate at 0.1 or 20 MPa at 5°C, although all of them germinated under 20 MPa at 30°C in sediment extract medium. Thus, it appears that more than elevated pressure, low temperature may impair the spore germination in mycelial fungi. Alternatively, their germination under deep-sea conditions may be substantially delayed (98). In contrast, when growth studies were carried out with mycelial fragments (and not spores) of deep sea-derived fungi or terrestrial fungi as the initial inoculum, substantial biomass was obtained at 20 MPa and 5°C (Table 3). Thus, the effects of pressure and low temperatures on spores and mycelia of the same fungi are totally different.

Using microarray technology, genes of heat shock proteins and protein degradation systems were observed in *S. cerevisiae* cells subjected to hydrostatic pressures of 1 to 180 MPa (86).

Numerous isolates of deep-sea-derived fungi did not grow at 20 MPa and 5°C, even when mycelial inocula were used, indicating that all fungi that reach the deep sea may remain viable but not actively grow under these conditions. A few of these fungi could be gradually adapted to grow at 30 MPa. The cultures were grown at 5 MPa and 30°C for 20 days. After depressurization, the cultures were checked for growth and viability. Cultures showing growth at 5 MPa were incubated at 10 MPa. This process was continued at 20, 30, and 40 MPa. A decreasing number of fungi showed viability and growth when they were gradually subjected to higher pressure. Only two strains of *Aspergillus terreus* and one unidentified yeast grew at 30 MPa. The yeast showed growth even at 40 MPa (18). Thus, pressure shock seems to have induced pressure tolerance in these three fungi. However, pressure pretreatment of tpsII mutant cells of *S. cerevisiae* (strain PF 9060) to a hydrostatic pressure of 50 MPa did not induce barotolerance (21). On the other hand, heat shock pretreatment increased barotolerance in *S. cerevisiae* (strain IFO-0224) cells, indicating that hydrostatic pressure and high temperature may have the same physiological effects on this yeast (31). Heat shock treatment also prevented yeast cells from freezing injury (42), suppressing the disruption of membranes and increasing the amount of unfreezable water (43). In addition, deuterium oxide and dimethyl sulfoxide treatment also protected *S. cerevisiae* cells against hydrostatic pressure and thermal damage. Thus, the mechanisms of cell injury by elevated temperature, hydrostatic pressure, and freezing appear to elicit similar physiological effects and responses in yeast cells. This was, however, not the case with fungal spores. Preliminary studies showed that pretreatment with dimethyl sulfoxide and elevated temperature did not induce barotolerance in spores of some of the test fungi. As fungal spores have a tougher wall, they do not appear to respond to these treatments in the same way as the yeast cells, or they may require higher concentrations and/or longer exposure times to respond.

In *S. cerevisiae* cells (strain IFO-0224), increasing trehalose content provided barotolerance to cells and not the heat shock proteins (23). *Trichosporon pullulans,* a psychrotrophic yeast, produced 26 cold shock proteins within 12 h when subjected to a temperature decrease from 21 to 5°C (35). Spores of *Aspergillus restrictus* isolated from a depth of 960 m in the Bay of Bengal showed increased percentages of germination with increasing concentrations of sucrose in the medium when grown under 10 MPa (73; Fig. 18). Increasing the osmophilic condition appeared to increase barotolerance. Several saccharides are

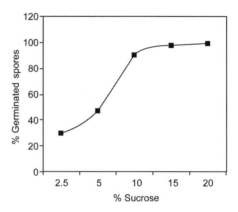

Figure 18. Effect of sucrose concentration on percentage of conidia germinated under 10 MPa.

known to provide protection against hydrostatic pressure damage in cells. In *Saccharomyces cerevisiae* strain IFO-10149, the survival rate after pressure stress of 150 MPa for 1 h without saccharide was very low (0.083%). Tolerance of the yeast increased with increasing concentration of saccharide. Added saccharides such as lactose, maltose, sucrose, and trehalose at concentrations of 0.5 and 1.0 M were effective in protection against high-pressure stress (7).

Substantial work has been reported with high-pressure-adapted (piezophilic) and 1-atm-pressure-adapted (mesophilic) yeasts (2–4). Whether these yeasts differ morphologically, physiologically, and ultimately at molecular level needs to be answered. Along a similar line, we do not know how similar are the common terrestrial fungi isolated by us from deep-sea sediments to the terrestrial ones. *Aspergillus ustus* (NIOCC20) isolated by us from deep-sea sediments produced cold-active alkaline serine protease, whereas *A. ustus* obtained from a terrestrial habitat did not (17). This might be due to differences in strains but may also indicate the adaptation that deep-sea fungi have undergone for their survival.

BIOTECHNOLOGICAL APPLICATIONS

The deep sea, the largest refrigerator unit of the earth, is a storehouse for novel low-temperature- and high-pressure-adapted microorganisms. Isolation, culturing, and characterization of culturable forms hold great potential for application in biotechnology. Understanding their physiology and adaptation to live in the primordial conditions of high temperature and pressure will provide clues to the origin of life. The fact that high-pressure-regulated genes are present in 0.1-MPa-adapted organisms is an exciting lead to the study of evolution (96). Studies on pressure resistance and sensitivity in fungi and their spores have direct application in food preservation technology. Pressure-adapted organisms can be used for high-pressure bioreactors for industrial production of enzymes and secondary metabolites. Although such processes will definitely be costly, they might be useful for production of expensive fine biochemicals and biotransformation reactions.

Fungi are generally used for industrial production of enzymes, as they produce them extracellularly, and raising large amounts of fungal biomass and detection of contaminants are relatively easy. Deep-sea-derived fungi will offer a good source for low-temperature active proteases, lipases, and amylases. Low-temperature-active proteases are in demand as additives to detergents for cold wash because cold wash not only saves energy but also increases the life and preserves the color of clothes in the long run. Low-temperature-active lipases and amylases are in demand by the food industry for retaining the freshness and fragrance of foods (65). Since cold-active cell wall-digestive enzymes such as cellulase and pectinase are suitable for protoplast preparation, they would be useful in plant biotechnology. The most successful example so far has been the lipase from the filamentous fungus *Thermomyces lanuginosus,* which is used in many standard detergents under the trade name Lipolase (64). We have isolated a serine protease from the deep-sea fungus *Aspergillus ustus* (strain NIOCC20). This protease is active in the pH range of 6.5 to 10.5 and retains 30% of its maximum activity at 15°C (17). It retains 80% of the maximum activity in the presence of 2 M NaCl. Under optimum conditions this fungus produces about 500 to 1,000 azocasein units ml^{-1}. Out of 221 fungal isolates obtained from deep-sea sediments of the Central Indian Basin, 73 produced protease active at pH 9.0 and a temperature of 5°C.

The most useful biomolecules produced by cold-adapted microorganisms are long-chain polyunsaturated fatty acids. Cold-adapted microorganisms might be the most likely candidates for production of polyunsaturated fatty acids, since it is well known that the proportion of unsaturated fatty acids in cold-adapted organisms is higher than in mesophilic organisms (27).

FUTURE DIRECTIONS

The topic of barobiology in the world of fungi offers many opportunities to researchers in fields of fundamental and applied research. Fungi offer great scope as eukaryotic model to study the effects of high pressure and low temperature, as do bacteria for prokaryotes. They can be easily cultured in the laboratory and thus are easy to use as model systems. Are there any nonculturable piezophilic fungi in the deep sea? The distribution of piezophilic genes in terrestrial fungi and adaptation mechanisms in piezotolerant fungi are some of the topics which need attention. The study of the effects of hydrostatic pressure on spores and mycelial phases of fungi is a challenging task. Screening culturable deep-sea fungi for cold-adaptive enzymes for biotechnological applications will be a spin-off from these studies. Study of the cold shock or stress proteins or genes produced in response to hydrostatic pressure shock in fungi using proteomics and microarray technology will help to understand the response in eukaryotic organisms to pressure. New techniques for retrieval of sediment samples with in situ pressure, isolation, and culture of the vast diversity of organisms from the deep sea will open new vistas in deep-sea biology.

Acknowledgments. C. Raghukumar is thankful to the Department of Biotechnology, New Delhi, India, for grant BT/PR1193/AAQ/03/102/2000, under which a substantial amount of work on deep-sea fungi was carried out. S. Damare acknowledges the Council of Scientific and Industrial Research, New Delhi, India, for a grant of Senior Research fellowship. This is NIO contribution no. 4308.

REFERENCES

1. **Abe, F., and K. Horikoshi.** 2000. Tryptophan permease gene *TAT*2 confers high-pressure growth in *Saccharomyces cerevisiae. Mol. Cell Biol.* **20:**8093–8102.

2. **Abe, F., and K. Horikoshi.** 2001. The biotechnological potential of pieozophiles. *Trends Biotechnol.* **19:**102–108.

3. **Abe, F., C. Kato, and K. Horikoshi.** 1999. Pressure-regulated metabolism in microorganisms. *Trends Microbiol.* **7:**447–453.

4. **Abe, F., C. Kato, and K. Horikoshi.** 2004. Extremophiles: pressure, p. 154–159. *In* T. A. Bull (ed.), *Microbial Diversity and Bioprospecting*. ASM Press, Washington, DC.

5. **Adu, J. K., and J. M. Oades.** 1978. Utilization of organic materials in soil aggregates by bacteria and fungi. *Soil Biol. Biochem.* **10:**117–122.

6. **Arao, T.** 1999. In situ detection of changes in soil bacterial and fungal activities by measuring ^{13}C incorporation into soil phospholipid fatty acids from ^{13}C acetate. *Soil Biol. Biochem.* **31:**1015–1020.

7. **Arao, T., Y. Suzuki, and K. Tamura.** 2002. Effects of saccharide in medium on stress tolerance of yeast. *Prog. Biotechnol.* **19:**331–342.

8. **Bailey, V. L., J. L. Smith, and H. Bolton, Jr.** 2002. Fungal-to-bacterial ratios in soils investigated for enhanced C sequestration. *Soil Biol. Biochem.* **34:**997–1007.

9. **Balser, T. C., K. K. Treseder, and M. Ekenler.** 2005. Using lipid analysis and hyphal length to quantify AM and saprophytic fungal abundance along a soil chronosequence. *Soil Biol. Biochem.* **37:**601–604.

10. **Banks, J. N., S. J. Cox, and B. J. Northway.** 1993. Polyclonal and monoclonal antibodies to field and storage fungi. *Int. Biodeterior. Biodegrad.* **32:**137–144.

11. **Bartlett, D. H.** 2002. Pressure effects on *in vivo* microbial processes. *Biochim. Biophys. Acta* **1595:**367–381.

12. **Bauer, H., A. Kasper-Giebl, M. Löflund, H. Giebl, R. Hitzenberger, F. Zibuschka, and H. Puxbaum.** 2002. The contribution of bacteria and fungal spores to the organic carbon content of cloud water, precipitation and aerosols. *Atmos. Res.* **64:**109–119.

13. **Bills, G. F., and J. D. Polishook.** 1994. Abundance and diversity of microfungi in leaf litter of a lowland rain forest in Costa Rica. *Mycologia* **86:**187–198.

14. **Buchalo, A. S., E. Nevo, S. P. Wasser, A. Oren, and H. P. Molitoris.** 1997. First records of fungal life in the extremely hypersaline Dead Sea, p. 178. *Abstr. 4th Int. Marine Biotechnol. Conf., 22–29 Sep. 1997, Sorrento, Paetsum, Otranto, Pugochiuso, Italy.*

15. **Corliss, J. B., J. A. Baross, and S. E. Hoffman.** 1981. An hypothesis concerning the relationship between submarine hot springs and the origin of life on Earth. *Oceanol. Acta* **4**(Suppl.):59–69.

16. **Damare, S., and C. Raghukumar.** 11 November 2007. Fungi and macroaggregation in deep-sea sediments. *Microb. Ecol.* doi:10.1007/S00248-007-9334-4.

17. **Damare, S., C. Raghukumar, U. D. Muraleedharan, and S. Raghukumar.** 2006. Deep-sea fungi as a source of alkaline and cold-tolerant proteases. *Enzyme Microb. Technol.* **39:**172–181.

18. **Damare, S., C. Raghukumar, and S. Raghukumar.** 2006. Fungi in deep-sea sediments of the Central Indian Basin. *Deep-Sea Res. Part I* **53:**14–27.

19. **Delong, E. F., and A. A. Yayanos.** 1985. Adaptation of the membrane lipids of a deep–sea bacterium to changes in hydrostatic pressure. *Science* **228:**1101–1103.

20. **Deming, J. W., and J. A. Baross.** 1993. The early diagenesis of organic matter: bacterial activity, p. 119–144. *In* M. H. Engel and S. Macko (ed.), *Organic Geochemistry: Principles and Applications*. Plenum Press, New York, NY.

21. **Fernandes, P. M. B., A. D. Panek, and E. Kurtenbach.** 1997. Effect of hydrostatic pressure on a mutant of *Saccharomyces cerevisiae* deleted in the trehalose-6-phosphate synthase gene. *FEMS Microbiol. Lett.* **152:**17–21.

22. **Friese, C. F., and M. F. Allen.** 1991. Tracking fates of exotic and local VA mycorrhizal fungi: methods and patterns. *Agric. Ecosyst. Environ.* **34:**87–96.

23. **Fujii, S., H. Iwahashi, K. Obuchi, T. Fujii, and Y. Komatsu.** 1996. Characterization of a barotolerant mutant of the yeast *Saccharomyces cerevisiae:* importance of trehalose content and membrane fluidity. *FEMS Microbiol. Lett.* **141:**97–101.

24. **Gessner, M. O., and S. Y. Newell.** 2001. Biomass, growth rate, and production of filamentous fungi in plant litter, p. 390–408. *In* C. J. Hurst, M. McInerney, L. Stetzenbach, G. Knudsen, and M. Walter (ed.), *Manual of Environmental Microbiology*, 2nd ed. ASM Press, Washington, DC.

25. **Golubic, S., G. Radtke, and T. L. Campion-Alsumard.** 2005. Endolithic fungi in marine ecosystems. *Trends Microbiol.* **13:**229–235.

26. **Hamamato, T., and K. Horikoshi.** 1993. Deep-sea microbiology research within the Deepstar program. *J. Mar. Biotechnol.* **1:**119–124.

27. **Hamamato, T., M. Kaneda, T. Kudo, and K. Horikoshi.** 1995. Characterization of a protease from a psychrophilic *Vibrio* sp. strain 5709. *J. Mar. Biotechnol.* **2:**219–222.

28. **Höhnk, W.** 1961. A further contribution to the oceanic mycology. *Rapp. P.-V. Reun. Cons. Int. Explor. Mer.* **149:**202–208.

29. **Höhnk, W.** 1969. Uber den pilzlichen Befall kalkiger Hartteile von Meerestieren. *Ber. Dtsch. Wiss. Komm. Meeresforsch.* **20:**129–140.

30. **Ittekot, V.** 1996. Particle flux in the ocean: introduction, p. 1–6. *In* V. Ittekot, P. Schafer, S. Honjo, and P. J. Depetris (ed.). *Particle Flux in the Ocean.* John Wiley & Sons Ltd., New York, NY.

31. **Iwahashi, H., S. C. Kaul, K. Obuchi, and Y. Komatsu.** 1991. Induction of barotolerance by heat shock treatment in yeast. *FEMS Microbiol. Lett.* **80:**325–328.

32. **Jahnke, R. A., and G. A. Jackson.** 1992. The spatial distribution of sea floor oxygen consumption in the Atlantic and Pacific oceans. *NATO ASI Ser. Ser. C* **160:**295–308.

33. **Jannasch, H. W., C. O. Wirsen, and K. W. Doherty.** 1996. A pressurized chemostat for the study of marine barophilic and oligotrophic bacteria. *Appl. Environ. Microbiol.* **62:**1593–1596.

34. **Jannasch, H. W., C. O. Wirsen, and C. L. Winget.** 1973. A bacteriological pressure-retaining deep-sea sampler and culture vessel. *Deep-Sea Res.* **20:**661–664.

35. **Julseth, C. R., and W. E. Inniss.** 1990. Induction of protein synthesis in response to cold shock in the psychrotrophic yeast *Trichosporon pullulans. Can. J. Microbiol.* **36:**519–524.

36. **Karl, D. M., and J. E. Dore.** 2001. Microbial ecology at sea: sampling, subsampling and incubation considerations, p. 13–51. *In* J. H. Paul (ed.), *Methods in Marine Microbiology.* Academic Press, London, United Kingdom.

37. **Kato, C., L. Li, Y. Nogi, Y. Nakamura, J. Tamoaka, and K. Horikoshi.** 1998. Extremely barophilic bacteria isolated from the Mariana Trench, Challenger Deep, at a depth of 11,000 meters. *Appl. Environ. Microbiol.* **64:**1510–1513.

38. **Kato, C., N. Masui, and K. Horikoshi.** 1996. Properties of obligately barophilic bacteria isolated from a sample of deep-sea sediment from the Izu-Bonin trench. *J. Mar. Biotechnol.* **4:**96–99.

39. **Klein, D. A., and M. W. Paschke.** 2000. A soil microbial community structural-functional index: the microscopy-based total active/active fungal bacterial (TA/AFB) biovolumes ratio. *Appl. Soil Ecol.* **14:**257–268.

40. **Kohlmeyer, J.** 1977. New genera and species of higher fungi from the deep sea (1615–5315m). *Rev. Mycol.* **41:**189–206.

41. **Kohlmeyer, J., and E. Kohlmeyer (ed.).** 1979. *Marine Mycology. The Higher Fungi.* Academic Press, New York, NY.

42. **Komatsu, Y., S. C. Kaul, H. Iwahashi, and K. Obuchi.** 1990. Do heat shock proteins provide protection against freezing? *FEMS Microbiol. Lett.* **72:**159–162.

43. **Komatsu, Y., K. Obuchi, H. Iwahashi, S. C. Kaul, M. Ishimura, G. M. Fahy, and W. F. Rall.** 1991. Deuterium oxide, dimethylsulfoxide and heat shock confer protection against hydrostatic pressure damage in yeast. *Biochem. Biophys. Res. Commun.* **174:**1141–1147.

44. **Kriss, A. E., and E. I. Zaichkin.** 1971. Distribution of barotolerant microorganisms in soil. *Mikrobiologii* **40:**558–562.

45. **Landau, J. V., and D. H. Pope.** 1980. Recent advances in the area of barotolerant protein synthesis in bacteria and implications concerning barotolerant and barophilic growth. *Adv. Aquat. Microbiol.* **2:**49–76.

46. **Lee, C., R. W. Howarth, and B. L. Howes.** 1980. Sterols in decomposing *Spartina alterniflora* and the use of Ergosterol in estimating the contribution of fungi to detrital nitrogen. *Limnol. Oceanogr.* **25:**290–303.

47. **Lopez-Garcia, P., F. Rodriguez-Valera, C. Pedros-Allo, and D. Moreira.** 2001. Unexpected diversity of small eukaryotes in deep-sea Antarctic plankton. *Nature* **409:**603–607.

48. **Lorenz, R.** 1993. Kultivierung Marine Pilze unter erhöhtem hydrostatischen Drück. Ph.D. thesis. University of Regensburg, Regensburg, Germany.

49. **Lorenz, R., and H. P. Molitoris.** 1992. High pressure cultivation of marine fungi: apparatus and method. *Colloq. INSERM/* **224:**537–539. John Libbey Eurotext Ltd.

50. **Lorenz, R., and H. P. Molitoris.** 1997. Cultivation of fungi under simulated deep-sea conditions. *Mycol. Res.* **11:**1355–1365.

51. **Margesin, R., and F. Schinner. (ed.).** 1999. *Biotechnological Applications of Cold-Adapted Organisms.* Springer-Verlag, Berlin, Germany.

52. **Mille-Lindblom, C., E. Wachenfeldt, and L. J. Tranvik.** 2004. Ergosterol as a measure of living fungal biomass: persistence in environmental samples after fungal death. *J. Microbiol. Methods* **59:**253–262.

53. **Miller, J. D., and N. J. Whitney.** 1981. Fungi of the Bay of Fundi. III. Geofungi in the marine environment. *Mar. Biol.* **65:**61–68.

54. **Moore, J. C., K. McCann, and P. C. Reutier.** 2005. Modeling trophic pathways, nutrient cycling and dynamic stability of soils. *Pedobiologia* **49:**499–510.

55. **Möttonen, M., E. Jarvinen, T. J. Hokkanen, T. Kuuluvainen, and R. Ohtonen.** 1999. Spatial distribution of Ergosterol in the organic layer of a mature Scots pine (*Pinus sylvestris* L.) forest. *Soil Biol. Biochem.* **31:**503–516.

56. **Mouzouras, R.** 1986. Patterns of timber decay caused by marine fungi, p. 341–354. *In* S. T. Moss (ed.), *The Biology of Marine Fungi.* Cambridge University Press, Cambridge, United Kingdom.

57. **Mudge, S. M., and C. E. Norris.** 1997. Lipid biomarkers in the Conwy Estuary (North Wales, UK): a comparison between fatty alcohols and sterols. *Mar. Chem.* **57:**61–84.

58. **Müller, V., and P. V. Sengbusch.** 1983. Visualization of aquatic fungi (Chytridiales) parasitizing on algae by means of induced fluorescence. *Arch. Hydrobiol.* **97:**471–485.

59. **Newell, S. Y.** 1989. Litterbags, leaf tags, and decay of nonabscised intertidal leaves. *Can. J. Bot.* **67:**2224–2227.

60. **Newell, S. Y.** 1996. Established and potential impacts of eukaryotic mycelial decomposers in marine/terrestrial ecotones. *J. Exp. Mar. Biol. Ecol.* **200:**187–206.

61. **Newell, S. Y.** 2000. Methods for determining biomass and productivity of mycelial marine fungi, p. 69–91. *In* K. D. Hyde and S. B. Pointing (ed.), *Marine Mycology—a Practical Approach.* Fungal Diversity Press, Hong Kong, Hong Kong.

62. **Newell, S. Y.** 2001. Fungal biomass and productivity. *Methods Microbiol.* **30:**357–372.

63. **Newell, S. Y., and R. D. Fallon.** 1991. Towards a method for measuring instantaneous fungal growth rates in field samples. *Ecology* **72:**1547–1559.

64. **Nielsen, T. B., M. Ishii, and O. Kirk.** 1999. Lipases A and B from the yeast *Candida antarctica,* p. 49–61. *In* R. Margesin and F. Schinner (ed.), *Biotechnological Applications of Cold-Adapted Organisms.* Springer-Verlag, Berlin, Germany.

65. **Ohgiya, S., T. Hoshino, H. Okuyama, S. Tanaka, and K. Ishizaki.** 1999. Biotechnology of enzymes from cold-adapted microorganisms, p. 17–34. *In* R. Margesin and F. Schinner (ed.), *Biotechnological Applications of Cold-Adapted Organisms.* Springer-Verlag, Berlin, Germany.

66. **Perkins, R. D., and S. D. Halsey.** 1971. Geologic significance of microboring fungi and algae in Carolina shelf sediments. *J. Sediment. Petrol.* **41:**843–853.

67. **Perkins, R. D., and C. I. Tsentas.** 1976. Microbial infestation of carbonate substrates planted on the St. Croix shelf, West Indies. *Bull. Geol. Soc. Am.* **87:**1615–1628.

68. **Poulicek, M., G. Goffinet, C. Jeuniaux, A. Simon, and M. F. Voss-Foucart.** 1988. Early diagenesis of skeletal remains in marine sediments: a 10 year study, p. 107–124. *In Recherches Oceanographiques en mer Mediterranee (Biologie, Chimie, Geologie, Physique).* Institute de Recherches Marines et d'Interactions Air-Mer.

69. **Poulicek, M., R. Machiroux, and C. Toussaint.** 1986. Chitin diagenesis in deep water sediments, p. 523–530. *In* R. Muzzarelli, C. Jeunix, and G. W. Gooday (ed.), *Chitin in Nature and Technology.* Plenum Press, New York, NY.

70. **Raghukumar, C.** 2005. Diversity and adaptations of deep-sea microorganisms, p. 53–70. *In* T. Satyanarayana and B. N. Joshi (ed.), *Microbial Diversity: Current Perspectives and Potential Applications.* I. K. International Pvt. Ltd., New Delhi, India.

71. **Raghukumar, C., P. A. Loka Bharathi, Z. A. Ansari, S. Nair, B. Ingole, G. Sheelu, C. Mohandass, B. N. Nath, and N. Rodrigues.** 2001. Bacterial standing stock, meiofauna and sediment-nutrient characteristics: indicators of benthic disturbance in the Central Indian Ocean. *Deep-Sea Res. Part II* **48:**3381–3399.

72. **Raghukumar, C., and S. Raghukumar.** 1998. Barotolerance of fungi isolated from deep-sea sediments of the Indian Ocean. *Aquat. Microb. Ecol.* **15:**153–163.

73. **Raghukumar, C., S. Raghukumar, S. Sharma, and D. Chandramohan.** 1992. Endolithic fungi from deep-sea calcareous substrate: isolation and laboratory studies, p. 3–9. *In* B. N. Desai (ed.), *Oceanography of the Indian Ocean.* Oxford & IBH, New Delhi, India.

74. **Raghukumar, C., S. Raghukumar, G. Sheelu, S. M. Gupta, B. Nagender Nath, and B. R. Rao.** 2004. Buried in time: culturable fungi in a deep-sea sediment core from the Chagos Trench, Indian Ocean. *Deep-Sea Res. Part I* **51:**1759–1768.

75. **Raghukumar, C., G. Sheelu, P. A. Lokabharathi, S. Nair, and C. Mohandass.** 2000. Microbial biomass and organic nutrients in the deep-sea sediments of the Central Indian Ocean Basin. *Mar. Geores. Geotechnol.* **19:**1–16.

76. **Raghukumar, S.** 1990. Speculation on niches occupied by fungi in the sea with relation to bacteria. *Proc. Indian Acad. Sci. Plant Sci.* **100:**129–138.

77. **Ravindran, J., C. Raghukumar, and S. Raghukumar.** 2001. Fungi in *Porites lutea*: association with healthy and diseased corals. *Dis. Aquat. Org.* **47:**219–228.

78. **Rossier, C., T. Ton-That, and G. Turian.** 1977. Microcyclic microconidiation in *Neurospora crassa. Exp. Mycol.* **1:**52–62.

79. **Roth, F. J., P. A. Orpurt, Jr., and D. G. Ahearn.** 1964. Occurrence and distribution of fungi in a subtropical marine environment. *Can. J. Bot.* **42:**375–383.

80. **Ruess, L., M. M. Haggblom, E. J. G. Zapata, and J. Dighton.** 2002. Fatty acids of fungi and nematodes—possible biomarkers in the soil food chain? *Soil Biol. Biochem.* **34:**745–756.

81. **Ruess, L., A. Tinuv, D. Haubert, H. H. Richnow, M. M. Haggblom, and S. Scheu.** 2005. Carbon stable isotope fractionation and trophic transfer of fatty acids in fungal based soil food chains. *Soil Biol. Biochem.* **37:**945–953.

82. **Sathe, V., and S. Raghukumar.** 1991. Fungi and their biomass in detritus of the seagrass *Thalassia hemprichii* (Ehrenberg) Ascherson. *Bot. Mar.* **34:**272–277.

83. **Saxena, R. K., N. Khurana, R. C. Kuhud, and R. Gupta.** 1992. D-Glucose soluble starch, a novel medium for inducing microcyclic conidiation in *Aspergillus. Mycol. Res.* **96:**490–494.

84. **Seiter, K., C. Hensen, J. Schroter, and M. Zabel.** 2004. Organic carbon content in surface sediments—defining regional provinces. *Deep-Sea Res. Part I* **51:**2001–2026.

85. **Seitz, L. M., D. B. Sauer, R. Burroughs, H. E. Mohr, and J. D. Hubbard.** 1979. Ergosterol as a measure of fungal growth. *Phytopathology* **69:**1202–1203.

86. **Shimizu, H., H. Iwahashi, and Y. Komatsu.** 2002. The stress response against high hydrostatic pressure in *Saccharomyces cerevisiae. Prog. Biotechnol.* **19:**265–270.

87. **Shinn, E. A., G. A. Smith, J. M. Prospero, P. Betzer, M. L. Hayes, V. Garrison, and R. T. Barber.** 2000. African dust and the demise of Caribbean coral reefs. *Geophys. Res. Lett.* **27:**3029–3032.

88. **Stahl, P. D., and T. B. Parkin.** 1996. Relationship of soil ergosterol concentration and fungal biomass. *Soil Biol. Biochem.* **28:**847–855.

89. **Takami, H.** 1999. Isolation and characterization of microorganisms from deep-sea mud, p. 3–26. *In* K. Horikoshi and K. Tsujii (ed.), *Extremophiles in Deep-Sea Environments.* Springer, Tokyo, Japan.

90. **Takami, H., A. Inoue, F. Fuji, and K. Horikoshi.** 1997. Microbial flora in the deepest sea mud of the Mariana Trench. *FEMS Microbiol. Lett.* **152:**279–285.

91. **Takami, H., K. Kobata, T. Nagahama, H. Kobayashi, and K. Horikoshi.** 1999. Biodiversity in deep-sea sites near the south part of Japan. *Extremophiles* **3:**97–102.

92. **Turley, C. M.** 1993. The effect of pressure on leucine and thymidine incorporation by free-living bacteria and by bacteria attached to sinking oceanic particles. *Deep-Sea Res. Part I* **40:**2193–2206.

93. **Turley, C. M., and J. L. Dixon.** 2002. Bacterial numbers and growth in surficial deep-sea sediments and phytodetritus in the NE Atlantic: relationships with particulate organic carbon and total nitrogen. *Deep-Sea Res. Part I* **49:**815–826.

94. **Van Veen, J. A., and E. A. Paul.** 1979. Conversion of biovolume measurements of soil organisms, grown under various moisture tensions, to biomass and their nutrient content. *Appl. Environ. Microbiol.* **37:**686–692.

95. **Welch, T. J., and D. H. Bartlett.** 1998. Identification of a regulatory protein required for pressure-responsive gene expression in the deep-sea bacterium *Photobacterium* species strain SS9. *Mol. Microbiol.* **27:**977.

96. **Welch, T. J., A. Farewell, F. C. Neidhart, and D. H. Bartlett.** 1993. Stress response of *Escherichia coli* to elevated hydrostatic pressure. *J. Bacteriol.* **175:**7170–7177.

97. **White, D. C., R. J. Bobbie, J. S. Nickels, J. D. King, and R. J. Bobbie.** 1980. Non–selective biochemical methods for the determination of fungal mass and community structure in estuarine detrital microflora. *Bot. Mar.* **23:**239–250.

98. **Zaunstöck, B., and H. P. Molitoris.** 1995. Germination of fungal spores under deep-sea conditions. *Abstr. VI Int. Mar. Mycol. Symp. Portsmouth.*

High-Pressure Microbiology
Edited by C. Michiels, D. H. Bartlett, and A. Aertsen
© 2008 ASM Press, Washington, DC

Chapter 16

Physiology and Biochemistry of *Methanocaldococcus jannaschii* at Elevated Pressures

Boonchai B. Boonyaratanakornkit and Douglas S. Clark

Methanocaldococcus jannaschii was originally isolated from the vicinity of a hydrothermal vent at a depth of 2,600 m, making it an ideal candidate for studies of pressure effects on the physiology of a deep-sea archaeon. In several different high-temperature-pressure bioreactors, *M. jannaschii* has exhibited piezophilic growth; for example, its growth rate at 86°C was nearly five times faster at 76 MPa than at 0.79 MPa. In addition, key enzymes of *M. jannaschii* are activated and stabilized by pressures up to 50 MPa, although pressure activation can be substrate dependent and pressure stabilization is not a general property of the organism's enzymes. The membrane composition of ether-derived polar lipids also shifted in response to pressure. At the genetic level, RNA expression has been profiled for cells grown at 50 MPa, lethally heat shocked at 50 MPa, and pressure shocked from 0.79 to 50 MPa. This review summarizes the effects of pressure on selected proteins, lipids, and gene expression levels of *M. jannaschii,* one of the few known hyperthermophilic piezophiles, and presents a few prospects for future research.

ECOLOGY AND ISOLATION OF *M. JANNASCHII*

The deep sea (>1,000 m deep) represents 75% of the total volume of the earth's oceans (28). With an average depth of 3,800 m, the mean oceanic temperature is 2°C. However, at deep-sea hydrothermal vents, hot water jets enriched with minerals are emitted at temperatures up to 380°C. As these fluids mix with cold surrounding water, precipitates arise, leading to the formation of statuesque mineral-derived chimneys (28). These vent environments represent one of most extreme habitats on the temperature-pressure plane where life is known to exist (Fig. 1). First generated by Aristides Yayanos (41), Fig. 1 indicates isobars and isotherms known to support life as of 1986. The figure shows that the upper temperature and pressure limits of life remain unknown. Recently, *Shewanella oneidensis* and *Escherichia coli* were reported to remain viable at gigapascal pressures (>1,000 MPa) (33). However, this result remains controversial (40).

Boonchai B. Boonyaratanakornkit and Douglas S. Clark • Department of Chemical Engineering, University of California at Berkeley, Berkeley, CA 94720.

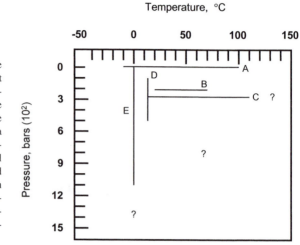

Figure 1. Life in the temperature-pressure plane. Horizontal solid lines represent three isobars where life is known to exist: atmospheric pressure (line A), the isobar in the depths of the Red Sea (line B), and the isobar near the deep-sea hydrothermal vents from which *M. jannaschii* was isolated (line C). Vertical lines represent the isotherms of the cold deep sea (line D) and the Mediterranean Sea (E). Question marks represent uncertainty with regard to the upper temperature and pressure limits of life. Reproduced from reference 20.

In April and May of 1982, members of an expedition led by Holger Jannasch manned the submersible ALVIN and collected effluent water and sedimentary material adjacent to hydrothermal vent chimneys at 21°N on the East Pacific Rise (EPR) (Fig. 2). Collected samples were transferred anaerobically into enrichment medium and stored at 20 to 45°C for 2 to 6 weeks. Under an atmosphere of 4:1 H_2-CO_2, samples were then incubated at 25,

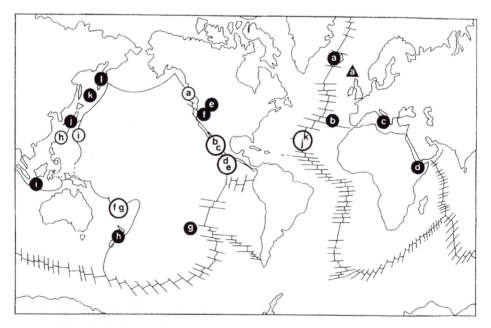

Figure 2. Distribution of hyperthermophilic archaea isolated from deep-sea hydrothermal vents (○), coastal marine hydrothermal vents and terrestrial hot springs (●), and an oil field reservoir (▲). *M. jannaschii* was isolated from the EPR at 21°N, indicated by "e" (○). Reprinted from reference 28 with permission of the publisher.

38, 60, 78, and 100°C and pH 4.0, 6.0, and 7.0 to identify new organisms. *M. jannaschii* was identified in a sample collected from a "white smoker" chimney on the EPR at 2,600 m (13), a depth that corresponds to a pressure of approximately 26 MPa (1 atm = 0.1013 MPa).

M. jannaschii cells are flagellated, irregular cocci up to 1.5 μm in diameter. Growth of *M. jannaschii* was originally described from 50 to 85°C on a substrate mixture of 80% H_2 and 20% CO_2 in 0.5 M NaCl, with a maximal doubling time of ca. 30 min near 85°C at 3 atm and pH 6.0 (13). Methane is produced through methanogenesis (35), which generates at least 1 mol of ATP per mol of CO_2. Methane has thus been used to measure growth rates of *M. jannaschii* at elevated pressures.

PRESSURE AND TEMPERATURE EFFECTS ON GROWTH AND METHANE PRODUCTION

M. jannaschii was cultivated at elevated pressures and temperatures in a sapphire pressure cell contained within a laboratory oven and capable of operation at temperatures up to 260°C and pressures up to 35 MPa (21). A magnetically operated pump was used to recirculate the 4:1 H_2-CO_2 substrate through the liquid phase to promote mixing and mass transfer in the vessel. This vessel was operational with a 43-ml volume at pressures up to 10 MPa and a 10-ml volume at pressures between 10 and 35 MPa. The reactor was filled with 0.79 MPa of the H_2-CO_2 substrate, and helium was added to achieve pressures above 0.79 MPa. At 90°C, the rate of methane production increased 3.5-fold when the pressure was raised from 0.79 to 10 MPa; however, below 88°C, methane production rates were similar at 0.79 and 10 MPa.

A stainless steel high-temperature, high-pressure bioreactor was constructed to replace the original prototype (23). The sapphire pressure vessel was replaced with a stainless steel 316 (SS316) vessel suitable for pressures up to 100 MPa with an internal volume of 167 ml and a working volume of 60 ml (Fig. 3). Helium was added for pressures above 0.79 MPa, and the gas phase was recirculated with a pneumatic pump. It was shown that growth was inhibited when Ar and H_2 were substituted for He as the pressurizing medium. When the dependence of their respective Henry's constants on pressure was calculated, the solubilities of H_2 and CO_2 were shown to vary little with pressure in the presence of He up to 76 MPa. Doubling times based on methane production decreased with increasing pressure at 86 and 90°C (Fig. 4). At 86°C, *M. jannaschii* grew nearly five times faster at 76 MPa than at 0.79 MPa. At 90°C, slower growth and longer lag times were observed than at 86°C. Protein production and methanogenesis proceeded in parallel at temperatures up to 90°C. The maximum temperature for methane production was extended from 92°C at 0.79 MPa to 98°C at 25 MPa. However, at temperatures above 90°C, protein production and methanogenesis were decoupled as protein production terminated but methanogenesis proceeded.

The 167-ml SS316 vessel was recently replaced with a 1.15-liter vessel capable of operating up to 200°C and 60 MPa with the gas phase recycled through a diaphragm compressor (Fig. 5). Construction and operation of this high-temperature and high-pressure bioreactor are described in a review by Park and coworkers (26). A glass liner is typically used in the vessel to minimize cell contact with the SS316, which may leach materials that are inhibitory to cell growth (24).

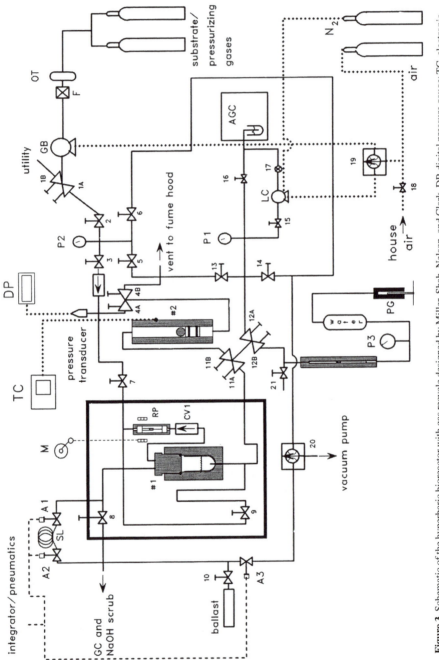

Figure 3. Schematic of the hyperbaric bioreactor with gas recycle designed by Miller, Shah, Nelson, and Clark. DP, digital pressure gauge; TC, electronic temperature controller; M, motor; OT, oxygen trap; GC, gas chromatograph; AGC, anaerobic glove chamber; PG, pressure generator; GB, gas booster; LC, liquid compressor; RP, recirculation pump; F, filter; SL, gas sample loop. Reproduced from the protocol by Nelson and Clark in reference 29.

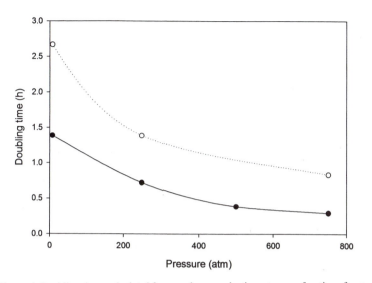

Figure 4. Doubling times calculated from methane production rates as a function of pressure at 86°C (●) and 90°C (○). Based on data from reference 23.

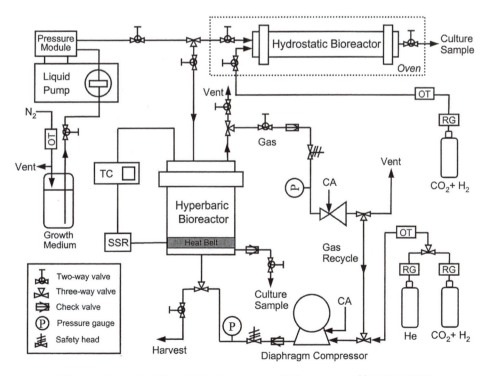

Figure 5. Schematic of the 1.15-liter high-pressure, high-temperature bioreactor system designed by Park and Clark. SSR, solid-state relay; TC, temperature controller; OT, oxygen trap; CA, compressed air; RG, regulator. Reproduced from reference 27.

The collection and transfer of deep-sea samples raise the question of how decompression prior to cultivation may affect organisms that originate from a high-pressure environment. In this connection, when samples were removed from the SS316 hyperbaric bioreactor through a microcontrol metering valve with a rapid decompression time (<1 s), increases in protein content indicated that *M. jannaschii* exhibited piezophilic growth at 26 MPa. However, there was no increase in turbidity of the culture. With a longer decompression time (5 min), the optical density at 660 nm after 7 h of growth increased more than sixfold compared to the sample that was depressurized in 1 s (24). Scanning electron microscopy images revealed that cultures grown at 26 MPa and decompressed in 1 s contained ruptured cells, while cultures that had been decompressed from 26 MPa over 5 min contained cells with mostly normal morphology (Fig. 6). In contrast, when a 65-ml hydrostatic bioreactor was decompressed in 1 s, cell lysis was not observed. Rapid decompression from the hyperbaric bioreactor apparently caused disruption of the cell envelope due to rapid expansion of pressurized gas dissolved within the cytoplasm.

Due to corrosion of seals on the diaphragm compressor caused by the recirculated gas phase, further operation of the 1.15-liter SS316 bioreactor continued without gas recirculation but with mixing provided by a magnetic stir bar. In this system, it was shown that piezophilic growth of *M. jannaschii* occurred when hydrogen consumption was physiologically limited but not when growth was mass transfer limited. When cell growth was limited by interfacial mass transfer of gaseous substrate into the liquid phase, growth was linear over time and was not accelerated at 50 MPa and 88°C (5). However, expression profiling with a whole-genome cDNA microarray revealed that a pressure-induced tran-

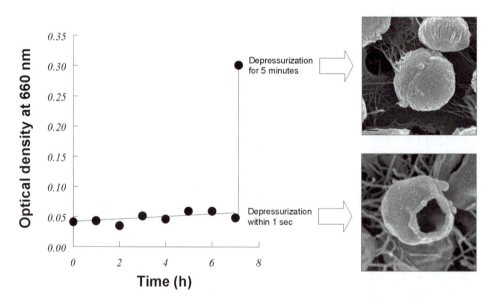

Figure 6. Scanning electron microscopy images of *M. jannaschii* after decompression for 5 min or 1 s. Cell rupture resulted in a negligible apparent turbidity increase when samples of *M. jannaschii* grown at 260 atm and 80°C were rapidly (<1 s) decompressed. However, after a 5-min decompression period, an increase in turbidity was evident. Adapted from reference 27.

scriptional response occurred at 50 MPa, regardless of whether piezophilic growth was observed. Furthermore, when growth was mass transfer limited at either 50 or 0.79 MPa, a stress response was evident (5). Thus, *M. jannaschii* exhibited a transcription level response to high pressure, even though the cells were stressed by low substrate availability and did not undergo accelerated growth (5). These experiments also showed that the possible effects of gas-liquid mass transfer limitations must be accounted for when determining whether gas-utilizing organisms are piezophilic based on measured growth rates.

PRESSURE EFFECTS ON PROTEINS

High pressure has been used in food sterilization to improve preservation of flavor and color. Pressure can also be used to stabilize certain enzymes and to control the products of an enzymatic reaction (1, 7, 20). For example, amyloglucosidase, an enzyme used in starch conversion at high temperatures, was stabilized by pressure when immobilized (30). In addition, the engineering of pressure-regulated promoters may allow pressure to be used to induce gene expression without the addition of new components into cells. Basal expression levels could then be restored upon decompression. It may thus be possible to engineer conventional hosts such as *Saccharomyces cerevisiae* into pressure-responsive organisms to facilitate industrial protein production through the application of pressure, although the economics of such a strategy would need to be evaluated.

Pressure effects on protein folding and reaction rates are based on Le Chatelier's principle. For an equilibrium process, application of pressure will cause the equilibrium to shift in favor of the state with the lowest overall volume. The entire system, which includes void volumes, bond lengths, and solvation volumes, must be taken into account when calculating this volume change (20). Although it is difficult to generalize, some trends have emerged from recent studies of pressure effects on protein stability, at least below 200 MPa. For example, hydrophobic interactions have been shown to contribute to pressure stabilization of proteins (7, 12), whereas pressure can promote disruption of protein salt bridges and induce protein denaturation (7). Pressure-induced stabilization can also occur if protein inactivation involves a transition between conformational substates of increasing volume, whereby pressure can favor the smaller, more compact substate(s). Such an effect was evident in the pressure stabilization of glutamate dehydrogenases described by Sun et al. (34).

In the case of *M. jannaschii*, the thermal half-life of partially purified hydrogenase at 90°C increased about fivefold at 50 MPa compared to that at 1 MPa (20) (Table 1). In addition, the thermal half-life of the *M. jannaschii* 20S proteasome at 125°C was 2.7-fold longer at 50 MPa than at 1 MPa (19). However, an adenylate kinase from *M. jannaschii* was not stabilized by pressure, illustrating that pressure stabilization is not a generic property of enzymes from the organism (15).

In addition to enhancing protein stability, pressure can also increase enzyme activity. Indeed, pressure activation of key enzymes may contribute to the piezophilic growth of *M. jannaschii*. Hydrogenase activity in crude extracts of *M. jannaschii* at 86°C increased more than threefold as pressure was increased from 0.76 MPa to 26 MPa (22). Moreover, hydrolytic activity of the *M. jannaschii* proteasome toward a tripeptide derivative at 130°C was fivefold higher at 50 MPa than at 1 MPa. However, when proteasome activity was measured at 35°C and pressures up to 100 MPa for a longer polypeptide substrate,

Table 1. Effect of elevated pressure on thermal half-lives and activity of enzymes from *M. jannaschii*[a]

Enzyme	Inactivation		Activity	
	Temp (°C)	$\dfrac{t_{1/2}(500\ \text{atm})}{t_{1/2}(10\ \text{atm})}$	Temp (°C)	Activity ratio[b]
Hydrogenase	90	4.8	86	3.2 (MV)[c]
Proteasome	125	2.7	130	5 (tripeptide)[d]
			35	0.2 (insulin)[e]
Adenylate kinase	83	Destabilized[f]		Destabilized[f]

[a]Hydrogenase data are from references 12 and 22. Proteasome data are from references 11 and 19. Adenylate kinase data are from reference 15.
[b]Substrates used in activity measurements are listed in parentheses.
[c]MV, methyl viologen. Activity ratio was calculated for 250 atm relative to 7.5 atm.
[d]Activity ratio for amidase activity was calculated for 500 atm relative to 10 atm.
[e]Activity ratio for proteolytic conversion was calculated for 500 atm relative to 1 atm.
[f]Residual activity of adenylate kinase was measured in the presence of 800 mM KCl.

oxidized insulin B-chain, overall conversion of the substrate was significantly lower at elevated pressures (11). Possible reasons for this reduced activity include inhibition of substrate entry and translocation through the catalytic core of the proteasome by pressure stabilization of hydrophobic interactions between the substrate and proteasome within the active-site chamber (11), and rigidification of the proteasome at 35°C.

The small heat shock protein (sHSP) from *M. jannaschii* also exhibited a favorable response to pressure. When the sHSP was prepressurized to 300 MPa for 15 min, chaperone activity against aggregated rhodanese at 48°C increased 2.8-fold compared to that in an unpressurized sample (36). The sHSP is a homomeric complex of 24 subunits; thus, pressure-induced penetration of water into the subunit interfaces may have loosened its oligomeric structure and exposed additional hydrophobic sites for binding denatured proteins (36).

PRESSURE EFFECTS ON LIPIDS

Archaeal lipids consist primarily of isoprenoid hydrocarbons and alkylglycerol ether-derived polar lipids not found in bacteria or eukaryotes. In addition to the archaeol and caldarchaeol lipids found in other archaea (Fig. 7), *M. jannaschii* contains a unique macrocyclic archaeol (macrocyclic diphytanyl glycerol ether) that is absent in several other *Methanococcus* strains. The proportion of this macrocyclic archaeol in *M. jannaschii* grown at 86°C and 25 or 50 MPa increased compared to that in cultures grown at 0.8 MPa, with corresponding reductions in the archaeol and caldarchaeol (Table 2) (14). Pressure affects membrane behavior by ordering lipids in a manner similar to that of low temperatures (16). Similarly, using a model core lipid system from *M. jannaschii*, Kaneshiro and Clark showed that pressure reduces membrane fluidity over a wide range of temperatures, shifting the temperature dependence of fluidity by ca. 10°C/50 MPa (14). Thus, in addition to activation of key enzymes by pressure, shifts in lipid composition and intrinsic membrane properties may also contribute to the barophilic growth of *M. jannaschii*.

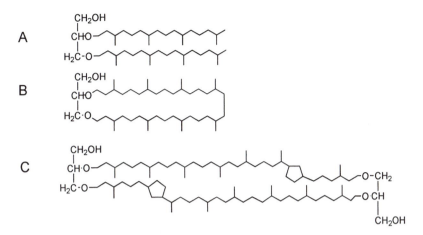

Figure 7. Structure of the archaeol (A), caldarchaeol (B), and the unique macrocyclic archaeol (macrocyclic diether) (C) lipids of *M. jannaschii.*

TRANSCRIPTION PROFILING OF *M. JANNASCHII* AT HIGH PRESSURE

In 1996, *M. jannaschii* became the first archaeon to have its complete genome sequenced. The 1.66-Mb genome, sequenced at The Institute for Genomic Research, contains 1,738 predicted protein-coding genes with an overall GC content of 31% (9). However, at that time, only 38% of the genes could be assigned a putative cellular role with high confidence. Since then, genomes from several methanogens have been sequenced *(Methanobacterium thermoautotrophicum, Methanococcus maripaludis, Methanopyrus kandleri, Methanosarcina acetivorans, Methanosarcina barkeri,* and *Methanosarcina mazei),* but only two are thermophiles *(M. thermoautotrophicum* and *M. kandleri)* and none have been identified as piezophiles. Expansion of the protein data banks has enabled about 88% of the protein-coding genes to be assigned to clusters of orthologous groups (10), although many of these clusters are not associated with specific functions. The genome of *M. jannaschii* revealed that genes related to energy production, cell division, and metabolism are similar to their bacterial counterparts, while genes involved in transcription, translation, and replication are similar to their eukaryotic counterparts (9), a trend also apparent in other archaeal genomes (17).

Recently, a cDNA microarray was fabricated from the whole genome of *M. jannaschii* (8). The array was used to profile gene expression in *M. jannaschii* under several growth conditions, including at 50 MPa and 88°C. No stress response was observed at 0.79 MPa

Table 2. Distribution of core polar lipids in *M. jannaschii* for different growth pressures at 86°C[a]

Pressure (atm)	Composition (%)		
	Archaeol	Macrocyclic archaeol	Caldarchaeol
8	7 ± 3	52 ± 5	41 ± 4
250	0–2	65 ± 5	35 ± 5
500	0–2	64 ± 6	36 ± 6

[a]Adapted from reference 14.

compared to 50 MPa, although a stress response has been observed in the deep-sea piezophiles *Thermococcus barophilus* and *Photobacterium profundum* SS9 when culti-vated at low pressures (18, 38). For *M. jannaschii,* gene expression ratios at 50 MPa rela-tive to 0.79 MPa indicated that several hypothetical proteins were differentially expressed, as were several genes encoding proteins that regulate glutamine synthesis (5).

Transcriptional studies were also performed on *M. jannaschii* cells that were lethally heat shocked at 50 MPa, and pressure shocked at 88°C from 0.79 to 50 MPa with helium. Preliminary results indicate that when cells were lethally heat shocked at 50 MPa, genes encoding the β subunit of the prefoldin chaperone and the α subunit of the 20S proteasome were down-regulated relative to expression at 0.79 MPa (6). Lethal heat shock at 50 MPa also stimulated the heat shock response observed at 0.79 MPa, while depressing the pressure effect observed at 50 MPa without heat shock. Although piezophilic growth was not evident, pressure upshift to 50 MPa at 88°C induced genes encoding the sHSP and the cold-induced chaperone, peptidyl-prolyl *cis/trans* isomerase (6).

FUTURE PROSPECTS FOR PRESSURE ADAPTATION STUDIES OF *M. JANNASCHII*

This review has summarized how pressure affects the growth, membrane composition, and gene expression profiles of *M. jannaschii,* as well as pressure effects on the activity and stability of several of the organism's key enzymes. These studies have thus expanded our understanding of how pressure affects the biochemistry and physiology of a hyperther-mophilic piezophile. However, many questions remain, particularly with regard to the mechanism(s) by which pressure regulates gene expression. One strategy for addressing this question is to examine pressure effects on the activity of transcriptional regulators. Two bacterial-type Lrp (leucine response regulatory protein) regulators have already been identified in *M. jannaschii* (25), and these regulators are good candidates for such studies. Additional pressure-controlled transcriptional regulators might be identified by perform-ing motif searches upstream of genes shown to be coregulated by pressure through expres-sion profiling. One such pressure-controlled transcriptional regulator, ToxR, was identified in the piezophile *P. profundum* SS9. At 28 MPa, the abundance of ToxR decreased about fourfold (39), and this protein was found to regulate genes whose products alter membrane structure and genes involved in the starvation response.

Pressure effects on protein-DNA interactions that underlie the activity of transcriptional regulators are also of interest. Protein-DNA interactions are likely to be destabilized by pressure due to packing effects (3). For example, 50% of the lactose repressor dissociated from operator DNA at 150 MPa (32). Whether any protein-DNA interactions are stabilized by pressure is unknown at this time, but a piezophile such as *M. jannaschii* represents a good candidate for exploratory studies of potential pressure stabilization.

Industrial biotechnology may also present opportunities for pressure regulation of mi-crobial processes. For example, piezotolerant host strains might be generated with the aim of using pressure as an operational variable to regulate recombinant protein production. Precedents for pressure-resistant mutants include a tryptophan permease-overexpressing variant of *S. cerevisiae* that was endowed with the ability to grow at pressures from 15 to 25 MPa (2). In addition, pressure-sensitive mutants of the piezophile *P. profundum* SS9 were used to identify the *recD* gene, which upon transfer into *E. coli* enabled cells to divide normally at 32 MPa (4).

It is conceivable that pressure-regulated genes identified through microarray studies of *M. jannaschii* could be engineered into *E. coli*, *S. cerevisiae,* and even pressure-sensitive methanococci to generate piezotolerant mutants. The genetic toolbox for methanogens currently includes a cassette with a strong promoter and ribosome binding site from *Methanococcus voltae* that confers puromycin resistance on *M. maripaludis* (37). Moreover, multiple mutations have been generated in *Methanosarcina acetivorans* using only one antibiotic resistance marker (31). Further examples of cassettes and markers for methanococci are described in the review by Tumbula and Whitman (37).

REFERENCES

1. **Abe, F., and K. Horikoshi.** 2001. The biotechnological potential of piezophiles. *Trends Biotechnol.* **19:**102–108.
2. **Abe, F., and K. Horikoshi.** 2000. Tryptophan permease gene *TAT2* confers high-pressure growth in *Saccharomyces cerevisiae. Mol. Cell Biol.* **20:**8093–8102.
3. **Bartlett, D. H.** 2002. Pressure effects on *in vivo* microbial processes. *Biochim. Biophys. Acta* **1595:**367–381.
4. **Bidle, K. A., and D. H. Bartlett.** 1999. RecD function is required for high-pressure growth of a deep-sea bacterium. *J. Bacteriol.* **181:**2330–2337.
5. **Boonyaratanakornkit, B., J. Cordova, C. B. Park, and D. S. Clark.** 2006. Pressure affects transcription profiles of *Methanocaldococcus jannaschii* despite the absence of barophilic growth under gas-transfer limitation. *Environ. Microbiol.* **8:**2031–2035.
6. **Boonyaratanakornkit, B. B., L. Y. Miao, and D. S. Clark.** 2007. Transcriptional responses of the deep-sea hyperthermophile *Methanocaldococcus jannaschii* under shifting extremes of temperature and pressure. *Extremophiles* **11:**495–503.
7. **Boonyaratanakornkit, B. B., C. B. Park, and D. S. Clark.** 2002. Pressure effects on intra- and intermolecular interactions within proteins. *Biochim. Biophys. Acta* **1595:**235–249.
8. **Boonyaratanakornkit, B. B., A. J. Simpson, T. A. Whitehead, C. M. Fraser, N. M. A. El-Sayed, and D. S. Clark.** 2005. Transcriptional profiling of the hyperthermophilic methanarchaeon *Methanococcus jannaschii* in response to lethal heat and non-lethal cold shock. *Environ. Microbiol.* **7:**789–797.
9. **Bult, C. J., O. White, G. J. Olsen, L. Zhou, R. D. Fleischmann, G. G. Sutton, J. A. Blake, L. M. FitzGerald, R. A. Clayton, J. D. Gocayne, A. R. Kerlavage, B. A. Dougherty, J. F. Tomb, M. D. Adams, C. I. Reich, R. Overbeek, E. F. Kirkness, K. G. Weinstock, J. M. Merrick, A. Glodek, J. L. Scott, N. S. Geoghagen, and J. C. Venter.** 1996. Complete genome sequence of the methanogenic archaeon, *Methanococcus jannaschii. Science* **273:**1058–1073.
10. **Ettema, T. J. G., W. M. de Vos, and J. van der Oost.** 2005. Discovering novel biology by *in silico* archaeology. *Nat. Rev. Microbiol.* **3:**859–869.
11. **Frankenberg, R. J., M. Andersson, and D. S. Clark.** 2003. Effect of temperature and pressure on the proteolytic specificity of the recombinant 20S proteasome from *Methanococcus jannaschii. Extremophiles* **7:**353–360.
12. **Hei, D. J., and D. S. Clark.** 1994. Pressure stabilization of proteins from extreme thermophiles. *Appl. Environ. Microbiol.* **60:**932–939.
13. **Jones, W. J., J. A. Leigh, F. Mayer, C. R. Woese, and R. S. Wolfe.** 1983. *Methanococcus jannaschii* sp. nov., an extremely thermophilic methanogen from a submarine hydrothermal vent. *Arch. Microbiol.* **136:**254–261.
14. **Kaneshiro, S. M., and D. S. Clark.** 1995. Pressure effects on the composition and thermal behavior of lipids from the deep-sea thermophile *Methanococcus jannaschii. J. Bacteriol.* **177:**3668–3672.
15. **Konisky, J., P. C. Michels, and D. S. Clark.** 1995. Pressure stabilization is not a general property of thermophilic enzymes—the adenylate kinases of *Methanococcus voltae, Methanococcus maripaludis, Methanococcus thermolithotrophicus,* and *Methanococcus jannaschii. Appl. Environ. Microbiol.* **61:**2762–2764.
16. **MacDonald, G. A.** 1987. The role of membrane fluidity in complex processes under high pressure, p. 207–223. *In* H. W. Jannasch, R. E. Marquis, and A. M. Zimmerman (ed.), *Current Perspectives in High Pressure Biology.* Academic Press, London, United Kingdom.

17. **Makarova, K. S., L. Aravind, M. Y. Galperin, N. V. Grishin, R. L. Tatusov, Y. I. Wolf, and E. V. Koonin.** 1999. Comparative genomics of the archaea (Euryarchaeota): evolution of conserved protein families, the stable core, and the variable shell. *Genome Res.* **9:**608–628.

18. **Marteinsson, V. T., A. L. Reysenbach, J. L. Birrien, and D. Prieur.** 1999. A stress protein is induced in the deep-sea barophilic hyperthermophile *Thermococcus barophilus* when grown under atmospheric pressure. *Extremophiles* **3:**277–282.

19. **Michels, P. C., and D. S. Clark.** 1997. Pressure-enhanced activity and stability of a hyperthermophilic protease from a deep-sea methanogen. *Appl. Environ. Microbiol.* **63:**3985–3991.

20. **Michels, P. C., D. Hei, and D. S. Clark.** 1996. Pressure effects on enzyme activity and stability at high temperatures. *Adv. Protein Chem.* **48:**341–376.

21. **Miller, J. F., E. L. Almond, N. N. Shah, J. M. Ludlow, J. A. Zollweg, W. B. Streett, S. H. Zinder, and D. S. Clark.** 1988. High pressure-temperature bioreactor for studying pressure temperature relationships in bacterial growth and productivity. *Biotechnol. Bioeng.* **31:**407–413.

22. **Miller, J. F., C. M. Nelson, J. M. Ludlow, N. N. Shah, and D. S. Clark.** 1989. High pressure-temperature bioreactor—assays of thermostable hydrogenase with fiber optics. *Biotechnol. Bioeng.* **34:**1015–1021.

23. **Miller, J. F., N. N. Shah, C. M. Nelson, J. M. Ludlow, and D. S. Clark.** 1988. Pressure and temperature effects on growth and methane production of the extreme thermophile *Methanococcus jannaschii*. *Appl. Environ. Microbiol.* **54:**3039–3042.

24. **Nelson, C. M., M. R. Schuppenhauer, and D. S. Clark.** 1991. Effects of hyperbaric pressure on a deep-sea archaebacterium in stainless-steel and glass-lined vessels. *Appl. Environ. Microbiol.* **57:**3576–3580.

25. **Ouhammouch, M., R. E. Dewhurst, W. Hausner, M. Thomm, and E. P. Geiduschek.** 2003. Activation of archaeal transcription by recruitment of the TATA-binding protein. *Proc. Natl. Acad. Sci. USA* **100:**5097–5102.

26. **Park, C. B., B. B. Boonyaratanakornkit, and D. S. Clark.** 2006. Toward the large scale cultivation of hyperthermophiles at high-temperature and high-pressure. *Methods Microbiol.* **35:**109–126.

27. **Park, C. B., and D. S. Clark.** 2002. Rupture of the cell envelope by decompression of the deep-sea methanogen *Methanococcus jannaschii*. *Appl. Environ. Microbiol.* **68:**1458–1463.

28. **Prieur, D., G. Erauso, and C. Jeanthon.** 1995. Hyperthermophilic life at deep-sea hydrothermal vents. *Planet. Space Sci.* **43:**115–122.

29. **Robb, F. T.** 1995. Archaea: a Laboratory Manual. Cold Spring Harbor Laboratory Press, Plainview, NY.

30. **Rohrbach, R. P., and M. J. Maliarik.** 15 November 1983. Increasing the stability of amyloglucosidase. U.S. patent 4,415,656.

31. **Rother, M., and W. W. Metcalf.** 2005. Genetic technologies for Archaea. *Curr. Opin. Microbiol.* **8:**745–751.

32. **Royer, C. A., A. E. Chakerian, and K. S. Matthews.** 1990. Macromolecular binding equilibria in the Lac repressor system—studies using high-pressure fluorescence spectroscopy. *Biochemistry* **29:**4959–4966.

33. **Sharma, A., J. H. Scott, G. D. Cody, M. L. Fogel, R. M. Hazen, R. J. Hemley, and W. T. Huntress.** 2002. Microbial activity at gigapascal pressures. *Science* **295:**1514–1516.

34. **Sun, M. M., N. Tolliday, C. Vetriani, F. T. Robb, and D. S. Clark.** 1999. Pressure-induced thermostabilization of glutamate dehydrogenase from the hyperthermophile *Pyrococcus furiosus*. *Protein Sci.* **8:**1056–1063.

35. **Thauer, R. K.** 1998. Biochemistry of methanogenesis: a tribute to Marjory Stephenson. *Microbiology* **144:**2377–2406.

36. **Tolgyesi, F., C. S. Bode, L. Smeller, D. R. Kim, K. K. Kim, K. Heremans, and J. Fidy.** 2004. Pressure activation of the chaperone function of small heat shock proteins. *Cell. Mol. Biol.* **50:**361–369.

37. **Tumbula, D. L., and W. B. Whitman.** 1999. Genetics of *Methanococcus*: possibilities for functional genomics in Archaea. *Mol. Microbiol.* **33:**1–7.

38. **Vezzi, A., S. Campanaro, M. D'Angelo, F. Simonato, N. Vitulo, F. M. Lauro, A. Cestaro, G. Malacrida, B. Simionati, N. Cannata, C. Romualdi, D. H. Bartlett, and G. Valle.** 2005. Life at depth: *Photobacterium profundum* genome sequence and expression analysis. *Science* **307:**1459–1461.

39. **Welch, T. J., and D. H. Bartlett.** 1998. Identification of a regulatory protein required for pressure-responsive gene expression in the deep-sea bacterium *Photobacterium* species strain SS9. *Mol. Microbiol.* **27:**977–985.

40. **Yayanos, A. A.** 2002. Are cells viable at gigapascal pressures? *Science* **297:**295.

41. **Yayanos, A. A.** 1986. Evolutional and ecological implications of the properties of deep-sea barophilic bacteria. *Proc. Natl. Acad. Sci. USA* **83:**9542–9546.

High-Pressure Microbiology
Edited by C. Michiels, D. H. Bartlett, and A. Aertsen
© 2008 ASM Press, Washington, DC

Chapter 17

Molecular Biology of the Model Piezophile, *Shewanella violacea* DSS12

Chiaki Kato, Takako Sato, Kaoru Nakasone, and Hideyuki Tamegai

The psychrophilic, moderately piezophilic bacterium *Shewanella violacea* strain DSS12 is a deep-sea isolate from a sediment sample collected from the Ryukyu Trench (depth, 5,110 m); it grows optimally at 30 MPa and 8°C but also grows at atmospheric pressure (0.1 MPa) and 8°C. This strain has been used as a model bacterium to elucidate the molecular basis for gene regulation as a function of pressure. In addition, with the completion of the whole genome sequence of this piezophilic bacterium, it is expected that many biotechnologically useful genes will be identified. In this chapter, we focus on the molecular characteristics of pressure adaptation in *S. violacea* DSS12 and recent advances in developing genetics and utilizing genomics.

PRESSURE-REGULATED PROMOTER OF *S. VIOLACEA* STRAIN DSS12

As stated above, the moderately piezophilic *S. violacea* strain DSS12 grows optimally at 30 MPa and 8°C but also grows well at atmospheric pressure (0.1 MPa) and 8°C (12, 24). Therefore, this piezophilic strain is useful as a model for comparison of various features of bacterial physiology under high and low hydrostatic pressure conditions. An operon identified as a pressure-regulated operon whose expression is activated by growth under high pressure was cloned and characterized from this strain. This operon, which has five transcription initiation sites, is controlled at the transcriptional level by elevated pressure (Fig. 1A; 18, 19). Moreover, transcriptional analysis showed that expression of the genes in the pressure-regulated operon is positively controlled at the transcriptional level by elevated pressure. At both low and high pressure, most transcripts arise from initiation site 2 (Fig. 1A). Upstream from this site is located a consensus sequence for the RNA polymerase sigma factor, sigma 54, and *S. violacea* sigma 54 has been shown to bind to this region (19). The sigma 54-containing RNA polymerase is responsible for the transcription

Chiaki Kato and Takako Sato • Extremobiosphere Research Center, Japan Agency for Marine-Earth Science and Technology, Yokosuka 237-0061, Japan. *Kaoru Nakasone* • Department of Chemistry and Environmental Technology, School of Engineering, Kinki University, Higashi-Hiroshima 739-2116, Japan. *Hideyuki Tamegai* • Department of Chemistry, College of Humanities and Sciences, Nihon University, Setagaya-ku, Tokyo 156-8550, Japan.

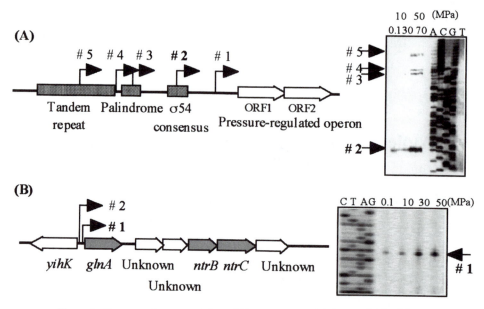

Figure 1. Diagrammatic representation of the pressure-regulated genes in *S. violacea* strain DSS12. (A) Pressure-regulated operon; (B) glutamine synthetase operon. Bold numbers (#2 in panel A and #1 in panel B) show the transcription start sites controlled by the sigma 54 factor.

of several genes, e.g., nitrogen metabolism genes such as *glnA* (encoding glutamine synthetase [17]). As shown in Fig. 1B, gene expression of the *glnA* operon is also induced at elevated pressure in *S. violacea*, particularly by sigma 54 (transcription initiation site 1 [10]). These results suggest that sigma 54 plays an important role in pressure-regulated transcription in *S. violacea*, although it should be noted that the expression of sigma 54 is not itself regulated by pressure (9, 20).

Our approach for understanding the molecular basis for high-pressure gene expression in *S. violacea* has been to undertake detailed characterization of the components of the transcriptional machinery and the accessory factors involved. This regulation is mediated by the alternative sigma factor sigma 54 and by a two-component regulatory system composed of the bacterial signal-transducing protein NtrB and the bacterial enhancer-binding protein NtrC in piezophilic *S. violacea*. In other bacteria, transcription from the sigma 54-dependent promoter, such as in the case of *glnAp2*, is regulated by enhancer-binding protein NtrC (23). The NtrC-binding sites are essential for the regulation of transcription by the sigma 54-containing RNA polymerase. Two NtrC-binding consensus sites are also present in the promoter region of *glnA* in *S. violacea*. Using electrophoretic mobility shift assay studies, we confirmed that the *S. violacea* NtrC protein specifically recognizes the element containing the NtrC consensus sequence on the *S. violacea glnA* operon (10, 11).

To reconstitute the two-component regulatory system and characterize the autophosphorylation of NtrB and transphosphorylation of phosphorylated NtrB to NtrC in vitro, we purified both of the proteins encoded by the *S. violacea ntrB* and *ntrC* genes using recombinant DNA techniques. The results of phosphorylation experiments performed at several

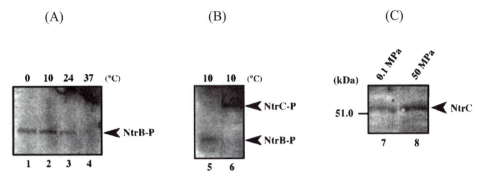

Figure 2. Autophosphorylation of the *S. violacea* NtrB protein, *trans*-phosphorylation of NtrB-P to the NtrC protein in vitro, and Western blot analysis of expression of the NtrC under different pressure conditions. (A) Autophosphorylation of NtrB incubated in the presence of [γ-^{32}P]ATP at several temperature conditions. Lane 1, 0°C; lane 2, 10°C; lane 3, 24°C; lane 4, 37°C. (B) *trans*-Phosphorylation to NtrC incubated with the phosphorylated NtrB-P at 10°C for 1 min. Lane 5, phosphorylated NtrB; lane 6, phosphorylated NtrC. (C) Lysates prepared from cells cultured at 0.1 or 50 MPa were fractionated by SDS–10% PAGE and then blotted onto a polyvinylidene fluoride membrane. The membrane was treated with antiserum against NtrC.

temperatures indicated that autophosphorylation of NtrB occurs only at low temperatures (0 to 10°C), whereas no activity was detected at 37°C (Fig. 2). Autophosphorylation activity of NtrB at such low temperatures was not detected in *Escherichia coli*. Furthermore, we have detected transcriptional activity at low temperatures in *S. violacea*. Therefore, this piezophilic bacterium adapts to the psychrosphere (low-temperature environment) and may have evolved a low-temperature-adapted system for function in the deep-sea environment. The phosphorelay between NtrB and NtrC in *S. violacea* could be detected particularly at low temperatures. This transphosphorylation might also occur under high-pressure conditions (20).

MODEL OF PRESSURE-REGULATED TRANSCRIPTION BY THE SIGMA 54 FACTOR

During transcription initiation at the sigma 54-dependent promoters, such as within the operator region of the pressure-regulated *glnA* operon, sigma 54-containing RNA polymerase holoenzyme transcription activation is facilitated by the activity of NtrC, which in turn is controlled by NtrB. Any of these *trans*-acting factors (sigma 54, NtrC, or NtrB) could play an important role in pressure-regulated transcription in this piezophilic bacterium. However, sigma 54 in *S. violacea* is expressed at a relatively constant level under both atmospheric and high-pressure conditions, indicating that changes to its levels are not likely to be a major factor in pressure-regulated gene expression. In contrast, Western blot analysis indicates that NtrC protein abundance is modulated by pressure. The levels of this factor expressed at high pressure are greater than that at atmospheric pressure (11).

A model for *glnA* pressure regulation by sigma 54 in the deep-sea piezophilic bacterium *S. violacea* is shown in Fig. 3. Because *S. violacea* sigma 54 is expressed at consistent levels at both atmospheric and high pressure, it is suggested that the intracellular levels of

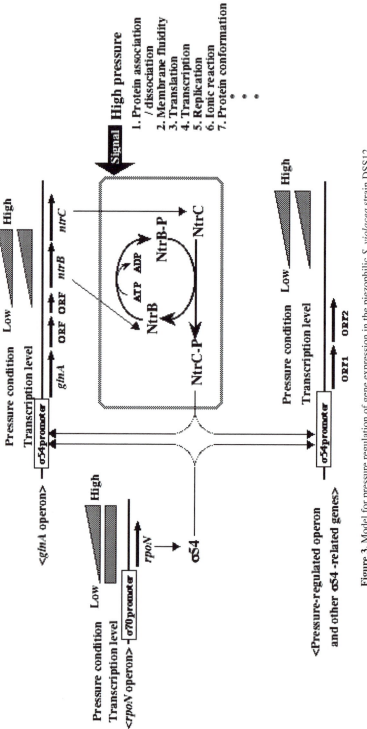

Figure 3. Model for pressure regulation of gene expression in the piezophilic *S. violacea* strain DSS12.

sigma 54-containing RNA polymerase holoenzyme under both conditions are constant. This observation also strongly suggests that the transcriptional activity at this sigma 54-dependent promoter is proportional to the amount of NtrC and that this rate-limiting factor regulates gene expression. This model also suggests that NtrB could function as a pressure sensor and that this protein might be preferentially autophosphorylated under high-pressure conditions. Subsequently, the NtrB-NtrC phospho form would result in phosphorylated NtrC activation of sigma 54-dependent promoters (Fig. 3). This model should be confirmed using molecular genetic approaches utilizing mutations in genes encoding components of this sensory transduction and transcription system.

RESPIRATORY SYSTEMS IN *S. VIOLACEA*

Downstream from the pressure-regulated operon described above is an open reading frame (ORF) homologous to the *cydD* gene of *E. coli*. This gene has been proposed to be important for bacterial growth under high pressure (13). The gene product of *cydD* in *E. coli* is thought to be required for the assembly of respiratory components (25–27). Further, the expression of the respiratory system is in fact regulated by hydrostatic pressure in *S. violacea* (22, 32, 35) as well as in another piezophilic bacterial strain, *Shewanella* sp. strain DB-172F (28, 29). These were the first reports addressing the role of respiratory systems in deep-sea bacteria and provided the first evidence that expression of genes for respiratory components is regulated by physical parameters, such as hydrostatic pressure. Generally, bacteria have branched respiratory chains. Specifically, *Shewanella* strains have many respiratory components for adaptation to environmental changes (8). These observations and the results of previous studies suggest that pressure regulation for expression of respiratory systems in *S. violacea* plays an important role in bacterial adaptation to high pressure.

Cytochrome *bd* is one of the members of the quinol oxidases, distinct from the heme-copper oxidase superfamily. In *E. coli*, two types of quinol oxidases, cytochrome *bo* and cytochrome *bd*, exist, and both have roles in respiration. Cytochrome *bo* is expressed in log phase, and cytochrome *bd* is expressed in stationary phase (15, 16). Cytochrome *bd* shows higher affinity for O_2 than does cytochrome *bo*, and the former acts as a terminal oxidase under low-oxygen-concentration conditions (16). For the biosynthesis of cytochrome *bd*, structural genes (encoded by the *cydAB* operon) and genes for assembly of the mature iron proteins (encoded by the *cydDC* operon) are required (5, 27). Expression of *cydAB* in *E. coli* is regulated by ArcA and Fnr, common O_2-regulated transcriptional regulators (2, 6), and that of *cydDC* is regulated by NarL (involved in the two-component regulatory system for nitrate respiration) as well as Fnr (1). However, in *S. violacea*, no cytochrome *bd* has been detected spectrophotometrically under atmospheric pressure even during the stationary phase. Surprisingly, cytochrome *bd* has been detected only under growth conditions of high hydrostatic pressure (Fig. 4 and Table 1; 32). Thus, transcriptional regulation of cytochrome *bd* in *S. violacea* may be different from that in other organisms, and this may be important for bacterial adaptation to high pressure. Transcriptional analysis of the *S. violacea cydAB* and *cydCD* operons indicates that transcription of the *cydDC* operon is strongly regulated by hydrostatic pressure (33). These results suggest that pressure regulation of the respiratory system in *S. violacea* plays an important role in bacterial adaptation to high hydrostatic pressure.

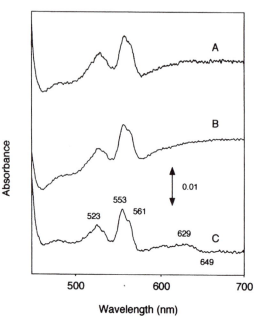

Figure 4. Reduced minus oxidized difference spectra of membrane fractions from *S. violacea*. Each fraction was obtained from the cells grown under a pressure of 0.1 MPa with shaking (A), grown under a pressure of 0.1 MPa under microaerobic conditions (B), and grown under a pressure of 50 MPa with microaerobic conditions (C). An absorption peak at 629 nm and a trough at 649 nm are specifically detected in the membrane fraction of cells grown under a pressure of 50 MPa. These spectral properties are typical of *d*-type cytochromes (16).

STABILITY OF THE RNA POLYMERASE FROM *S. VIOLACEA* UNDER PRESSURE

Another aspect of transcription in piezophiles is the stability of the quaternary structure of RNA polymerase. This has been addressed using a high-pressure electrophoresis apparatus (HPEA), developed by a modification of the method previously reported by Erijman and Clegg (3; see also reference 14) (Fig. 5). Using this system, pressures up to 200 MPa can be applied within 1 min using silicon oil and a hand pump. The HPEA has been utilized to investigate the relative stability of the *S. violacea* RNA polymerase (svRNAP) and *Escherichia coli* RNA polymerase (ecRNAP) as a function of pressure. Native polyacrylamide gel electrophoresis (PAGE) was carried out in glass capillary tubes. Following electrophoresis and decompression, the gel was removed and equilibrated in sodium dodecyl sulfate (SDS) buffer. Next, the gel was overlaid onto an SDS gel and subjected to SDS-PAGE using 10% polyacrylamide gels at atmospheric pressure, and proteins were visualized by silver staining.

Table 1. Composition of cytochromes in *S. violacea*

Fraction	Cytochrome(s) in:	
	Cells grown at 0.1 MPa	Cells grown at 50 MPa
Soluble	*c*-type (greater amount)	*c*-type (lesser amount)
Membrane	*b*- (and/or *o*-) type	*b*- (and/or *o*-) type
	c-type	*d*-type
		c-type

(A)

(B)

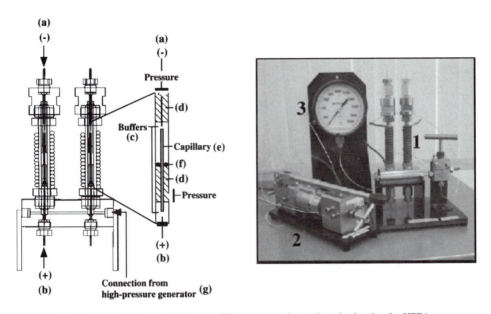

Figure 5. Diagram of the HPEA. (A) High-pressure electrophoresis chamber for HPEA. (a) Connection to the power supply (anode); (b) connection to the power supply (cathode); (c) buffers; (d) silicone oil KF-96-1.5CS; (e) glass microcapillary tube; (f) O-ring to partition a space into the upper and lower spaces; (g) connection to a high-pressure pump. (B) Photograph of the HPEA. 1, high-pressure electrophoresis chamber; 2, high-pressure hand pump; 3, pressure gauge.

These experiments revealed that a pressure of 140 MPa caused dissociation of ecRNAP but not that of svRNAP (Fig. 6A and B). On the other hand, the core enzyme of svRNAP, which lacked the sigma factor, was dissociated at 140 MPa (Fig. 6C; 14). These results suggest that the sigma factor is required for stabilization of svRNAP under high-pressure conditions. The sigma subunit is known to change the quaternary structure of ecRNAP (7, 34). It is likely that the *S. violacea* sigma subunit stabilizes the core enzyme through alteration of the quaternary structure of RNA polymerase, resulting in piezotolerance. In this context, the predicted β-sheet domain, which is not observed in the *E. coli* or *S. oneidensis* sigma 70 subunit, may have a role in stabilization of RNA polymerase at high pressure (14). Further experimentation is required to determine the significance of the β-sheet domain by comparing the structure with that of other mesophilic *Shewanella* strains and by analysis of the effects of mutations within the domain on the piezotolerance of RNA polymerase in terms of transcriptional activity and subunit association. For investigations of the molecular adaptation of proteins to high hydrostatic pressure, the HPEA is a powerful tool in combination with techniques based on molecular biology and bioinformatics.

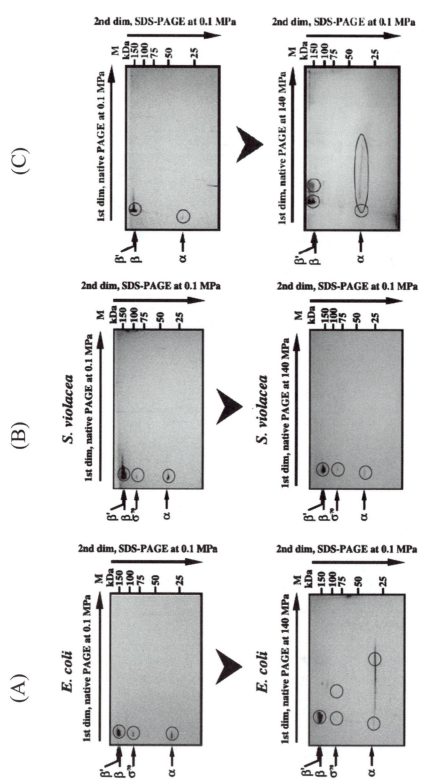

Figure 6. Effects of high hydrostatic pressure on subunit association in RNA polymerase in *E. coli* (A), *S. violacea* (B), and *S. violacea* without sigma factor (C). Native PAGE was performed at 0.1 MPa (upper panels) or 140 MPa (lower panels), followed by SDS-PAGE at 0.1 MPa. Proteins were visualized by silver staining. Each subunit of RNA polymerase is shown by an open circle. M, molecular mass markers; 1st dim, one-dimensional separation; 2nd dim, two-dimensional separation.

312

HOST VECTOR SYSTEM OF *S. VIOLACEA*

Several gene clusters from *S. violacea* have been studied with respect to their expression at high pressure. In addition, control mechanisms for gene expression in vivo and in comparison to *E. coli* have also been examined as described above. However, how piezophile genes are actually regulated under high pressure in this piezophilic bacterium is not yet understood in detail. Therefore, we developed a gene transfer system in a rifampin-resistant *S. violacea* strain, DSS12R, to further expand upon the genetic studies of the piezophiles (4). This system utilizes a biparental method for bacterial mating, resulting in transformation. In this biparental system, the donor is *E. coli* strain S17-1, which contains a *tra* gene in its chromosome (31). The plasmid to be transferred into the recipient cells is transformed into the donor. Early-log-phase cells of the donor harboring the plasmid and late-log-phase cells of the recipient are mixed and cultivated on nonselective agar. After mating, the cells are cultivated on selective agar plates at 8°C to identify transconjugants. Using this method, plasmids have been successfully introduced into *S. violacea*. This has been verified by direct plasmid isolation from the transconjugants and by PCR analyses. The results of this approach are summarized in Table 2 as well as in Fig. 7, which display electrophoretic characterization of the plasmids in the transconjugants. DSS12R was successfully transconjugated with the broad-host-range vectors pKT231 and pTS4 (30) as well as plasmid pACYC184, a plasmid often used as a control in *E. coli* transformation experiments. The efficiency of conjugation (transconjugants/recipient cell) using pKT231 was 4.2×10^{-7}, which is much lower than that reported for mesophile bacterial conjugations. One possible explanation for the low conjugation frequency is that the optimal mating temperature for DSS12R is 20°C, which is much lower than the optimal temperature for *E. coli*. This study provided the first demonstration of gene transfer in the piezophilic bacterium DSS12. The study of the mechanisms for adaptation to high-pressure environments, including gene regulatory systems, may now also proceed in vivo using genetic approaches in this piezophile.

S. VIOLACEA GENOME ANALYSIS

To better understand pressure regulation in deep-sea bacteria, we performed genome analysis of the piezophilic bacterium *S. violacea* strain DSS12 as a model deep-sea bacterium (21). The genome size of this bacterium is about 4.9 Mbp based on the results of pulse-field gel electrophoresis analysis, and 12 rRNA operons have been identified. The

Table 2. Comparison of *S. violacea* DSS12 conjugation efficiencies with different plasmids[a]

Plasmid	Size (kb)	Replication origin	Antibiotic resistance phenotype	Conjugation efficiency (transconjugants/recipient cell)
Broad-host-range vector				
pKT231*	13.0	RSF1010	Km Sm	4.2×10^{-7}
pRK415	10.5	RK2	Tc	—
Vector for *E. coli*				
pTS4**	2.7	pUC	Amp Cm	1×10^{-8}
pACYC184**	4.2	p15A	Tc Cm	2×10^{-8}

[a]—, no transformants; *, plasmids detected following direct isolation; **, predicted-length fragments from plasmids were detected by PCR amplification.

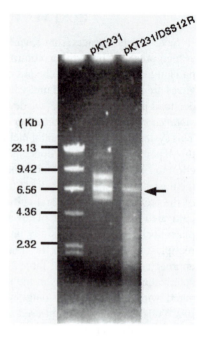

Figure 7. Images of 1% agarose electrophoresis of plasmid extracted from DSS12R transconjugants.

high number of ribosomal operons might correspond to extreme environmental conditions, and it is likely that the ability of *S. violacea* to grow under high pressure depends on the improved translational efficiency conferred by large numbers of ribosomes. From the results of genome sequencing, approximately 4,600 ORFs have been identified. As shown in Fig. 8,

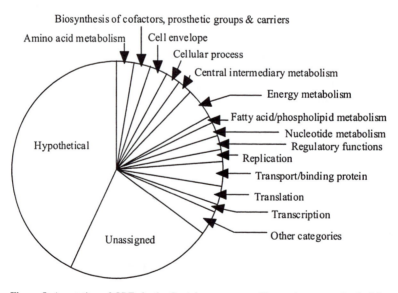

Figure 8. Annotation of ORFs in the *S. violacea* genome. The total genome size is 4.9 Mbp, and approximately 4,600 ORFs are predicted.

preliminary analyses of the annotation suggested that almost 70% of the ORFs are hypothetical proteins and unassigned. This result indicates that many genes from deep-sea microbes could be novel and might also have novel functions. Therefore, genomic analysis of marine extremophiles may lead to the discovery of new functions for genes. In the *S. violacea* genome, the following interesting genes were identified: those encoding haloalkane dehalogenase, extracellular proteases, chitinase, eicosapentaenoic acid synthesis clusters, thiamine biosynthesis clusters, lipoproteins, vitamin B_6 biosynthesis protein, biotin synthetase, cellulases, organic solvent tolerance proteins, etc.

Acknowledgments. We are very grateful to Koki Horikoshi for his continued support of our studies of extremophiles. We also thank the crews of the research ships and members of the submersible operation division at JAMSTEC for their efforts in collecting samples from the deep-sea environment as well as Howard K. Kuramitsu for assistance in editing the manuscript.

REFERENCES

1. **Cook, G. M., J. Membrillo-Hernandez, and R. K. Poole.** 1997. Transcriptional regulation of the *cydDC* operon, encoding a heterodimeric ABC transporter required for assembly of cytochromes *c* and *bd* in *Escherichia coli* K-12: regulation by oxygen and alternative electron acceptors. *J. Bacteriol.* **179:**6525–6530.

2. **Cotter, P. A., S. B. Melville, J. A. Albrecht, and R. P. Gunsalus.** 1997. Aerobic regulation of cytochrome *d* oxidase (*cydAB*) operon expression in *Escherichia coli:* roles of Fnr and ArcA in repression and activation. *Mol. Microbiol.* **25:**605–615.

3. **Erijman, L., and R. M. Clegg.** 1995. Heterogeneity of *E. coli* RNA polymerase revealed by high pressure. *J. Mol. Biol.* **253:**259–265.

4. **Fukuchi, J., T. Sato, C. Kato, M. Ito, and K. Horikoshi.** 2002. The host-vector system for deep-sea piezophilic bacteria, *Shewanella violacea* DSS12 and *Moritella japonica* DSK1. *JAMSTECR* **46:**157–161. (In Japanese.)

5. **Georgiou, C. D., H. Fang, and R. B. Gennis.** 1987. Identification of the *cydC* locus required for expression of the functional form of the cytochrome *d* terminal oxidase complex in *Escherichia coli. J. Bacteriol.* **169:**2107–2112.

6. **Govantes, F., J. A. Albrecht, and R. P. Gunsalus.** 2000. Oxygen regulation of the *Escherichia coli* cytochrome *d* oxidase (*cydAB*) operon: roles of multiple promoters and the Fnr-1 and Fnr-2 binding sites. *Mol. Microbiol.* **37:**1456–1469.

7. **Greiner, D. P., K. A. Hughes, A. H. Gunasekera, and C. F. Meares.** 1996. Binding of the σ^{70} protein to the core subunits of *Escherichia coli* RNA polymerase, studied by iron-EDTA protein footprinting. *Proc. Natl. Acad. Sci. USA* **93:**71–75.

8. **Heidelberg, J. F., I. T. Paulsen, K. E. Nelson, E. J. Gaidos, W. C. Nelson, T. D. Read, J. A. Eisen, R. Seshadri, N. Ward, B. Methe, R. Clayton, T. Meyer, A. Tsapin, J. Scott, M. Beanan, L. Brinkac, S. Daugherty, R. T. DeBoy, R. J. Dodson, A. S. Durkin, D. H. Haft, J. F. Kolonay, R. Madupu, J. D. Peterson, L. A. Umayam, O. White, A. M. Wolf, J. Vamathevan, J. Weidman, M. Impraim, K. Lee, K. Berry, C. Lee, J. Mueller, H. Khouri, J. Gill, T. R. Utterback, L. A. McDonald, T. V. Feldblyum, H. O. Smith, J. C. Venter, K. H. Nealson, and C. M. Fraser.** 2003. Genome sequence of the dissimilatory metal ion-reducing bacterium *Shewanella oneidensis. Nat. Biotechnol.* **20:**1118–1123.

9. **Ikegami, A., K. Nakasone, M. Fujita, S. Fujii, C. Kato, R. Usami, and K. Horikoshi.** 2000. Cloning and characterization of the gene encoding RNA polymerase sigma factor σ^{54} of deep-sea piezophilic *Shewanella violacea. Biochim. Biophys. Acta* **1491:**315–320.

10. **Ikegami, A., K. Nakasone, C. Kato, Y. Nakamura, I. Yoshikawa, R. Usami, and K. Horikoshi.** 2000. Glutamine synthetase gene expression at elevated hydrostatic pressure in a deep-sea piezophilic *Shewanella violacea. FEMS Microbiol. Lett.* **192:**91–95.

11. **Ikegami, A., K. Nakasone, C. Kato, R. Usami, and K. Horikoshi.** 2000. Structural analysis of *ntrBC* genes of deep-sea piezophilic *Shewanella violacea. Biosci. Biotechnol. Biochem.* **64:**915–918.

12. **Kato, C., T. Sato, and K. Horikoshi.** 1995. Isolation and properties of barophilic and barotolerant bacteria from deep-sea mud samples. *Biodivers. Conserv.* **4:**1–9.

13. **Kato, C., H. Tamegai, A. Ikegami, R. Usami, and K. Horikoshi.** 1996. Open reading frame 3 of the barotolerant bacterium strain DSS12 is complementary with *cydD* in *Escherichia coli: cydD* functions are required for cell stability at high pressure. *J. Biochem.* **120:**301–305.

14. **Kawano, H., K. Nakasone, M. Matsumoto, R. Usami, C. Kato, and F. Abe.** 2004. Differential pressure resistance in the activity of RNA polymerase isolated from *Shewanella violacea* and *Escherichia coli. Extremophiles* **8:**367–375.

15. **Kita, K., K. Konishi, and Y. Anraku.** 1984. Terminal oxidases of *Escherichia coli* aerobic respiratory chain. I. Purification and properties of cytochrome b_{562}-o complex from cells in the early exponential phase of aerobic growth. *J. Biol. Chem.* **259:**3368–3374.

16. **Kita, K., K. Konishi, and Y. Anraku.** 1984. Terminal oxidases of *Escherichia coli* aerobic respiratory chain. II. Purification and properties of cytochrome b_{558}-d complex from cells grown with limited oxygen and evidence of branched electron-carrying systems. *J. Biol. Chem.* **259:**3375–3381.

17. **Merrick, M. J., and R. A. Edwards.** 1995. Nitrogen control in bacteria. *Microbiol. Rev.* **59:**604–622.

18. **Nakasone, K., A. Ikegami, C. Kato, R. Usami, and K. Horikoshi.** 1998. Mechanisms of gene expression controlled by pressure in deep-sea microorganisms. *Extremophiles* **2:**149–154.

19. **Nakasone, K., A. Ikegami, C. Kato, R. Usami, and K. Horikoshi.** 1999. Analysis of *cis*-elements upstream of the pressure-regulated operon in the deep-sea barophilic bacterium *Shewanella violacea* strain DSS12. *FEMS Microbiol. Lett.* **176:**351–356.

20. **Nakasone, K., A. Ikegami, H. Kawano, R. Usami, C. Kato, and K. Horikoshi.** 2002. Transcriptional regulation under pressure conditions by the RNA polymerase σ^{54} factor with a two component regulatory system in *Shewanella violacea. Extremophiles* **6:**89–95.

21. **Nakasone, K., H. Mori, T. Baba, and C. Kato.** 2003. Whole-genome analysis of piezophilic and psychrophilic microorganism. *Kagaku to Seibutsu* **41:**32–39. (In Japanese.)

22. **Nakasone, K., M. Yamada, M. H. Qureshi, C. Kato, and K. Horikoshi.** 2001. Piezoresponse of the *cyo*-operon coding for quinol oxidase subunits in a deep-sea piezophilic bacterium, *Shewanella violacea. Biosci. Biotechnol. Biochem.* **65:**690–693.

23. **Ninfa, A. J., L. J. Reitzer, and B. Magasanik.** 1987. Initiation of transcription at the bacterial *glnAp2* promoter by purified *E. coli* components is facilitated by enhancers. *Cell* **50:**1039–1046.

24. **Nogi, Y., C. Kato, and K. Horikoshi.** 1998. Taxonomic studies of deep-sea barophilic *Shewanella* species, and *Shewanella violacea* sp. nov., a new barophilic bacterial species. *Arch. Microbiol.* **170:**331–338.

25. **Poole, R. K., F. Gibson, and G. Wu.** 1994. The *cydD* gene product, component of a heterodimeric ABC transporter, is required for assembly of periplasmic cytochrome *c* and of cytochrome *bd* in *Escherichia coli. FEMS Microbiol. Lett.* **117:**217–224.

26. **Poole, R. K., L. Hatch, M. W. J. Cleeter, F. Gibson, G. B. Cox, and G. Wu.** 1993. Cytochrome *bd* biosynthesis in *Escherichia coli:* the sequences of the *cydC* and *cydD* genes suggest that they encode the components of an ABC membrane transporter. *Mol. Microbiol.* **10:**421–430.

27. **Poole, R. K., H. D. Williams, A. Downie, and F. Gibson.** 1989. Mutations affecting the cytochrome *d*-containing oxidase complex of *Escherichia coli* K12: identification and mapping of a fourth locus, *cydD. J. Gen. Microbiol.* **135:**1865–1874.

28. **Qureshi, M. H., C. Kato, and K. Horikoshi.** 1998. Purification of a *ccb* type quinol oxidase specifically induced in a deep-sea barophilic bacterium, *Shewanella* sp. strain DB-172F. *Extremophiles* **2:**93–99.

29. **Qureshi, M. H., C. Kato, and K. Horikoshi.** 1998. Purification of two pressure-regulated *c*-type cytochromes from a deep-sea bacterium, *Shewanella* sp. strain DB-172F. *FEMS Microbiol. Lett.* **161:**301–309.

30. **Sato, T., C. Kato, and K. Horikoshi.** 1995. Effect of high pressure on gene expression by *lac* and *tac* promoters in *Escherichia coli. J. Mar. Biotechnol.* **3:**89–92.

31. **Simon, R., U. Priefer, and A. Puhler.** 1983. A broad host range mobilization system for *in vivo* genetic engineering. *Bio/Technology* **1:**784–791.

32. **Tamegai, H., C. Kato, and K. Horikoshi.** 1998. Pressure-regulated respiratory system in barotolerant bacterium, *Shewanella* sp. strain DSS12. *J. Biochem. Mol. Biol. Biophys.* **1:**213–220.

33. **Tamegai, H., H. Kawano, A. Ishii, S. Chikuma, K. Nakasone, and C. Kato.** 2005. Pressure-regulated biosynthesis of cytochrome *bd* in piezo- and psychrophilic deep-sea bacterium *Shewanella violacea* DSS12. *Extremophiles* **9:**247–253.

34. **Wu, F. Y., L. R. Yarbrough, and C. W. Wu.** 1976. Conformational transition of *Escherichia coli* RNA polymerase induced by the interaction of sigma subunit with core enzyme. *Biochemistry* **15:**3254–3258.

35. **Yamada, M., K. Nakasone, H. Tamegai, C. Kato, R. Usami, and K. Horikoshi.** 2000. Pressure-regulation of soluble cytochromes *c* in a deep-sea piezophilic bacterium, *Shewanella violacea*. *J. Bacteriol.* **182:**2945–2952.

High-Pressure Microbiology
Edited by C. Michiels, D. H. Bartlett, and A. Aertsen
© 2008 ASM Press, Washington, DC

Chapter 18

Adaptations of the Psychrotolerant Piezophile *Photobacterium profundum* Strain SS9

Douglas H. Bartlett, Gail Ferguson, and Giorgio Valle

Strains of *Photobacterium profundum* and *Shewanella violacea* have become the workhorses for examining the adaptations of cold-adapted deep-sea microbes to high pressure. The latter is described by Kato and colleagues in the preceding chapter. Here the focus is on *P. profundum,* with an emphasis on strain SS9 (Fig. 1). SS9 grows from atmospheric pressure to 90 MPa, with a broad pressure optimum at ~28 MPa (32). Its temperature optimum is 15°C, and its temperature range extends from below 0 to ~25°C. Much of the progress that has been made with SS9 has relied on genetics. Transposon mutagenesis, insertional mutagenesis, *lacZ*-based transcriptional reporter studies, allelic exchange, and complementation are all possible (2–4, 12, 15, 26, 27, 62, 102). SS9 is both the most psychrotolerant and the most piezophilic microorganism for which such studies have been reported. More recently its entire genome sequence has been determined, and transcriptome and comparative genomic analyses have been undertaken (23, 98).

The utility of strain SS9 for laboratory investigations of life at depth is reflected in the environment from which it was obtained. SS9 was obtained in the early 1980s from the Sulu Sea, located in the western Pacific Ocean (32). The Sulu Sea is a basin flanked by a number of shallow sills. As a result it has relatively warm, deep waters and an extremely low oxygen content (0.57 ml liter^{-1}). It extends to depths of about 5,000 m; a temperature of 10°C has been recorded at this depth. SS9 was collected using a thermally insulated free vehicle trap to collect amphipod crustaceans from a depth of 2,551 m. These were subsequently homogenized and used as an inoculum in a nutrient-rich marine broth at 26 MPa and 7°C. SS9 is a member of a group of piezophiles which are tolerant of higher temperatures, up to and in some cases exceeding 20°C, which were isolated by DeLong and Yayanos from the warmer marine basins of the Sulu Sea, the Celebes Sea, or the Mediterranean Sea (33).

Douglas H. Bartlett • Mail code 0202, Marine Biology Research Division, Center for Marine Biotechnology and Biomedicine, Scripps Institution of Oceanography, University of California, San Diego, La Jolla, CA 92093-0202. ***Gail Ferguson*** • School of Medicine, Department of Medicine and Therapeutics, Institute of Medical Sciences, University of Aberdeen, Foresterhill, Aberdeen AB25 2ZD, United Kingdom. ***Giorgio Valle*** • Department of Biology, University of Padua, 35131 Padua, Italy.

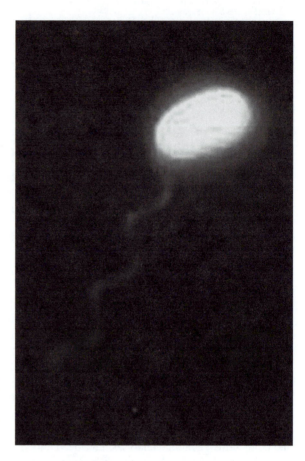

Figure 1. NanoOrange fluorescent image of a *P. profundum* strain SS9 cell from a culture grown at 30 MPa. The body of the cell is approximately 2.5 μm in length. A single unsheathed polar flagellum is present.

SS9 was originally selected for further investigation from among a group of piezophiles in the Yayanos culture collection because it demonstrated a vigorous response to changes in pressure. This was first determined by looking at protein banding patterns via one-dimensional sodium dodecyl sulfate polyacrylamide gel electrophoresis (9). Since then SS9 has been shown to exhibit pressure-responsive changes in gene expression, membrane fatty acid species, and osmolyte levels (4, 9, 23, 27, 67, 98, 102). It is interesting in this regard that within the Sulu Sea there are a variety of organisms which undergo vertical migrations of 1 to 5 km (36). These include shrimp *(Acanthephyra)*, squid *(Leachia)*, and fish (Macrouridae and *Bregmaceros*). Thus, it is conceivable that as a result of its association with vertically migrating fauna SS9 evolved its pressure-sensing capabilities and adaptation to a broad range of pressures.

Subsequent to the isolation and taxonomic characterization of SS9, additional *P. profundum* strains and closely related species have been uncovered (Fig. 2). These have been isolated from sediments and water samples from depths as great as 5.1 km (14, 21, 23, 75, 82). *P. profundum* is very closely related to the species *Photobacterium indicum* and *Photobacterium frigidiphilum* (containing more than 98% rRNA sequence identity), which are also common in deep-sea sediment environments (88). Indeed, the characterized

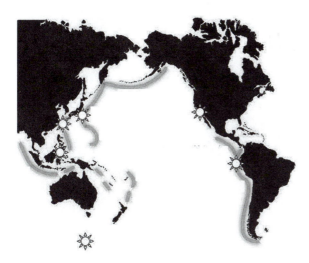

Figure 2. Locations in which *P. profundum* isolates have been reported (14, 21, 23, 75, 82). Stars indicate eastern Antarctica, the northwestern Pacific Ocean, the Peru Margin, the Ryukyu Trench, the San Diego Bay, and the Sulu Sea.

P. frigidiphilum strain is also a moderate piezophile. Another close relative of *P. profundum* has been associated with petroleum degradation (accession number EF628004).

In addition to SS9, two other *P. profundum* strains have also been studied beyond the taxonomic level (23). They are DSJ4 and 3TCK. DSJ4 is the type strain for the species and was recovered from a Ryukyu Trench (Japan) shallow-sediment sample obtained from a water depth of 5,110 m (75). It is a moderate piezophile with a pressure optimum of 10 MPa and a temperature optimum of 10°C. 3TCK was isolated from a shallow-sediment sample obtained in shallow water in the San Diego Bay (California) (23). Its pressure optimum is atmospheric pressure (0.1 MPa), and its temperature range for growth is from below 0°C to above 25°C.

The following sections describe the roles of the membrane, membrane-associated signaling systems, and DNA recombination in SS9 piezoadaptation, and provide recent insights obtained from the SS9 genome sequence and functional and comparative genomics.

MEMBRANE LIPIDS

A number of membrane components are known or postulated to be important for low-temperature and high-pressure growth, the best understood of which are unsaturated fatty acids (Fig. 3). The ability of many organisms, including bacteria, to adapt their phospholipid unsaturated fatty acid content in response to changes in temperature has been known for many years (66). This process, traditionally known as homeoviscous adaptation, acts to maintain membrane fluidity within a certain range, as a function of temperature (90). It has also been suggested that these and other phospholipid alterations are part of a homeostatic process for the maintenance of optimal ion permeability or membrane curvature elastic stress (7, 96). In the 1980s, DeLong and Yayanos discovered that the piezophilic marine bacterium CNPT3 was also able to adapt its unsaturated fatty acid content in response to changes in pressure (33). Specifically, CNPT3 increased the ratio of unsaturated to saturated fatty acids when the pressure was increased from atmospheric (0.1 MPa) to 70 MPa. Since the changes in the fatty acid content of CNPT3

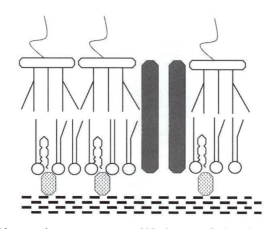

Figure 3. Many membrane components could be important for integrity and function at high pressure. Shown in schematic format is the outer membrane of a gram-negative bacterial cell. The outer leaflet is composed of LPS with its lipid A (vertical lines, some with branches), core oligosaccharide (horizontal oval), and O-antigen polysaccharide (wavy line) regions. The inner leaflet contains phospholipids containing head groups (shown as open circles) attached to different types of fatty acids at their *sn*-1 and *sn*-2 positions. Straight lines are used to depict saturated fatty acids, lines with one bend are used to depict MUFAs such as *cis*-vaccenic acid, and lines with multiple bends are used to depict the PUFA EPA. A porin protein is shown spanning the entire outer membrane. Lipoprotein is shown connecting the outer membrane to the cell wall below. Phospholipid composition is also important for inner membrane function (not shown). Many lipid-protein interactions, and thus many essential cellular processes, could be subject to strong influences by pressure.

were similar to the low-temperature-induced fatty acid changes observed in other bacteria (76), these findings suggested that low temperature and high pressure exerted similar effects on bacterial membranes.

Subsequently, *P. profundum* SS9 was found to adapt its fatty acid content in response to both low temperature and high pressure (4). Thus, when *P. profundum* SS9 was grown at either high pressure (28 to 50 MPa, 15°C) or low temperature (4°C, 0.1 MPa), the amounts of monounsaturated fatty acids (MUFAs) and the omega-3 polyunsaturated fatty acid (PUFA) eicosapentaenoic acid (EPA), 20:5, were increased compared to growth at atmospheric pressure (15°C). However, although both growth conditions led to an increase in MUFAs, high pressure and low temperature led to the greatest increase in 18:1 and 16:1 levels, respectively, suggesting that these conditions exert slightly different effects on *P. profundum* SS9 fatty acids. Since mesophilic bacteria such as *Escherichia coli* adapt their fatty acid composition in response to temperature changes but *not* to pressure changes (2, 66), this suggested that the ability to adapt fatty acids in response to pressure changes may be a unique feature of piezophilic bacteria and could be necessary for optimum growth under these conditions.

MUFAs Are Required for High-Pressure-Adapted Growth

To investigate the relative importance of MUFAs and PUFAs in the psychrophilic and piezophilic growth of *P. profundum* SS9, a number of studies have been carried out either by adding inhibitors of fatty acid biosynthesis or by constructing defined mutants with disruptions in genes involved in either MUFA or PUFA biosynthesis (2, 4, 66). For example, cerulenin is an inhibitor of β-ketoacyl-acyl carrier protein (ACP) synthases I and II (KAS I and II) and therefore preferentially inhibits MUFA biosynthesis (77). Since EPA production is not inhibited by cerulenin (77), this antibiotic provided a useful tool to investigate the importance of MUFAs in *P. profundum* SS9 growth under different conditions (4). Treatment of *P. profundum* SS9 with cerulenin resulted in a pronounced growth inhibition at both high pressure and low temperature compared to untreated control cultures (4). Thus, this provided the first piece of evidence that MUFAs could be important for high-pressure- and low-temperature-adapted growth of *P. profundum* SS9. Additionally, supplementation of the growth medium of the cerulenin-treated cultures with 18:1 noticeably enhanced the growth deficiency of *P. profundum* SS9 at high pressure, and also had a slight stimulatory effect on growth at low temperature (4). Therefore, combined, these data provide further support that an increase in 18:1 levels is important for the growth of *P. profundum* SS9 at high pressure, and they also suggest that it could play some role in growth at low temperature.

The *E. coli* KAS II enzyme catalyzes the elongation of 16:1-ACP to 18:1-ACP and is required for the increase in 18:1 levels observed during the growth of *E. coli* at low temperatures (44, 66). Since *P. profundum* SS9 KAS II is 88% similar and 76% identical to the *E. coli* KAS II protein, it was hypothesized that the *P. profundum* SS9 KAS II protein could also be involved in the cold-induced, and possibly also the high-pressure-induced, increase in 18:1 levels (2). To investigate this, a *P. profundum* SS9 mutant with an insertion in the *fabF* gene, which encodes the KAS II protein, was constructed (2). Consistent with this hypothesis, the KAS II enzyme of *P. profundum* SS9 was required for the increase in 18:1 levels at both low temperature and high pressure, since this increase was not observed in the *fabF* mutant lacking KAS II. Additionally, growth of the *P. profundum* SS9 *fabF* mutant was significantly reduced at high pressure, suggesting that KAS II and the increase in 18:1 levels appear to be required for piezophilic growth. In contrast, like the *E. coli* KAS II protein (45), the *P. profundum* SS9 KAS II protein was not required for growth at low temperatures (2), despite being required for the 18:1 increase under these conditions. Thus, it may be that some additional change in the fatty acid content of the *E. coli* and *P. profundum* SS9 *fabF* mutants at low temperatures is able to compensate for the absence of the 18:1 increase. However, these data suggest that although low temperature and high pressure exert some similar effects on bacterial cells, there are clearly differences by which bacteria perceive and respond to these conditions.

KAS II Regulation Model

Although KAS II is required for the low-temperature and high-pressure-induced increase in 18:1 levels in *P. profundum* SS9, the question remained as to the molecular mechanism of this process. In *E. coli,* the low-temperature-induced increase in 18:1 levels was found to be due to an increase in KAS II activity, rather than an increase in either *fabF* gene transcription or KAS II protein stability (44). Thus, lowering the growth temperature increased the affinity of the *E. coli* KAS II protein for its substrate (16:1-ACP) and also resulted in a higher V_{max} (44). Similarly, it was hypothesized that low temperature, and possibly high pressure, could also affect *P. profundum* SS9 KAS II activity (2). Consistent with such a model, preliminary studies have found that transcription of the *P. profundum* SS9 *fabF* gene is unaltered by growth at either high pressure or low temperature (2). Furthermore, introduction of the SS9 gene into *E. coli* confers pressure-inducible 18:1 production (E. E. Allen and D. H. Bartlett, unpublished results).

Polyunsaturated Fatty Acids

As mentioned above, when *P. profundum* SS9 is grown at either low temperature or high pressure, there is also an increase in the omega-3 PUFA EPA (4). Since omega-3 PUFAs have been found in a number of deep-sea bacteria, it was suggested that these fatty acids could be important for piezophilic growth (34, 39, 75). However, a *P. profundum* SS9 EPA-deficient mutant is unaffected in either high-pressure- or low-temperature-adapted growth (4), suggesting that EPA was not required for growth under these conditions. Since the EPA-deficient mutant produced elevated levels of MUFAs at high pressure and low temperature, the increase in MUFAs may compensate for the absence of EPA in this mutant. Biophysical studies using phosphatidylcholine containing a mixture of 18:0 with either 18:1 or 20:4 fatty acids showed that the mixture with the MUFA was far more effective at lowering the gel-to-liquid-crystalline-phase transition temperature than the mixture containing the PUFA (30). Thus, an increase in MUFA could potentially have a profound effect of membrane fluidity, possibly masking the loss of EPA.

Another possibility is that EPA and related molecules are important only under certain physiological or symbiotic conditions. Valentine and Valentine (95) have suggested that omega-3 PUFAs are used for respiration under conditions of proton-driven bioenergetics. They can also protect cells from oxidative stress (74). Since omega 3-PUFAs are essential components of animal membrane lipids and are also precursors of hormones (6, 59, 87), another possibility is that deep-sea bacteria may produce PUFAs as a food source for deep-sea animals. Intriguingly, *P. profundum* SS9 was originally isolated from a deep-sea amphipod homogenate (32), suggesting that in the deep sea it may form an interaction with amphipods. Consistent with this hypothesis, the genome of *P. profundum* SS9 encodes a potential type IV secretion system (98), which is often present in bacteria that form interactions with eukaryotic hosts.

It should also be stressed that not all piezophiles produce PUFAs, and hence PUFAs are not necessarily essential to piezophily. The recently discovered gram-positive deep-sea piezophiles *Carnobacterium* strains AT7 and AT12 (61) do not produce PUFAs (F. M. Lauro and D. H. Bartlett, unpublished results).

Role of MUFAs

It has been known for many years that the activity of membrane proteins can be affected by altering the fluidity of the membrane (94). Thus, since MUFAs have lower melting temperatures and can be less tightly packed than saturated fatty acids (48), it may be that MUFAs are required to maintain the correct membrane fluidity of *P. profundum* SS9 at high pressure. Additionally, as described earlier, MUFAs may be better able to maintain optimum membrane fluidity over a range of conditions than PUFAs. However, to confirm this, biophysical measurements of membrane fluidity during high-pressure growth and in different fatty acid biosynthetic mutants would be essential. Additionally, it is also possible that membrane proteins require specific MUFAs for their function under different growth conditions. Previous research has determined that certain lipids could be important for the proper functioning of membrane proteins (85).

Role of Surface Polysaccharides

Since *P. profundum* SS9 is a gram-negative bacterium, lipopolysaccharide (LPS) comprises the outer leaflet of its outer membrane. LPS consists of three subunits: O antigen, the core, and lipid A. The lipid A component of LPS is modified with fatty acids. In other bacteria, the LPS fatty acid composition was found to have a dramatic effect on growth under certain environmental conditions (41, 47). Additionally, bacteria have also been found to adapt their LPS composition, including their LPS fatty acids, in response to environmental changes (19, 55). Thus, it seems possible that *P. profundum* SS9 LPS may undergo structural changes during growth at low temperature and/or high pressure, and these changes may be important for optimum growth under these conditions. Consistent with this hypothesis, a number of cold-sensitive mutants of *P. profundum* SS9 with transposon insertions in either putative LPS or capsular polysaccharide (CPS) genes have recently been isolated (62a).

MEMBRANE PROTEINS

ToxR Is a Pressure Sensor

In SS9, ToxR has been revealed to be a major "one-component" signaling system for pressure-regulated gene expression. It was first demonstrated to determine the inverse pressure regulation of the expression of two outer membrane protein genes, the low-pressure-inducible *ompL* gene and the high-pressure-inducible *ompH* gene (102). A subsequent study indicated that it controls the expression of a number of genes involved in membrane structure and starvation adaptation (16). Overproduction of ToxR and its associated protein ToxS compromises SS9 growth ability at high pressure (T. J. Welch and D. H. Bartlett, unpublished results), but the genes under ToxR control which are responsible for this phenotype have yet to be identified.

Because SS9 ToxR is largely homologous to *Vibrio cholerae* ToxR, which has been the subject of more intensive study, much of the hypothesized SS9 ToxR function has been inferred from knowledge gained during *V. cholerae* ToxR research. The name ToxR is in fact a misnomer arising from the fact that in *V. cholerae* and in some other pathogenic vibrios ToxR has been co-opted to direct the expression of laterally transferred toxin genes and other virulence factors (29, 99). More recent *V. cholerae* transcriptome studies have demonstrated that it controls the expression of more than 150 genes, mostly genes not directly related to virulence, including many hypothetical and conserved hypothetical genes and genes whose products influence envelope structure, transport, motility and chemotaxis, and energy metabolism (18). Seventy-nine percent of these genes are also present in SS9. Although *toxR*-related genes are found outside the family, when considering overall identities above 40%, *toxR* and its companion gene *toxS* are two of only a handful of distinguishing ancestral genes for members of the *Vibrionaceae*. It would appear most likely that a primary function for ToxR is to homeostatically control membrane structure and energy flow under diverse environmental conditions.

The *V. cholerae* transcription factor ToxR is anchored within the cytoplasmic membrane via a single 16- to 19-amino-acid (aa) transmembrane segment with its 180-aa amino-terminal cytoplasmic region containing an OmpR-type winged helix-turn-helix domain for DNA binding and its 100-aa carboxy-terminal region extending into the periplasmic space, where it influences dimerization (50, 51). *V. cholerae* ToxR forms both homodimers and heterodimers with ToxS. ToxR dimerization appears to be required for full virulence function. Among the genes activated by *V. cholerae* ToxR, the *ompU* gene (a homolog of the SS9 *ompL* gene) confers resistance to membrane-perturbing bile salts and antimicrobial peptides produced in the intestines of many organisms (68, 80).

A common feature with both the *V. cholerae* and SS9 ToxR proteins is that they are responsive to chemical treatments which influence membrane structure. However, only SS9 ToxR appears to sense changes in hydrostatic pressure. This could reflect changes in the more divergent periplasmic protein domains of the two proteins or perhaps result from differences in the membrane environments of the two organisms. Considering the compressibility of lipids and the sensitivity of the quaternary structure of many proteins to pressure, the presumed location and multimeric structure of SS9 ToxR would seem to lend themselves well to piezometer function (84, 100). The pressure responsiveness of SS9 ToxR can be titrated back to the low-pressure state, even while the cells are at high pressure, with the addition of increasing amounts of either of the membrane-fluidizing anesthetics procaine and phenethyl alcohol at concentrations which are known to produce measurable reductions in lipid chain order (102). These results do more than only suggest a connection between membrane structure and ToxR function in both *V. cholerae* and SS9: they also provide another example of the opposing effects of anesthetics and pressure. This is well documented in high-pressure medicine (42). In this case high pressure brought about by deep diving increases the excitability of the vertebrate central nervous system through changes in both synaptic and intrinsic membrane properties (72). One dramatic example of the inverse effects of pressure and anesthetics on this phenomenon is that elevated pressure has the remarkable property of being able to reawaken anesthetized tadpoles (54).

The abundance as well as activity of SS9 ToxR decreases with increasing pressure. This was demonstrated by monitoring the abundance of the ToxR protein in wild-type cells and in cells engineered to overexpress the *toxRS* operon (102). The gene encoding heat shock protein 90, designated *htpG,* is immediately upstream of the SS9 *toxRS* operon and oriented in the opposite direction. This could be significant if *htpG* induction occurred at high pressure, and if the RNA polymerase directing *htpG* transcription sterically prevented RNA polymerase access to the *toxRS* promoter. Such a model is analogous to one previously proposed for temperature regulation of *V. cholerae toxR* expression. However, more recent SS9 transcriptomic data (see below) indicate that *htpG* transcription, like that of many other heat shock genes, is higher at atmospheric pressure than at high pressure (i.e., 28 MPa). A more plausible scenario is that SS9 ToxR regulates its own expression, as previously documented for *Vibrio anguillarum* ToxR (99).

RseC

As indicated above, *P. profundum* SS9 regulates the amount of the outer membrane protein OmpH in response to changes in pressure (102). Thus, at low pressure, OmpH levels are low, whereas at high pressure, the amount of OmpH is substantially increased. ToxR is not the only regulatory protein which influences OmpH abundance. Mini-Mu transposon insertions that reduced OmpH abundance at atmospheric pressure were isolated and characterized (27). Of the eight mutants that were isolated, six of these had insertions in an *rpoE*-like locus. The *rpoE* operon of *E. coli* has been extensively investigated. *rpoE* encodes an alternative RNA polymerase sigma factor, sigma E, involved in altering gene expression in response to elevated temperatures (38). Since *rpoE* loci in a number of different bacteria deal with extracytoplasmic functions, they are known also as ECF sigma factors (71). In general, ECF sigma factors respond to some type of extracytoplasmic stimuli and are involved in regulating the expression of genes whose products result in extracytoplasmic changes (71). The *rpoE* operon is comprised of *rpoE,* two downstream genes whose products modulate RpoE activity in response to environmental cues, and the *rseC* gene, whose product has a negligible effect on RpoE activity. RseC and related proteins are believed to function as inner transmembrane proteins which anchor protein complexes to the membrane for the transfer of electrons to key enzymes, such as those involved in thiamine biosynthesis (13). RseC also influences the reduction of the reactive oxygen species sensor SoxR (56). Mutations in SS9 which decrease *rseC* expression result in both high-pressure-sensitive and cold-sensitive growth phenotypes, perhaps because of low-temperature and high-pressure effects on membrane-based electron transport. Inactivation of either of the anti-sigma factors RseA and RseB, or of RseC, leads to a decrease in *ompH* transcript levels. However, psychrophilic and piezophilic growth appears to be influenced only by RseC, independent of any effects on RpoE activity.

NUCLEIC ACID FUNCTION AT HIGH PRESSURE

Piezoadaptation is not restricted to the membrane. For example, as described in the following section, recent evidence in *P. profundum* and in other bacteria suggests that rRNA structure is modified for growth ability at high pressure (61). Another component of high-pressure adaptation which is related to nucleic acid-protein interactions concerns DNA recombination in general and the *recD* gene in particular. *recD* was perhaps the most unexpected gene discovered to be critical for growth at high pressure. Its product is a component of the RecBCD helicase/nuclease. This enzyme processes DNA flanking double-stranded-DNA breaks, which can occur under a variety of physiological and environmental conditions, for repair by DNA recombination (35, 57). And while DNA replication is one of the most pressure-sensitive essential cell processes yet uncovered in *Escherichia coli* (104), and DNA supercoiling is also subject to modification at elevated pressure (28, 93), DNA recombination was one aspect of DNA metabolism which had not previously been implicated in any pressure effects on microbes prior to the SS9 *recD* findings.

Interest in the SS9 *recD* gene began when a pressure-sensitive SS9 acridine mutagen-derived chemical mutant, designated strain EC1002, was found to harbor a stop codon within the *recD* gene (15). Piezophilic growth was restored by the reintroduction of the wild-type *recD* gene. EC1002 also exhibited an altered cell morphology, being transformed from rod-shaped cells to larger, more irregularly shaped cells upon a shift to high pressure. This phenotype too was reversed by the introduction of a wild-type *recD* gene into the mutant on integrating or autonomously replicating plasmids. Construction of SS9 strains possessing insertion mutations within *recD* further established the role of this gene in growth and cell division at high pressure, although the extent of debilitation at high pressure depended on the amount of *recD* remaining 5' to the insertion site.

The basis of pressure sensitivity in *recD* mutants could be linked to an increased frequency of DNA recombination. *E. coli recD* mutants are hyperrecombinogenic owing to decreased RecBCD nuclease activity (5, 17, 25). As a result, plasmids in *recD* mutants exhibit excessive intermolecular recombination, leading to unequal partitioning among daughter cells and eventual plasmid loss. Indeed, plasmid stability in strain EC1002 is dramatically decreased relative to that in the parental strain (15). SS9 strains harbor an endogenous plasmid (23, 98), but it is not essential and its loss (at least in laboratory culture) appears to be inconsequential. However, if the defect in DNA partitioning of SS9 *recD* mutants extends to one of its two chromosomes, then this could be reflected in both the growth and cell division defects, since chromosome partitioning is a prerequisite for septation (63). While we find this model attractive, no evidence linking RecD function to interchromosomal recombination has yet been obtained in any bacterium. However, it should be noted that SS9, like all other members of the *Vibrionaceae* examined, has two chromosomes, one of which has replication properties in common with certain plasmids (37).

Other possibilities exist. For example, cell growth and division could be compromised in the SS9 *recD* mutant as a result of DNA damage or an SOS response (43). One of the consequences of inducing this stress regulon is the inhibition of cell division. The *recD* gene has been found to be essential for growth at low temperature in the Antarctic bacterium *Pseudomonas syringae* Lz4W (83). *P. syringae recD* mutants produce increased levels of DNA fragments at low temperature compared to *recD*⁺ cells, which could lead to an SOS response. Also consistent with this model is the fact that *E. coli recD* mutants experience particularly high mutation rates in stationary phase (22).

Another clue to SS9 *recD* function is the effect this gene has on *E. coli* cells. Surprisingly, an *E. coli recD* mutant harboring a plasmid containing SS9 *recD* gains the ability to divide normally at high pressure under conditions in which the mutant containing the vector alone forms long filaments (15). The latter phenotype is characteristic of other *E. coli* strains incubated at elevated pressure (86, 105). The *E. coli* cells do not become piezophilic but do exhibit piezoadapted cell division. These results are significant for two reasons. First, they provide further evidence for the direct or indirect role of SS9 RecD in cell division at high pressure, and second, they suggest that *E. coli* filamentation at high pressure is not the result of an inherent defect in the cell division apparatus, but rather that it is the consequence of the presence or absence of a regulatory signal that precedes septation. This signal is now known to be DNA damage. Aertsen and Michiels have obtained compelling evidence for the high-pressure induction of an SOS response in *E. coli* at elevated pressure, and further that it is the result of the formation of double-stranded-DNA breaks whose repair and SOS signal generation require the RecB component of the RecBCD exonuclease (see chapter 5).

Additional work is needed to clarify the role of SS9 RecD in piezophilic growth. Is the pressure-sensitive phenotype of an SS9 *recD* mutant due to a defect in chromosome partitioning or because of increased DNA damage? The partitioning of both chromosomes should be examined in *recD*⁺ and *recD* mutant strains. And while SS9 transcriptome studies suggest greater DNA damage at low pressure than at high pressure (see below), it would nevertheless be useful to directly measure DNA damage and mutation rates in SS9 parental and *recD* mutants at low and high pressure. Analysis of an SS9 double mutant containing a defective *recD* gene and a deletion in a second gene required for SOS signal generation *(recA)* or for the SOS inhibition of cell division *(sulA)* should also be performed to assess the possible involvement of the SOS response. Finally, it is noteworthy that SS9 *recD* mutants readily give rise to suppressor mutants after 60 h of incubation at high pressure. Curiously, *P. syringae* Lz4W *recD* mutants readily give rise to extragenic cold-adapted mutants as well (83). The genetic alterations underlying some of these SS9 suppressor mutants should be identified, and any RecD interactions outside of its exonuclease subunit contacts should be clarified.

Another issue that arises is the apparent connection between SS9 RecD function and *ompH* gene expression. The pressure-sensitive *recD* mutant (strain EC1002) was originally obtained as a chemical mutant altered in β-galactosidase expression in an *ompH::lacZ* transcriptional fusion strain (27). This is consistent with the observation that some chromosome partitioning mutants have altered membrane properties (101). One pertinent example of this is that *ompF* expression is considerably reduced in an *E. coli mukA* mutant (8). However, the wild-type *recD* gene does not restore β-galactosidase expression to wild-type levels in EC1002, and so it appears that this strain harbors two unlinked mutations, the characterized *recD* mutation and a second, unknown genetic alteration affecting *ompH* gene expression. The isolation of both mutations in one strain was most likely simply a fortunate coincidence.

GENOMICS

Advances in genomic approaches have dramatically altered the understanding of the evolution, diversity, biochemistry, and physiology of life. These technologies have also been applied to two piezophiles: *P. profundum* strain SS9 (98) and *Shewanella violaceae* strain DSS12 (described in the preceding chapter), and additional sequencing projects are in the works (60). The genome of SS9 is divided into three replicons: a 4.1-Mbp chromosome, a 2.2-Mbp chromosome, and an 80-kbp circular plasmid. Its genome size is larger than in all other members of the *Vibrionaceae* and includes many transposable elements, as many rRNA operons as any microbe yet characterized, plus many unique genes, particularly on chromosome 2. Many of the 16S rRNA copies contain certain elongated helices, a feature which correlates well with the extent of adaptation to growth at elevated pressure in many bacteria (61). These changes are presumed to facilitate ribosome function at high pressure. In addition to rRNA operons, SS9 also contains a large number of other types of gene duplications. This includes extra sets of genes for cytochromes, the F_0F_1 ATP synthase, and motility, all of which are known to be pressure sensitive in mesophilic bacteria (65, 69, 92). It is possible that SS9 contains multiple sets of certain genes to adapt to both shallow-water and deep-sea pressure conditions. SS9 and other piezophiles also tend to have large intergenic distances, a property which suggests the presence of extensive regulatory networks.

An SS9 microarray was used to monitor gene expression under different conditions of pressure (0.1, 28, and 45 MPa) and temperature (4 and 16°C) (23; Color Plate 3). Among the 260 differentially expressed genes identified, many genes for amino acid or ion transport or protein folding were turned on at atmospheric pressure, whereas genes for complex polymer degradation were up-regulated at high pressure. Surprisingly, genes for a Stickland amino acid fermentation pathway, previously found only in certain strictly anaerobic bacteria (46), were found to be induced at elevated pressure as well. Thus, transcriptional profiling indicated specific, large-scale, and complex changes in gene expression and physiology as a function of pressure, changes that were dramatically different from the stress response of mesophilic bacteria to similar pressures (52, 103).

HORIZONTAL GENE TRANSFER

While horizontal gene transfer (HGT) was once thought to play a minor role in microbial evolution, current estimates of HGT-acquired DNA range from 0.5% for an obligate intracellular symbiont up to 25% for the methanogen *Methanosarcina acetivorans,* with 14% being the average among 116 sequenced prokaryotic genomes examined (73). There are many examples of HGT involving members of the *Vibrionaceae* (24, 40, 64, 78, 81, 97). Members of this family can take up DNA by transduction, conjugation, or transformation (10, 70). It may be noteworthy when considering HGT in deep-sea microbes that extracellular DNA can be present at high concentrations in deep-sea sediments (31). The value of extracellular DNA as a source of genetic or nutritional material has been recently reviewed (10).

HGT was examined in SS9 using several methods (23). Microarray hybridization was used to compare the genome content of *P. profundum* strain SS9 with that of another piezophilic *P. profundum* isolate (strain DSJ4 from the Ryuku Trench) and the piezosensitive *P. profundum* strain 3TCK (obtained from San Diego Bay). Genes present in SS9 but of variable distribution in the other *P. profundum* strains were further examined for a possible HGT origin in SS9. This included examining their GC content, tetranucleotide composition, and codon bias relative to that of the average gene at the codon site 3 position.

From these analyses it was found that large clusters of HGT DNA (5 to 145 kbp) are located within 28 regions, representing more than 13% of the SS9 genome. A greater percentage of the HGT DNA is present on chromosome 2 than on chromosome 1, a bias also present in the genomes of other *Vibrionaceae* (49, 97). Much of the HGT DNA within SS9 appears to consist of phage, integrated or freely replicating plasmids, transposases, and flagellum and pilus genes. A complete collection of genes for assembling what appears to be a lateral flagellum is specific only to the two piezophilic *P. profundum* strains. This is particularly intriguing considering the exquisite pressure sensitivity of flagellum assembly and rotation from mesophilic bacteria (69).

Also noteworthy is that much of the DNA acquired by HGT encodes functions associated with the cell envelope. This includes possible LPS O-antigen biosynthesis genes, a number of sets of phosphotransferase system genes, iron and oligopeptide transport, and a phage-associated tryptophan transport and utilization system. As noted above, the membrane is highly compressible, and because of this, the observation that many membrane components respond to pressure is consistent with their localization. Also significant to the above results is that tryptophan utilization has previously been implicated in limiting mesophilic microbial growth at elevated pressure (1).

When combined with the transcriptome results, only six genes are found to be present exclusively in the piezophilic *P. profundum* strains and transcriptionally activated at high pressure. These correspond to five transport proteins and the hypothetical sensor protein TorS. In *Escherichia coli* TorS responds to trimethylamine-*N*-oxide (TMAO) levels to regulate TMAO reductase activity in anticipation of alkaline stress (20). Since no TMAO was added to the SS9 cultures employed for transcriptome profiling, it may be that the SS9 sensory protein is responding to an as-yet-unidentified signal which accumulates at high pressure.

The results of these analyses provide a hint that HGT may have contributed to SS9 piezophily or other aspects of its adaptations to depth. *P. profundum* piezophile strain-specific genes such as those for motility, LPS O antigen, phosphotransferase system components, iron and oligopeptide transport, and the phage-associated tryptophan transport and utilization system should all be considered in this context. Genetic manipulations are needed to establish which genes actually influence growth ability at high pressure. Additional important issues are which piezophile-specific genes were acquired by HGT and which were lost by the mesophilic strain 3TCK, which genes 3TCK might have needed to acquire for adaptation to shallow-water and atmospheric pressure, and whether the last common ancestor to the *P. profundum* strains was piezophilic or mesophilic. It will also be important to establish how much of a role HGT has played in high-pressure adaptation versus modification of preexisting genes.

OVERALL PROTEIN ADAPTATION

A central issue to address is the general complexity of piezophilic adaptation and the nature of its accompanying changes. Some evidence suggests that a few genes are critical for growth at high pressure; however, there is also evidence supporting the idea that global changes in protein structure are also required.

A useful piece of information comes from the microarray analysis of gene expression in SS9, using cells grown under different pressure conditions (98). It was shown that genes involved in protein folding and response to stress conditions, including *htpG, dnaK, dnaJ,* and *groEL,* increase their expression levels at atmospheric pressure, indicating that the proteins of this piezophilic bacterium are optimized for high pressure and require the help of these chaperones to fold correctly at atmospheric pressure. Interestingly, in *E. coli,* the corresponding orthologous genes are overexpressed at high pressure (52). These data support the hypotheses that the pressures employed have an influence on protein structure and that adaptation to high pressure requires the optimization of many proteins.

The next question, then, is whether there are classes of proteins that may be more affected by high pressure than other classes. A partial answer to this question came from *E. coli,* for which it was shown that cell motility, cell division, and many substrate transport processes are among the processes most sensitive to high pressure (53, 58, 69, 79, 105). Similarly, the same processes are affected when SS9 is grown at low pressure (11; E. A. Eloe, F. M. Lauro, V. F. Vogel, and D. H. Bartlett, unpublished results). It is interesting that these processes are all involved in the movement of large macromolecules. An indirect confirmation that some of the proteins involved in these processes are not fully functional when operating at a nonoptimal pressure came once again from microarray data. It was shown that amino acid transporters are generally overexpressed at low pressure in SS9 (98). A simple explanation could be that the overexpression is used to compensate for the reduced activity of these proteins, but other explanations are also possible.

It is plausible that most structural adaptations of proteins to high pressure do not follow general rules, but are rather specific to the peculiarity of each particular protein. However, some general trends of amino acid substitution have been observed in other extremophiles, such as thermophiles (89, 91). To address this issue with regard to piezophily, a comparative analysis of amino acid composition of SS9 versus *Vibrio vulnificus* and *Vibrio parahaemolyticus* was performed (N. Vitulo, A. Vezzi, S. Campanaro, and G. Valle, unpublished data). These three free-living marine bacteria are phylogenetically related and share similar GC contents, but only SS9 is a piezophile, allowing the detection of possible differences due to high-pressure adaptation. Although a statistical analysis of the overall amino acid composition did not reveal any significant bias, the analysis of the substitutions in aligned positions of orthologous genes disclosed several highly significantly unbalanced substitutions, namely, Ala, Glu, and Val decreasing in SS9 and Cys, Asp, Ile, Asn, Ser, and Thr increasing, compared to both *V. vulnificus* and *V. parahaemolyticus*.

Two hypotheses were considered to explain the selective pressure responsible for driving these amino acid substitutions in SS9: (i) high pressure may change the metabolic cost of some amino acids, or (ii) some amino acids may be more suitable to optimize structural adaptation to high pressure. To distinguish between these hypotheses, three subsets of ortholog proteins were selected: (i) proteins with a high codon adaptation index (CAI); (ii) proteins belonging to functional classes particularly susceptible to high pressure, such as transporters and flagella; and (iii) control proteins. The rationale was that highly expressed proteins (CAI-high subset) should be the most susceptible to amino acid cost, while the other subset should be the most receptive to adopt amino acids that are structurally suitable to high pressure. Only the CAI-high subset showed an unbalanced rate of substitution of amino acids (with Ala and Glu decreasing and Ser increasing). Thus, the data favor the view that the SS9 proteome amino acid differences with related bacteria reflect metabolic adjustment rather than global adaptations to protein function. Many genes may require modification for the evolution of piezophily, but no generalizable pattern of amino acid substitutions is apparent.

FUTURE PERSPECTIVES

The large-scale isolation of pressure-sensitive mutants, whole-genome sequencing and comparative genomics of additional strains, and application of additional postgenomic approaches (i.e., proteomics and metabolomics) hold great promise for future insights into pressure adaptation in *Photobacterium profundum*. In addition, the large platform of information and technology available also makes strain SS9 a useful system for the introduction of other scientific approaches, such as biophysical studies of its membrane properties, structure-function examination of ToxR, or suppressor mutant mapping to further assess the functions of piezoadaptation genes such as *fabF, rseC,* and *recD*.

REFERENCES

1. **Abe, F., and K. Horikoshi.** 2000. Tryptophan permease gene *TAT2* confers high-pressure growth in *Saccharomyces cerevisiae. Mol. Cell. Biol.* **20:**8098–8102.
2. **Allen, E. E., and D. H. Bartlett.** 2000. FabF is required for piezoregulation of *cis*-vaccenic acid levels and piezophilic growth of the deep-sea bacterium *Photobacterium profundum* strain SS9. *J. Bacteriol.* **182:**1264–1271.
3. **Allen, E. E., and D. H. Bartlett.** 2001. Structure and regulation of the omega-3 polyunsaturated fatty acid synthase from the deep-sea bacterium *Photobacterium profundum* strain SS9. *Microbiology* **148:**1903–1913.
4. **Allen, E. E., D. Facciotti, and D. H. Bartlett.** 1999. Monounsaturated but not polyunsaturated fatty acids are required for growth at high pressure and low temperature in the deep-sea bacterium *Photobacterium profundum* strain SS9. *Appl. Environ. Microbiol.* **65:**1710–1720.

5. **Amundsen, S. K., A. F. Taylor, A. M. Chaudhury, and G. R. Smith.** 1986. *recD:* the gene for an essential third subunit of exonuclease V. *Proc. Natl. Acad. Sci. USA* **83:**5558–5582.

6. **Angerer, P., and C. von Schacky.** 2000. n-3 polyunsaturated fatty acids and the cardiovascular system. *Curr. Opin. Clin. Nutr. Metab. Care* **3:**439–445.

7. **Attard, G. S., R. H. Templer, W. S. Smith, A. N. Hunt, and S. Jackowski.** 2000. Modulation of CTP:phosphocholine cytidylyltransferase by membrane curvature elastic stress. *Proc. Natl. Acad. Sci. USA* **97:**9032–9036.

8. **Bahloul, A., J. Meury, R. Kern, J. Garwood, S. Guha, and M. Kohiyama.** 1996. Coordination between membrane *oriC* sequestration factors and a chromosome partitioning protein, TolC (MukA). *Mol. Microbiol.* **22:**275–282.

9. **Bartlett, D., M. Wright, A. Yayanos, and M. Silverman.** 1989. Isolation of a gene regulated by hydrostatic pressure. *Nature* **342:**572–574.

10. **Bartlett, D. H., and F. Azam.** 2005. Chitin, cholera, and competence. *Science* **310:**1775–1777.

11. **Bartlett, D. H., and E. Chi.** 1994. Genetic characterization of *ompH* mutants in the deep-sea bacterium *Photobacterium* species strain SS9. *Arch. Microbiol.* **162:**323–328.

12. **Bartlett, D. H., and T. J. Welch.** 1995. *ompH* gene expression is regulated by multiple environmental cues in addition to high pressure in the deep-sea bacterium *Photobacterium* species strain SS9. *J. Bacteriol.* **177:**1008–1016.

13. **Beck, B. J., L. E. Connolly, A. De Las Peñas, and D. M. Downs.** 1997. Evidence that *rseC*, a gene in the *rpoE* cluster, has a role in thiamine synthesis in *Salmonella typhimurium. J. Bacteriol.* **179:**6504–6508.

14. **Biddle, J. F., C. H. House, and J. E. Brenchley.** 2004. Enrichment and isolation of psychrophilic microorganisms from sediment collected at ocean drilling program site 1230, p. 105. *Abstr. 104th Gen. Meet. Am. Soc. Microbiol.*

15. **Bidle, K. A., and D. H. Bartlett.** 1999. RecD function is required for high-pressure growth in a deep-sea bacterium. *J. Bacteriol.* **181:**2330–2337.

16. **Bidle, K. A., and D. H. Bartlett.** 2001. RNA arbitrarily primed PCR survey of genes regulated by ToxR and ToxS in the deep-sea bacterium *Photobacterium profundum* strain SS9. *J. Bacteriol.* **183:**1688–1693.

17. **Biek, D. P., and S. N. Cohen.** 1986. Identification and characterization of *recD*, a gene affecting plasmid maintenance and recombination in *Escherichia coli. J. Bacteriol.* **167:**594–603.

18. **Bina, J., J. Zhu, M. Dziejman, S. Faruque, S. Calderwood, and J. J. Mekalanos.** 2003. ToxR regulon of *Vibrio cholerae* and its expression in vibrios shed by cholera patients. *Proc. Natl. Acad. Sci. USA* **100:**2801–2806.

19. **Bishop, R. E., H. S. Gibbons, T. Guina, M. S. Trent, S. I. Miller, and C. R. Raetz.** 2000. Transfer of palmitate from phospholipids to lipid A in outer membranes of gram-negative bacteria. *EMBO J.* **19:**5071–5080.

20. **Bordi, C., L. Theraulaz, V. Mejean, and C. Jourlin-Castelli.** 2003. Anticipating an alkaline stress through the Tor phosphorelay system in Escherichia coli. *Mol. Microbiol.* **48:**211–223.

21. **Bowman, J. P., S. A. McCammon, J. A. Gibson, L. Robertson, and P. D. Nichols.** 2003. Prokaryotic metabolic activity and community structure in Antarctic continental shelf sediments. *Appl. Environ. Microbiol.* **69:**2448–2462.

22. **Bull, H. J., G. J. McKenzie, P. J. Hastings, and S. M. Rosenberg.** 2000. Evidence that stationary-phase hypermutation in the *Escherichia coli* chromosome is promoted by recombination. *Genetics* **154:**1427–1437.

23. **Campanaro, S., A. Vezzi, N. Vitulo, F. M. Lauro, M. D'Angeo, F. Simonato, A. Cestaro, G. Malacrida, G. Bertoloni, G. Valle, and D. H. Bartlett.** 2005. Laterally transferred elements and high pressure adaptation in *Photobacterium profundum* strains. *BMC Genomics* **6:**122.

24. **Charbit, A., and N. Autret.** 1998. Horizontal transfer of chromosomal DNA between the marine bacterium *Vibrio furnissii* and *Escherichia coli* revealed by sequence analysis. *Microb. Comp. Genomics* **3:**119–132.

25. **Chaudhury, A. M., and G. R. Smith.** 1984. A new class of *Escherichia coli recBC* mutants: implications for the role of RecBC enzyme in homologous recombination. *Proc. Natl. Acad. Sci. USA* **81:**7850–7854.

26. **Chi, E., and D. H. Bartlett.** 1995. An *rpoE*-like locus controls outer membrane protein synthesis and growth at cold temperatures and high pressures in the deep-sea bacterium *Photobacterium* SS9. *Mol. Microbiol.* **17:**713–726.

27. **Chi, E., and D. H. Bartlett.** 1993. Use of a reporter gene to follow high-pressure signal transduction in the deep-sea bacterium *Photobacterium* sp. strain SS9. *J. Bacteriol.* **175:**7533–7540.

28. **Chilukuri, L. N., P. A. G. Fortes, and D. H. Bartlett.** 1995. High pressure modulation of DNA gyrase activity. *Biochem. Biophys. Res. Commun.* **239:**552–556.

29. **Conejero, M. J., and C. T. Hedreyda.** 2003. Isolation of partial *toxR* gene of *Vibrio harveyi* and design of *toxR*-targeted PCR primers for species detection. *J. Appl. Microbiol.* **95:**602–611.

30. **Coolbear, K. P., C. B. Berde, and K. M. Keough.** 1983. Gel to liquid-crystalline phase transitions of aqueous dispersions of polyunsaturated mixed-acid phosphatidylcholines. *Biochemistry* **22:**1466–1473.

31. **Dell'Anno, A., and R. Danovaro.** 2005. Extracellular DNA plays a key role in deep-sea ecosystem functioning. *Science* **309:**2179.

32. **DeLong, E. F.** 1986. Adaptations of deep-sea bacteria to the abyssal environment. Ph.D. dissertation. University of California, San Diego.

33. **DeLong, E. F., and A. A. Yayanos.** 1985. Adaptation of the membrane lipids of a deep-sea bacterium to changes in hydrostatic pressure. *Science* **228:**1101–1103.

34. **DeLong, E. F., and A. A. Yayanos.** 1986. Biochemical function and ecological significance of novel bacterial lipids in deep-sea prokaryotes. *Appl. Environ. Microbiol.* **51:**730–737.

35. **Dillingham, M. S., M. Spies, and S. C. Kowalczykowski.** 2003. RecBCD enzyme is a bipolar DNA helicase. *Nature* **423:**893–897.

36. **Dolar, M. L. L., W. A. Walker, G. L. Kooyman, and W. F. Perrin.** 2003. Comparative feeding ecology of spinner dolphins *(Stenella longirostris)* and Fraser's dolphins *(Lagenodelphis hosei)* in the Sulu Sea. *Mar. Mamm. Sci.* **19:**1–19.

37. **Egan, E. S., M. A. Fogel, and M. K. Waldor.** 2005. Divided genomes: negotiating the cell cycle in prokaryotes with multiple chromosomes. *Mol. Microbiol.* **56:**1129–1138.

38. **Erickson, J. W., and C. A. Gross.** 1989. Identification of the sigma E subunit of Escherichia coli RNA polymerase: a second alternate sigma factor involved in high-temperature gene expression. *Genes Dev.* **3:**1462–1471.

39. **Fang, J., O. Chan, C. Kato, T. Sato, T. Peeples, and K. Niggemeyer.** 2003. Phospholipid FA of piezophilic bacteria from the deep sea. *Lipids* **38:**885–887.

40. **Faruque, S. M., D. A. Sack, R. B. Sack, R. R. Colwell, Y. Takeda, and G. B. Nair.** 2003. Emergence and evolution of *Vibrio cholerae* O139. *Proc. Natl. Acad. Sci. USA* **100:**1304–1309.

41. **Ferguson, G. P., A. Datta, R. W. Carlson, and G. C. Walker.** 2005. Importance of unusually modified lipid A in Sinorhizobium stress resistance and legume symbiosis. *Mol. Microbiol.* **56:**68–80.

42. **Finch, E. D., and L. A. Kiesow.** 1979. Pressure, anesthetics, and membrane structure: a spin-probe study. *Undersea Biomed. Res.* **6:**41–53.

43. **Friedberg, E. C., G. C. Walker, and W. Siede.** 1995. *DNA Repair and Mutagenesis.* ASM Press, Washington, DC.

44. **Garwin, J. L., and J. E. Cronan, Jr.** 1980. Thermal modulation of fatty acid synthesis in *Escherichia coli* does not involve de novo enzyme synthesis. *J. Bacteriol.* **141:**1457–1459.

45. **Gelmann, E. P., and J. E. Cronan, Jr.** 1972. Mutant of *Escherichia coli* deficient in the synthesis of *cis*-vaccenic acid. *J. Bacteriol.* **112:**381–387.

46. **Graentzdoerffer, A., A. Pich, and J. R. Andreesen.** 2001. Molecular analysis of the grd operon coding for genes of the glycine reductase and of the thioredoxin system from *Clostridium sticklandii.* *Arch. Microbiol.* **175:**8–18.

47. **Guo, L., K. B. Lim, C. M. Poduje, M. Daniel, J. S. Gunn, M. Hackett, and S. I. Miller.** 1998. Lipid A acylation and bacterial resistance against vertebrate antimicrobial peptides. *Cell* **95:**189–198.

48. **Hazel, J. R., and E. E. Williams.** 1990. The role of alterations in membrane lipid composition in enabling physiological adaptation of organisms to their physical environment. *Prog. Lipid Res.* **29:**167–227.

49. **Heidelberg, J. F., J. A. Eisen, W. C. Nelson, R. A. Clayton, M. L. Gwinn, R. J. Dodson, D. H. Haft, E. K. Hickey, J. D. Peterson, L. Umayam, S. R. Gill, K. E. Nelson, T. D. Read, H. Tettelin, D. Richardson, M. D. Ermolaeva, J. Vamathevan, S. Bass, H. Qin, I. Dragoi, P. Sellers, L. Mcdonald, T. Utterback, R. D. Fleishmann, W. C. Nierman, O. White, S. L. Salzberg, H. O. Smith, R. R. Colwell, J. J. Mekalanos, J. C. Venter, and C. M. Fraser.** 2000. DNA sequence of both chromosomes of the cholera pathogen *Vibrio cholerae. Nature* **406:**477–483.

50. **Hennecke, F., A. Muller, R. Meister, A. Strelow, and S. Behrens.** 2005. A ToxR-based two-hybrid system for the detection of periplasmic and cytoplasmic protein-protein interactions in *Escherichia coli*: minimal requirements for specific DNA binding and transcriptional activation. *Protein Eng. Des. Sel.* **18:**477–486.

51. **Hung, D. T., and J. J. Mekalanos.** 2005. Bile acids induce cholera toxin expression in *Vibrio cholerae* in a ToxT-independent manner. *Proc. Natl. Acad. Sci. USA* **102:**3028–3033.

52. **Ishii, A., T. Oshima, T. Sato, K. Nakasone, H. Mori, and C. Kato.** 2005. Analysis of hydrostatic pressure effects on transcription in *Escherichia coli* by DNA microarray procedure. *Extremophiles* **9:**65–73.

53. **Ishii, A., T. Sato, M. Wachi, K. Nagai, and C. Kato.** 2004. Effects of high hydrostatic pressure on bacterial cytoskeleton FtsZ polymers *in vivo* and *in vitro*. *Microbiology* **150:**1965–1972.

54. **Johnson, F. H., and E. A. Flagler.** 1950. Hydrostatic pressure reversal of narcosis in tadpoles. *Science* **112:**91–92.

55. **Kannenberg, E. L., and R. W. Carlson.** 2001. Lipid A and O-chain modifications cause Rhizobium lipopolysaccharides to become hydrophobic during bacteroid development. *Mol. Microbiol.* **39:**379–391.

56. **Koo, M. S., J. H. Lee, S. Y. Rah, W. S. Yeo, J. W. Lee, K. L. Lee, Y. S. Koh, S. O. Kang, and J. H. Roe.** 2003. A reducing system of the superoxide sensor SoxR in *Escherichia coli*. *EMBO J.* **22:**2614–2622.

57. **Kowalczykowski, S. C.** 2000. Initiation of genetic recombination and recombination-dependent replication. *Trends Biochem. Sci.* **26:**156–165.

58. **Landau, J. V.** 1967. Induction, transcription, and translation in *Escherichia coli:* a hydrostatic pressure study. *Biochem. Biophys. Acta* **149:**506–512.

59. **Lauritzen, L., H. S. Hansen, M. H. Jorgensen, and K. F. Michaelsen.** 2001. The essentiality of long chain n-3 fatty acids in relation to development and function of the brain and retina. *Prog. Lipid Res.* **40:**1–94.

60. **Lauro, F. M., and D. H. Bartlett.** 17 January 2007. Prokaryotic lifestyles in deep-sea habitats. *Extremophiles* doi:10.1007/s00792-006-0059-5.

61. **Lauro, F. M., R. A. Chastain, L. E. Blankenship, A. A. Yayanos, and D. H. Bartlett.** 8 December 2006. The unique 16S rRNA genes of piezophiles reflect both phylogeny and adaptation. *Appl. Environ. Microbiol.* doi:10.1128/AEM.01726-06. (Subsequently published, *Appl. Environ. Microbiol.* **73:**838–845.)

62. **Lauro, F. M., E. A. Eloe, and D. H. Bartlett.** 2005. Conjugal vectors for cloning, expression, and insertional mutagenesis in Gram-negative bacteria. *BioTechniques* **38:**708–710.

62a. **Lauro, F. M., K. Tran, A. Vezzi, N. Vitulo, G. Valle, and D. H. Bartlett.** Large-scale transposon mutagenesis of *Photobacterium profundum* SS9 reveals new genetic loci important for growth at low temperature and high pressure. *J. Bacteriol.*, in press.

63. **Leonard, T. A., J. Moller-Jensen, and J. Lowe.** 2005. Towards understanding the molecular basis of bacterial DNA segregation. *Philos. Trans. R. Soc. Lond. B* **360:**523–535.

64. **Li, M. R., T. Shimada, J. G. Morris, A. Sulakvelidze, and S. Sozhamannan.** 2002. Evidence for the emergence of non-O1 and non-O139 *Vibrio cholerae* strains with pathogenic potential by exchange of O-antigen biosynthesis regions. *Infect. Immun.* **70:**2441–2453.

65. **Marquis, R. E., and G. R. Bender.** 1987. *Barophysiology of Prokaryotes and Proton-Translocating ATPases*. Academic Press, London, United Kingdom.

66. **Marr, A. G., and J. L. Ingraham.** 1962. Effect of temperature on the composition of fatty acids in *Escherichia coli*. *J. Bacteriol.* **84:**1260–1267.

67. **Martin, D. D., D. H. Bartlett, and M. F. Roberts.** 2002. Solute accumulation in the deep-sea bacterium *Photobacterium profundum*. *Extremophiles* **6:**507–514.

68. **Mathur, J., and M. K. Waldor.** 2004. The *Vibrio cholerae* ToxR-regulated porin OmpU confers resistance to antimicrobial peptides. *Infect. Immun.* **72:**3577–3583.

69. **Meganathan, R., and R. E. Marquis.** 1973. Loss of bacterial motility under pressure. *Nature* **246:**526–527.

70. **Meibom, K. L., X. B. Li, A. T. Nielsen, C.-Y. Wu, S. Roseman, and G. K. Schoolnik.** 2004. The *Vibrio cholerae* chitin utilization program. *Proc. Natl. Acad. Sci. USA* **101:**2524–2529.

71. **Missiakas, D., and S. Raina.** 1998. The extracytoplasmic function sigma factors: role and regulation. *Mol. Microbiol.* **28:**1059–1066.

72. **Mulkey, D. K., R. A. Henderson III, R. W. Putnam, and J. B. Dean.** 2003. Pressure ($<$ or $=$ 4 ATA) increases membrane conductance and firing rate in the rat solitary complex. *J. Appl. Physiol.* **95:**922–930.

73. **Nakamura, Y., T. Itoh, H. Matsuda, and T. Gojobori.** 2004. Biased biological functions of horizontally transferred genes in prokaryotic genomes. *Nat. Genet.* **36:**760–766.

74. **Nishida, T., Y. Orikasa, Y. Ito, R. Yu, A. Yamada, K. Watanabe, and H. Okuyama.** 2006. *Escherichia coli* engineered to produce eicosapentaenoic acid becomes resistant against oxidative damages. *FEBS Lett.* **580:**2731–2735.

75. **Nogi, Y., N. Masui, and C. Kato.** 1998. *Photobacterium profundum* sp. nov., a new moderately barophilic bacterial species isolated from a deep-sea sediment. *Extremophiles* **2:**1–7.

76. **Oliver, J. D., and R. R. Colwell.** 1973. Extractable lipids of gram-negative marine bacteria: fatty acid composition. *Int. J. Syst. Bacteriol.* **23:**442–458.

77. **Omura, S.** 1981. Cerulenin. *Methods Enzymol.* **72:**520–532.

78. **O'Shea, Y. A., S. Finnan, F. J. Reen, J. P. Morrissey, F. O'Gara, and E. F. Boyd.** 2004. The *Vibrio* seventh pandemic island-II is a 26.9 kb genomic island present in *Vibrio cholerae* El Tor and O139 serogroup isolates that shows homology to a 43.4 kb genomic island in *V. vulnificus. Microbiology* **150:**4053–4063.

79. **Pope, D. M., and L. R. Berger.** 1973. Inhibition of metabolism by hydrostatic pressure: what limits microbial growth? *Arch. Mikrobiol.* **93:**367–370.

80. **Provenzano, D., D. A. Schuhmacher, J. L. Barker, and K. E. Klose.** 2000. The virulence regulatory protein ToxR mediates enhanced bile resistance in *Vibrio cholerae* and other pathogenic *Vibrio* species. *Infect. Immun.* **68:**1491–1497.

81. **Purdy, A., F. Rohwer, R. Edwards, F. Azam, and D. H. Bartlett.** 2005. A glimpse into the expanded genome content of *Vibrio cholerae* through identification of genes present in environmental strains. *J. Bacteriol.* **187:**2992–3001.

82. **Radjasa, O. K., H. Urakawa, K. Kita-Tsukamoto, and K. Ohwada.** 2001. Characterization of psychrotrophic bacteria in the surface and deep-sea waters from the northwestern Pacific Ocean based on 16S ribosomal DNA analysis. *Mar. Biotechnol.* **3:**454–462.

83. **Regha, K., A. K. Satapathy, and M. K. Ray.** 2005. RecD plays an essential function during growth at low temperature in the antarctic bacterium *Pseudomonas syringae* Lz4W. *Genetics* **170:**1473–1484.

84. **Royer, C. A.** 1995. Application of pressure to biochemical equilibria: the other thermodynamic variable. *Methods Enzymol.* **259:**357–377.

85. **Sandermann, H., Jr.** 1978. Regulation of membrane enzymes by lipids. *Biochim. Biophys. Acta* **515:**209–237.

86. **Sato, T., A. Ishii, C. Kato, M. Wachi, K. Nagai, and K. Horikoshi.** 2000. High hydrostatic pressure represses FtsZ-ring formation and chromosomal DNA condensation in Escherichia coli, p. 171–172. *Third Int. Cong. Extremophiles Abstr.*

87. **Sauer, L. A., R. T. Dauchy, and D. E. Blask.** 2001. Polyunsaturated fatty acids, melatonin, and cancer prevention. *Biochem. Pharmacol.* **61:**1455–1462.

88. **Seo, H. J., S. S. Bae, J.-H. Lee, and S.-J. Kim.** 2005. *Photobacterium frigidiphilum* sp. nov., a psychrophilic, lipolytic bacterium isolated from deep-sea sediments of Edison Seamount. *Int. J. Syst. Evol. Microbiol.* **55:**1661–1666.

89. **Siddiqui, K. S., and R. Cavicchioli.** 2006. Cold-adapted enzymes. *Annu. Rev. Biochem.* **75:**403–433.

90. **Sinensky, M.** 1974. Homeoviscous adaptation—a homeostatic process that regulates the viscosity of membrane lipids in Escherichia coli. *Proc. Natl. Acad. Sci. USA* **71:**522–525.

91. **Singer, G. A., and D. A. Hickey.** 2003. Thermophilic prokaryotes have characteristic patterns of codon usage, amino acid composition and nucleotide content. *Gene* **317:**39–47.

92. **Tamegai, H., C. Kato, and K. Horikoshi.** 1998. Pressure-regulated respiratory system in barotolerant bacterium *Shewanella* sp. strain DSS12. *J. Biochem. Mol. Biol. Biophys.* **1:**213–220.

93. **Tang, G.-Q., N. Tanaka, and S. Kunugi.** 1998. In vitro increases in plasmid DNA supercoiling by hydrostatic pressure. *Biochim. Biophys. Acta* **1443:**364–368.

94. **Thilo, L., H. Trauble, and P. Overath.** 1977. Mechanistic interpretation of the influence of lipid phase transitions on transport functions. *Biochemistry* **16:**1283–1290.

95. **Valentine, R. C., and D. L. Valentine.** 2004. Omega-3 fatty acids in cellular membranes: a unified concept. *Prog. Lipid Res.* **43:**383–402.

96. **Van de Vossenberg, J. L. C. M., T. Ubbink-Kok, M. G. L. Elferink, A. J. M. Driessen, and W. N. Konings.** 1995. Ion permeability of the cytoplasmic membrane limits the maximal growth temperature of bacteria and archaea. *Mol. Microbiol.* **18:**925–932.

97. **van Passel, M. W., A. Bart, H. H. Thygesen, A. C. Luyf, A. H. van Kampen, and A. van der Ende.** 2005. An acquisition account of genomic islands based on genome signature comparisons. *BMC Genomics* **6:**163.

98. **Vezzi, A., S. Campanaro, M. D'Angelo, F. Simonato, N. Vitulo, F. M. Lauro, A. Cestaro, G. Malacrida, B. Simionati, N. Cannata, C. Romualdi, D. H. Bartlett, and G. Valle.** 2005. Life at depth: *Photobacterium profundum* genome sequence and expression analysis. *Science* **307:**1459–1461.

99. **Wang, S. Y., J. Lauritz, J. Jass, and D. L. Milton.** 2002. A ToxR homolog from *Vibrio anguillarum* serotype O1 regulates its own production, bile resistance, and biofilm formation. *J. Bacteriol.* **184:**1630–1639.

100. **Weber, G., and H. G. Drickamer.** 1983. The effect of high pressure upon proteins and other biomolecules. *Q. Rev. Biophys.* **16:**89–112.

101. **Wegrzyn, A., B. Wrobel, and G. Wegrzyn.** 1999. Altered biological properties of cell membranes in *Escherichia coli dnaA* and *seqA* mutants. *Mol. Gen. Genet.* **261:**762–769.

102. **Welch, T. J., and D. H. Bartlett.** 1998. Identification of a regulatory protein required for pressure-responsive gene expression in the deep-sea bacterium *Photobacterium* species strain SS9. *Mol. Microbiol.* **27:**977–985.

103. **Welch, T. J., A. Farewell, F. C. Neidhardt, and D. H. Bartlett.** 1993. Stress response of *Escherichia coli* to elevated hydrostatic pressure. *J. Bacteriol.* **175:**7170–7177.

104. **Yayanos, A. A., and E. C. Pollard.** 1969. A study of the effects of hydrostatic pressure on macromolecular synthesis in *Escherichia coli*. *Biophys. J.* **9:**1464–1482.

105. **ZoBell, C. E., and A. B. Cobet.** 1963. Filament formation by *Escherichia coli* at increased hydrostatic pressures. *J. Bacteriol.* **87:**710–719.

INDEX